Approaches to Human Geography

Approaches to Human Geography

Edited by
Stuart Aitken and Gill Valentine

First published 2006

SAGE Publications Ltd
1 Oliver's Yard
55 City Road
London EC1Y 1SP

SAGE Publications Inc.
2455 Teller Road
Thousand Oaks, California 91320

SAGE Publications India Pvt Ltd
B-42, Panchsheel Enclave
Post Box 4109
New Delhi 110 017

British Library Cataloguing in Publication data

A catalogue record for this book is available from the British Library

ISBN13 978 0 7619 4262 7
ISBN10 0 7619 4262 9

ISBN13 978 0 7619 4263 4 (pbk)
ISBN10 0 7619 4263 7 (pbk)

Library of Congress Control Number available

Typeset by C&M Digitals (P) Ltd., Chennai, India
Printed on paper from sustainable resources
Printed in Great Britain by The Cromwell Press Ltd, Trowbridge, Wiltshire

Contents

List of Contributors

Stuart Aitken is Professor of Geography in the Department of Geography at San Diego State University.

Clive Barnett is a Lecturer in Human Geography at the Open University.

Fernando J. Bosco is Assistant Professor of Geography in the Department of Geography at San Diego State University.

Vera Chouinard is Professor of Geography in the School of Geography and Geology at McMaster University.

David B. Clarke is Professor in the Department of Geography, University of Wales, Swansea.

Deborah P. Dixon is a Senior Lecturer in Geography at the Institute of Geography and Earth Sciences at Aberystwyth, University of Wales.

Isabel Dyck is Reader in Geography, Queen Mary, University of London.

Kim England is Associate Professor of Geography in the Department of Geography at the University of Washington.

J. Nicholas Entrikin is Professor of Geography in the Department of Geography at the University of California, Los Angeles.

A. Stewart Fotheringham is Science Foundation Ireland Research Professor and Director of National Centre for Geocomputation, National University of Ireland, Maynooth.

Reginald G. Golledge is Professor in the Department of Geography at the University of California, Santa Barbara.

Michael F. Goodchild is Professor in the Department of Geography at the University of California, Santa Barbara.

Paul Harrison is a Lecturer in the Department of Geography at the University of Durham.

David Harvey is Professor of Geography in the Department of Geography and Environmental Engineering, City University of New York.

George Henderson is Associate Professor in the Department of Geography at the University of Minnesota.

John Paul Jones III is Professor of Geography in the Department of Geography and Regional Development at the University of Arizona.

Robin A. Kearns is Associate Professor in the School of Geography and Environmental Science at the University of Auckland.

Rob Kitchin is Director of the National Institute of Regional and Spatial Analysis at the National University of Ireland, Maynooth.

Lawrence Knopp is Professor of Geography and Urban Studies in the Department of Geography at the University of Minnesota, Duluth.

David Ley is Professor of Geography in the Department of Geography at the University of British Columbia.

Linda McDowell is Professor of Human Geography in the School of Geography and the Environment at the University of Oxford.

Janice Monk is Professor of Geography and Regional Development and Research Social Scientist Emerita in Women's Studies at the University of Arizona.

Richa Nagar is Associate Professor in Women's Studies at the University of Minnesota.

Paul Robbins is Associate Professor of Geography and Regional Development, University of Arizona.

Paul Rodaway is Director of the Centre for Learning and Teaching at the University of Paisley.

Gerard Rushton is Professor of Geography in the Department of Geography at the University of Iowa.

Michael Samers is a Senior Lecturer in the School of Geography at the University of Nottingham.

Andrew Sayer is Professor of Sociology in the Department of Sociology at Lancaster University.

Eric Sheppard is Professor of Geography in the Department of Geography at the University of Minnesota.

John H. Tepple is a researcher in the Department of Geography at the University of California, Los Angeles.

Gill Valentine is Professor of Human Geography in the School of Geography at the University of Leeds.

John W. Wylie is a Lecturer in Human Geography at the University of Sheffield.

Acknowledgements

A project this large is almost always a long time in coming together. Many of the chapter authors were long suffering through several rounds of edits and unforeseen delays. We would like to thank those who were on board from the beginning, who believed in what we were trying to do, and who stuck with us. Others came on board later in the project and we'd like to thank them for the swiftness with which they worked. We would also like to acknowledge the support of faculty and student seminar and reading groups at San Diego State University, and colleagues and students at Universities of Leeds and Sheffield. Special thanks go to Fernando Bosco, who allowed some of the ideas in this book to be shared in his 'Philosophy in Geography' seminar; and to Charlotte Kenten for all her hard work reformatting chapters and chasing missing information.

There is only one Robert Rojek! We owe a huge debt to him for commissioning and supporting the development of this manuscript through its long gestation process. The quality of the final product is due to the efforts of David Mainwaring, Brian Goodale the copyeditor, and Vanessa Harwood the production editor.

Finally, we want to acknowledge the continuing inspiration and energy we gain from our graduate students, and we want to thank them for challenging our ways of knowing.

1 WAYS OF KNOWING AND WAYS OF DOING GEOGRAPHIC RESEARCH

Stuart Aitken and Gill Valentine

This book is intended as an accessible introduction to the diverse ways of knowing in contemporary geography with the purpose of demonstrating important and strategic links between philosophies, theories, methodologies and practices. As such it builds on the other books in this series: *Key Concepts* (Holloway, Rice and Valentine, 2003); *Key Methods* (Clifford and Valentine, 2003); and *Key Thinkers* (Hubbard, Kitchin and Valentine, 2004). Our intention is to guide beginning students in the sometimes complex and convoluted links between ways of knowing and ways of doing geographical research. It is a philosophical reader designed to be a practical and usable aid to establishing a basis for researcher projects, theses and dissertations. It is an attempt to lift the seemingly impenetrable veil that sometimes shrouds philosophical and theoretical issues, and to show how these issues are linked directly to methodologies and practices. The book highlights some intensely serviceable aspects of a diverse array of philosophical and theoretical underpinnings – what we are calling ways of knowing. It makes a case for embracing certain ways of knowing in terms of how they inform methods and practices. We believe that ways of knowing drive not only individual research projects but also the creative potential of geography as a discipline. Philosophies and theories, as ways of knowing, are not simply academic pursuits with little bearing on how we work and how we live our lives.

The book avoids jargon-laden, impenetrable language and concepts while not sacrificing the rigour and complexity of the ideas that underlie geographic knowledge and the ways that it is conflicted and contested. It is written for students who have not encountered philosophical or theoretical approaches before and, as such, we see the book as a beginning guide to geographic research and practice. We believe that grounding research in philosophy and theory is essential for human geography research because it provides a hook for empirical work, it contextualizes literature reviews, it elaborates a corpus of knowledge around which the discipline grows, it energizes ideas, and it may legitimate social and political activism. In addition, and importantly, an understanding of philosophy and practice directs the discipline of geography conceptually and practically towards progressive social change by elaborating clearer understandings of the complexity of our spatial world.

The book is split into three parts: philosophies, people and practices. In the first part, leading academics make special and partial 'cases for' particular philosophies, and illustrate their argument with short examples. Although it is far from comprehensive, the part covers a large swathe of philosophical perspectives and highlights some of the tensions between various ways of knowing. It is not intended to offer the student an all-inclusive guide to philosophies in geography (this is better achieved by more specialist texts such as

Johnston, 1991; Cloke et al., 1991; Unwin, 1992) but rather it offers practical insight into how philosophies inform work and how research questions are always based on assumptions and choices between different ways of knowing. The chapters do not resolve philosophical debates; instead they lead students to consider what choices and assumptions must be made when beginning a research project, and when choosing methodologies. The second part of the book places geographic thought amidst the complexity and struggle of people contextualized in places. Within contemporary human geography there is an emphasis on situated or contextual knowledges – which has its roots in the feminist belief that 'the personal is political' and critical feminist science's challenge to traditional conceptions of scientific practice as objective and disembodied (Haraway, 1991; Rose, 1997). Thus personal writing is seen by many as an important strategy to challenge the disembodied and dispassionate nature of previous academic writing (e.g. Moss, 2001). In the second part, several prominent geographers write about the people, places and events that shaped their personal ways of knowing. Finally, philosophy is often taught separately from methodology, which means that students sometimes fail to recognize the connections between theories and practices. The final part outlines some of these relationships and illustrates them with examples from a range of geographical studies.

Students beginning a research project in geography encounter a mind-boggling array of methodologies and practices. These methodologies and practices are linked in complex ways to theories and philosophies. Geographical research comprising a cloudy web of methodologies, theories, philosophies and practices ultimately elaborates geographical knowledge. We have tried to represent this complexity in Figure 1.1, and yet this diagram structures and represents our concerns too simply.

Ways of doing are not attached to static ways of knowing but rather are changing as one set of ideas is challenged and informed by others. How we come to approach the world through theories and philosophies – our ways of knowing – is constantly refined, challenged, rejected and/or transformed. Customarily, theoretical traditions (positivism, humanism, Marxism, feminism, etc.) have been understood to emerge and dominate geographical thinking at particular times for a particular period. In other words, they have become what Kuhn (1962) termed 'dominant paradigms'. As such, some writers have mapped out the development and adoption of different philosophic approaches within the discipline of geography (e.g. Johnston, 1991; Unwin, 1992) highlighting paradigm shifts – when new philosophical approaches emerge to challenge previous ways of thinking. Johnston (1996) suggests that paradigm shifts are a result of generational transitions. Thus new ways of thinking are taken up at first by younger academics; as this generation becomes established, and takes on editing journals and writing textbooks, so their ways of thinking come to the fore. A paradigmatic approach to geography begins in the 1950s when positivistic spatial science emerged to challenge and supersede the regional tradition in geography. In turn the positivist paradigm is understood to have been overturned in the 1970s by other approaches such as behaviourial geography, humanistic geography and radical approaches including Marxism and feminism. In the 1990s a paradigmatic perspective would understand poststructuralism as displacing these ways of thinking.

Yet, while sometimes a whole set of ideas is thrown out in light of perceived shortcomings, usually part of the thinking continues in one form or another (see Figure 1.2). The institutional framework of geography – professional organizations, journals and departmental cultures – may privilege or reinforce

Figure 1.1 Ways of knowing and ways of doing

particular fashionable ways of thinking, but there are always dissenting voices. In reality, most ways of knowing are partial and are in flux; they continue to change as geographers examine and re-examine their strengths and weaknesses and as new ideas come along as a challenge. The discipline always includes a range of generations, and scholars who don't act their age! The linear narrative of the development of unified paradigms thus falsely creates a sense of sequential progress when consensus is rarely complete or stable. Although the chapters in this book are loosely ordered in relation to the genealogy of their emergence in the discipline, it is not our intention to suggest that one displaced another. Rather, our intention is to show how each approach to geography (positivistic geography, humanistic geography, Marxism, feminism and so on) contains within it multiple trajectories of thought and how each has continued to evolve whatever its paradigmatic status. Part of the excitement of doing geographical research is the continual struggle to make sense of these changing perspectives and their connections.

When writing a research proposal, choices must be made about appropriate ways of knowing and doing. Students must be aware of the assumptions of particular ways of knowing, how they help raise appropriate questions and their adequacy for addressing those questions. Ultimately, all researchers must be able to justify the answers they give to their research questions and that justification

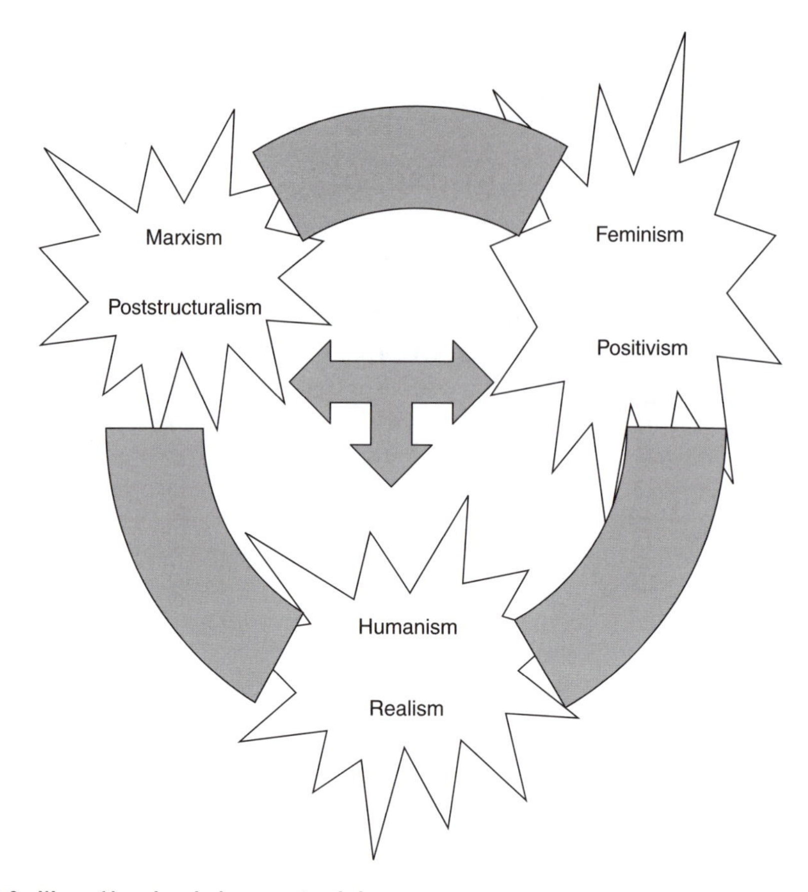

Figure 1.2 Ways of knowing clash, connect and change

cannot avoid philosophical and theoretical ways of knowing. In this sense, philosophy is a form of communicating not only what we know but also how we know it. Understanding philosophical processes as forms of communication suggests an important pedagogical metaphor. Elspeth Graham argues that 'philosophy is to research as grammar is to language ... just as we cannot speak a language without certain grammatical rules, so we cannot conduct a successful piece of research without making certain philosophical choices' (1997: 8). Philosophy helps contextualize and justify the answers to our research questions in ways that communicate what we know. We can still speak and write without awareness of grammar, but it is always there. Grammar is a useful metaphor for understanding the role of philosophy in research projects because it suggests that the more we know about philosophical underpinnings the better we appreciate how influential they are to our work. If doing research is like the grammatical foundations of a language then, Graham (1997) notes, pushing the metaphor further, the beginning researcher must learn the appropriate vocabulary and terms. This involves reading and learning the vocabulary and the grammar and syntax of the speech community you wish to join. Just as Mexican Spanish and the practice of Mexican culture are intimately tied together, and are quite different from Scottish English and the practice of Scottish culture, then so too are philosophies differentiated. Marxist geographers use terms like production, social reproduction, class, superstructure and dialectics; positivist geographers use terms like paradigms, hypotheses, laws and verifiability; feminists and queer theorists use terms like patriarchy, bodies, sexualities and performativity; humanistic and experiential geographers use terms like essences, taken-for-grantedness and nihilism (these terms and others are defined and explained in Johnston et al., 2000 and McDowell and Sharp, 1999). Built around these language differences are systems of meaning, and so the beginning researcher must master more than just the terms: she must also engage associated cultures and practices. A positivist researcher engaging the practice of falsification, for example, might follow the rules of hypothesis testing; a feminist researcher engaging in the practice of positionality might want to understand fully her own personal politics and situatedness. And just as aspects of Scottish and Mexican cultures and practices collide and meld, so too do aspects of humanism, Marxism, feminism, queer theory and positivism. The connections and conflicts are at once daunting and exhilarating. Exhilarating because this is the stuff of creative debates and purposeful practices; daunting because students reading this book are being asked to gain a working knowledge of many languages at once.

Ways of knowing are, of course, quite different from grammar in that they are at once more fundamental, and they are often more convoluted. Philosophy as a way of knowing elaborates the structures and essences of our existence. This is known as ontology. Ontology comprises theories, or sets of theories, which seek to answer questions about what the world must be like for knowledge to be possible. Philosophy also investigates the origin, methods and limits of our knowledge about existence. That is, it establishes what is accepted as valid knowledge. This is known as epistemology.

In the tradition of Greek Enlightenment, logic and reason are touted as the basis for all epistemologies. From this western perspective, it is assumed that minds are essentially rational and have similar experiences of the world (Peet, 1998: 5). It is also assumed that ideas can be abstracted from the material world, and it is the purpose of philosophy to organize these ideas into coherent patterns and then evaluate the knowledge derived from those ways of knowing. Once thought of, these patterns are spoken of and written about so that they may be understood as axioms around which aspects of existence revolve, or they may be criticized and rejected. In its strictest form, the assumption that all minds work in the same way suggests that there can be one, unitary and all-encompassing philosophy. An alternative set of philosophical traditions hold that how we think is a social construction rather than derived from some innate, universal logic. From this social constructivist perspective the distinctions between different philosophies are derived from different political and cultural milieux and then imposed upon the minds of those who are part of that context. This position accepts that ontology is grounded in epistemology and that all epistemologies are embedded in social practice.

Most of the authors in this book do not view philosophy as a basis of knowledge that is completely abstracted from people and the places they work. Rather, they assume it to be the driving force that connects us with others, and that contextualizes who we are, what we know and what we do. Nor do most of the authors believe that philosophy and theory need to employ only logic and reasoning to organize knowledge into formal systems of understanding. Some believe that knowledge comes also from less reasoned and less representable ways of knowing derived from emotions such as anger, passion, love, joy and fear. Ways of knowing are at least in part derived from these and other emotions that are sometimes difficult to write about and represent in a logical form. Philosophy as outlined in the chapters in this book is seen as a social, political, and cultural construction that contains elements of rationality and irrationality. And so, some of the authors argue that the rationality so valued by Greek Enlightenment thinkers is influenced by irrational beliefs and meanings derived from our bodies and our emotions as well as cultural meanings and the places where we work and live.

Theories as Ways of Knowing and Being

Theory can be less heady than philosophy but it is equally important as a way of knowing. If philosophy encompasses larger ways of knowing that connect us to the beliefs, values and meanings of others (sometimes known as metaphysics) and systematize what we know, then theory extends this to the experiences of everyday life. As Richard Peet (1998: 5) points out, theory 'has a more direct contact with the occurrences, events, and practices of lived reality' than philosophy. He argues that theory is derived inductively (working from the specific to the general) and primarily from empirical sources (those derived directly from experience). He goes on to suggest that theory looks 'for commonalities or similarities, but also (perhaps) systems of difference or, maybe, just difference'. Theories are also deductive (working from the general to the specific) because they often speculate from one aspect of difference and uniqueness to others.

Whereas philosophy engages larger systems and webs of meaning, theory engages a more specific sphere of understanding and being in the world. In the field of the empirical sciences, hypotheses are constructed as systems of theories that are tested against experience by observation and experiment. In the humanities and social sciences, social or critical theories deal directly with understanding social, political and cultural perspectives and characteristics as they relate to transformations within societies and the day-to-day lives of people.

Practices as Ways of Knowing, Being and Doing

Practices are ways of knowing in action. Academics are engaged in the production of knowledge and its dissemination. Philosophies help articulate the ontological and epistemological bases of that production. Theories help elaborate the production of knowledge from experience and experimentation, and they sometimes challenge conventional wisdom. As such, theories are not impartial or neutral but, rather, they are instruments of persuasion backed by experience. For some, they suggest action. This practice may play out in day-to-day lives or it may take the form of social and political activism. Teaching and research practices are also modes of doing, and are charged with political will and intent that are sometimes explicit and sometimes veiled. For some academics, doing is not just about teaching and writing, it is also about taking their values and beliefs, their philosophies and theories, out into the world from which they are derived in an attempt to transform that world for the better.

Research, like social and political activism, is almost always intensely political. It reacts to, and informs, the larger contexts of societal crisis, injustice and wellbeing. Within this realm, disciplines and subdisciplines clash and contend with each other in their attempts to respond to social crises and injustices. These internal struggles within academia can become vitriolic given limited access to finite resources and money. While touting a quest for truth or a better world, academic debate is also about status, power, and control of resources. These struggles sometimes delimit boundaries between different discourses and sometimes transgress them; they often inflame passionate struggles between seemingly rival ways of knowing. In sum, teaching, researching, writing and practising geography open a myriad of different ways of knowing that often clash.

We believe that diverse ways of knowing and practising geography are the basis of the discipline. When they collide and lock horns, as they often do, a creative energy is unleashed that questions assumptions and pushes thinking forward, often in intriguing, innovative and exciting ways.

A number of years ago a panel at the annual meetings of the Association of American Geographers positioned advocates of two seemingly competing philosophies – humanism and positivism – in formal debate. The auditorium was packed with geographers anxious to see some intellectual giants do battle. Battle was not the intention of the organizers of the plenary session who, in the session abstract, elaborated the possibility of a common ground between humanism and positivism. The debate began politely enough as the moderator articulated her desire to use this forum as a basis for moving a common ground forward towards synthesis. While accepting the possibility of a basis for discussion, the protagonists presented diverse cases for their respective philosophical leanings in very particular ways. In making their respective cases, the speakers either used rhetoric that politely accepted alternative ways of knowing but only as perspectives that could be subsumed within the practice of their particular philosophical leaning, or attacked the premises of their opponents as untenable. Humanistic philosophies, for example, were positioned as the basis of being and consciousness from which mathematical analysis and logical deduction were derived as merely abstract ways of knowing. Alternatively, positivism and scientific perspectives were seen as the logical end point of humanistic assessments that merely provided qualitative data from which quantitative categories could be built. After the presentations a debate ensued that was quite vitriolic. Scholars who had built their careers on a particular philosophy were loath to accept the possibility that their way of knowing was either subservient to or less practical than another way of knowing, and they definitely did not accept the possibility that their way of knowing was flawed. In the last innumerable years other conflicts have arisen between diverse philosophies in most of the major geography meetings around the world and also in published work. Using a variety of rhetorical devices, structuralism has been pitted against poststructuralism; Marxism against poststructuralism or feminism; ideography against nomotheticism; postcolonialism against environmentalism; environmentalism against feminism; possibilism against probabilism; relational approaches against theories of structuration; and so forth. Sometimes the debates become intensely myopic and perhaps a little impenetrable when, for example, queer theory challenges feminism or behaviouralism admonishes behaviourism. And yet, in each interaction of ideas and practices there is the creative potential for change.

Although the rhetoric changes, the terms of these clashes often revolve around what a set of philosophies and theories proposes as a basis for geographic knowledge and how practical those philosophies and theories are in delivering that knowledge. We purposefully list some 'isms' above without definition because we argue that the meat is in the process of debate: that is where the passion lies! This is not to suggest that intensely practical ways of knowing set the tone for subsequent scholarship. Nor is this about philosophical fads and the current 'ism' of the day. For example, the debates between particularity (ideography) and generality (nomotheticism) that popularly smattered the pages of academic geography in the 1950s returned in different forms throughout the last half-century with critiques of metanarratives, discussions about the merits of humanistic, poststructural and relativistic approaches, and so forth. The context of the discussion changes in different times and in difference places. The point is not just what is contested, but that there is contestation that is creatively adopted and used to propel geographical ways of knowing.

Geographical Ways of Knowing

When first confronted with the literature on how human geographers construct their world intellectually, the new student is faced with a bewildering set of apparent alternatives. As a named discipline, geography is an ancient form of intellectual inquiry, predating Greek classicism and its notions of rational thinking. And yet there is little agreement about how the discipline is constituted, what it studies and how it should go about that study. Certainly what is thought of as geographic inquiry has changed significantly over the millennia, and the last half-century in particular has resulted in an increasingly conflicted and contradictory set of arguments for how the discipline is constituted and practised.

This book attempts to uncover ways of knowing geography (how it is thought about) and the practice of geography (thought expressed in action) without sacrificing people and places as an important part of that practice. It attempts to capture contemporary geography as a known and practised discipline that is internally differentiated and contested. Knowledge is always partial and practice is often infused with passion. The book does not attempt to elaborate the entire corpus of knowledge that comprises contemporary human geographic thought, but rather it brings to light the contested and hotly debated nature of diverse ways of knowing.

Disciplinary boundaries are not cast in stone; they are fuzzy and chameleon-like, changing before our eyes as we focus deeper. Subdisciplinary boundaries are even more difficult to tie down, and yet each embraces an accepted body of knowledge that legitimizes practice. Embracing a particular way of knowing distinguishes a thesis or dissertation, enabling some degree of classification. It is what examiners and reviewers focus on as they try to place the work; the success or failure of a particular study often resides with its ability to contextualize itself in a larger corpus of knowledge. For example, thesis or dissertation abstracts that announce respectively a postcolonial approach to the development of squatter settlement, a humanistic appraisal of belonging and being-at-homeness, an econometric appraisal of regional housing demand, or a feminist critique of suburban spatial entrapment, suggest diverse and perhaps contradictory ways of establishing academic credibility. Postcolonialism, humanism, econometrics and feminism are three sets of methods and practices with their own assumptions, values and ways of proceeding. Each are legitimate geographic ways of knowing that leave a new student struggling to place them amongst dozens of others and to get a sense of how they might relate to each other as well as to the student's own interests and passions. There is nothing absolute or sacred about any particular way of knowing; each is elaborated upon and argued about, and there is no single set of criteria by which one way of knowing legitimizes itself over another. The clash of knowledge, the lack of boundaries and absolutes, the tension between ways of knowing are at once confusing and exhilarating. They are confusing because each philosophy presents a laudable case for its own existence, leaving difficult choices for students seeking to legitimize their own interests; and exhilarating because the creative tension between different ways of knowing engenders passion amongst adherents. And passion is always stimulating.

Constructing Geographical Knowledge and Practice

The passion of academic debate is sometimes disregarded as the synthesizers of geographic knowledge tackle through simplification the myriad arguments and accounts that make up the discipline. Traditionally, geographic knowledge has been constructed in five ways.

First, confusion is bypassed and underlying philosophies are disregarded simply by suggesting that geography is primarily what geographers do (Gould, 1985; Johnson, 1991). This perspective relies on geographers' self-definitions and focuses on disciplinary practices. Referring to actions and activities rather than underpinning structures of knowledge emphasizes output, productivity, utility and problem-solving above all else. From this perspective, academic geographers attract students to their departments by teaching something that is seen as useful and of some interest to those who study it. It has been argued that they also are inclined to do research that is of interest to, and is tied in with, the agendas of financial sponsors (Unwin, 1992: 6). It might be argued further that constructing the corpus of geographic knowledge in this way ties it most successfully to societal needs, but this argument presupposes that 'doing' and productivity through problem-solving are always useful and can be divorced from larger ways of knowing. It neglects the fundamental issues of how problem-solving and utility are constructed and for whom.

The second way of synthesizing geographic ways of knowing is methodological (see Clifford and Valentine, 2003 for a guide to methods in human and physical geography). Many geography degree programmes offer methodological and technical options as tracks or even as full-blown diplomas. A unique set of tools – such as those comprised in and defined by spatial analysis or environmental modelling – delimits and justifies disciplinary boundaries (see chapters in Part 3). The tools can be learnt and applied to different spatial and environmental phenomena. It may be argued that a large part of the recent success of geography in technological societies may be attributed to geographical information systems, which manage and analyse spatially referenced data through sophisticated computer software programs (see Chapter 23). The recent change in name and orientation from geographic information *systems* to geographic information *science* suggests an appreciation of the limitations of technological systems that are not energized by ideas and frameworks of knowledge.

A third attempt to tie down human geography is by identifying a subject matter around what the discipline studies and how it studies it. Such definitions delimit certain objects as legitimately geographic and others that are not. For example, in a famous and influential essay, Norman Fenneman (1919) described the circumference of geography as best defined by the region, arguing that its use would serve to focus the discipline and prevent its absorption by other sciences. And, at around the same time, American cultural geographer Carl Sauer stated simply that 'we are not concerned in geography with energy, customs or beliefs of man [*sic*] but with man's [*sic*] record on the landscape' (1928: 342). Key concepts (see Holloway et al., 2003) and terms such as landscape, region, environment, space, place, culture, scale and so forth are often adhered to specific categories of knowledge in various ways, changing and transforming as the ideas about them are tugged in different directions by different philosophical bents (cf. Earle et al., 1996). These objects of geographical analysis are often uncritically accepted as part of a particular way of knowing comprising uniform categories, sometimes referred to as stable referents within a particular philosophy. Geographic knowledge produced for a particular audience constitutes these categories. Language operates to establish social and natural worlds through signifying or discursive practices that generate and organize signs or discourse into particular geographical knowledge or 'ways of seeing' such as those proposed by Fenneman and Sauer. In its attempt to sort out the complexity, this framework provides a seemingly neutral way of engaging geographic knowledge.

A fourth strategy may acknowledge other ways of knowing but usually positions them as less consequential or subsumes them as precursors to a dominant way of knowing. For example, coming from a positivist and quantitative perspective, Brian Berry (1964) argued that all geographic patterns and processes could be accessed through establishing a huge matrix of variables across time. Alternatively, in the 1980s Larry Ford (1984) argued that geography has its origins in how the landscape is observed and all other methods and practices follow. Michael Goodchild and Don Janelle (1988) used multidimensional scaling techniques on data from speciality group membership amongst members of the Association of American Geographers to argue a practical and dynamic core for the discipline in those speciality groups that were most connected. And later that decade, Michael Dear (1988) defined a core of human geography quite differently in terms of social theory development. He argued for the discipline's pivotal role in the social sciences with its focus on three primary processes that structure what he calls the fabric of time–space: the political, the economic and the social. These strategies are important to the extent that they gain favour with geographers, and all are agenda based. Most of the authors cited above are willing to acknowledge those agendas, but with nonetheless convincing arguments they also provide a singular way forward that smooths out or disregards tensions and conflicts.

A fifth way of coming to terms with complex and divergent ways of knowing also is inclined to smooth out tensions and conflicts. This strategy offers a synthesis that relies on understandings that change through time (Johnston, 1991; Livingston, 1992). This way of approaching philosophy in geography attempts to provide a linear and relatively objective and impartial appraisal of how knowledge is built and transformed. There is what might be thought of as a patterned sequence to how geographers have come to know the world. In this formulation, the discipline's so-called paradigms or 'isms' stretch back over time and help define what comes after. This way of structuring knowledge is essentially about lumping philosophies into categories that may begin, for example, with environmental determinism in the early twentieth century and then flow through possibilism, regionalism, the quantitative revolution, structuralism, realism, humanism, Marxism, feminism, queer geographies and postcolonialism to end, perhaps, with poststructuralism or the latest intellectual fad. It is a common practice of textbook writers to smooth out and generalize the connections between different philosophies in this way because it is deemed too hard for beginning students to get their minds around all these debates. Too often texts on geographic thought neglect the contested nature of the world and our knowledge of it by supplying a relatively linear set of approaches melding into each other and ending with a professor's preferred way of knowing. No wonder students are put off by this plethora of 'isms' and the challenges that they hold out to each other.

The 'isms' suggest abstract knowledge that is extracted and simplified from a very complex set of interactions between people, places and intellectual movements (see Part 2). For today's students, they often suggest a way of structuring knowledge that has little bearing on research projects and is, rather, an interpretation of dead or barely alive geographers' ways of thinking that has only a remote connection with today's world. The fact that most of the existing books and articles on philosophy and human geography are either written by a single author or presented to the reader in one voice means that the outline of each philosophy is very balanced, neutral and even. As such students often fail to grasp the contested nature of the discipline and regard the approaches as pick 'n' mix alternatives rather than recognizing the tensions between those who adopt different philosophical positions or the possibilities of collaboration between those who have different ways of thinking. Those tensions often arise from a body of literature that is adopted and elaborated by geographers. Particular people writing from particular places at particular times also often spur them (see Moss, 2001 or Gould and Pitts, 2002 for autobiographical accounts of the intellectual development of geographers; or Hubbard et al., 2004 for a biographical approach to understanding key thinking on space and place). The energy of a social movement or an individual's ideas, or the culture of a specific academic department, will enhance certain ways of knowing over others. Johnston (2004), for example, highlights the significance of individuals' networks and the career trajectories from which geography develops by tracing the path taken by David Smith – the connections he forged, and the influences on his decision-making as he made the switch from a spatial analysis tradition to other paradigms. Thus instead of assuming a geographic imaginary that organizes itself around an ordered timeline of ideas, what happens if we say it is ordered around different sets of people, places and contexts for the ideas? What if we openly acknowledge the political and moral connections, and the personal and social stories, that give the ideas life? What if we probe the ways that philosophical approaches are energized by conflict, critique and career advancement? What kinds of lessons do we glean from documenting encounters between scholarship and practice? How does the way we live our lives, the way we connect with social and political struggles and the seemingly random opportunities that come our way, affect our geographical imagination? These questions drive the chapters in this book. The chapter authors do not try to explain or smooth out tensions between their preferred way of knowing and others.

The chapters in this book provide accessible accounts of the ways different philosophies and theories intersect with and scrunch against each other. Rather than searching for a common ground, we accept that knowledge is contested, controversial and partial; that it is about power and career enhancement as much as it is about a search for enlightenment; that it is about moral integrity and a need to understand more fully social and spatial injustices; but that it is also about the academic culture of particular places and particular times. Further, this book provides a new way of encountering geographical thought because it ties it intimately with methodologies and practices. We dismiss past pedagogies that abstract thought from people, places and their practices. We do not disengage from the conflict that arises between ideas and factions that compete for control of geography as an intellectual resource that helps make sense of the world. Rather, we engage intellectual conflict and tension as the harbingers of change and social engagement through practice. Ultimately geography, like all academic pursuits, is about changing the world for the better and, as such, it is not a neat and ordered practice.

References

Berry, B. (1964) 'Approaches to regional analysis: a synthesis', *Annals of the Association of American Geographers*, March: 2–11.

Clifford, N. and Valentine, G. (eds) (2003) *Key Methods in Geography*. London: Sage.

Cloke, P., Philo, C. and Sadler, D. (eds) (1991) *Approaching Human Geography*. London: Chapman.

Earle, C., Mathewson, K. and Kenzer, M.S. (eds) (1996) *Concepts in Human Geography.* Lanham, MD: Rowman and Littlefield.

Fenneman, N. (1919) 'The circumference of geography', *Annals of the Association of American Geographers*, 9: 3–11.

Ford, L. (1984) 'A core of geography: what geographers do best', *Journal of Geography*, 2: 102–6.

Goodchild, M.F. and Janelle, D.G. (1988) 'Specialization in the structure and organization of geography', *Annals of the Association of American Geographers*, 78 (1): 1–28.

Gould, P.R. (1985) *The Geographer at Work.* London: Routledge.

Gould, P.R. and Pitts, F.R. (eds) (2002) *Geographical Voices: Fourteen Autobiographical Essays.* Syracuse, NY: Syracuse University Press.

Graham, E. (1997) 'Philosophies underlying human geography research', in Robin Flowerdew and David Martin (eds), *Methods in Human Geography: A Guide for Students Doing Research Projects.* Edinburgh: Longman, pp. 6–30.

Haraway, D.J. (1991) *Simians, Cyborgs, and Women: The Reinvention of Nature.* New York: Routledge.

Holloway, S.L., Rice, S. and Valentine, G. (eds) (2003) *Key Concepts in Geography.* London: Sage.

Hubbard, P.J., Kitchin, R. and Valentine, G. (eds) (2004) *Key Thinkers on Space and Place.* London: Sage.

Johnston, R.J. (1991). *Geography and Geography: Anglo-American Human Geography since 1945*, 4th edn. London: Arnold.

Johnston, R.J. (1996) 'Paradigms and revolution or evolution?', in J. Agnew, D. Livingstone and A. Rogers (eds), *Human Geography.* Oxford: Blackwell.

Johnston, R.J. (2004) 'Disciplinary change and career paths', in R. Lee and D.M. Smith (eds), *Geographies and Moralities.* Oxford: Blackwell.

Johnston, R.J., Gregory, D., Pratt, G. and Watts, M. (eds) (2000) *The Dictionary of Human Geography.* Oxford: Blackwell.

Kuhn, T.S. (1962) *The Structure of Scientific Revolutions*, 1st edn (2nd edn 1970). Chicago: Chicago University Press.

Livingston, D. (1992) *The Geographical Tradition.* Oxford: Blackwell.

McDowell, L. and Sharp, J. (eds) (1999) *A Feminist Glossary of Human Geography.* London: Arnold.

Moss, P. (2001) *Placing Autobiography.* Syracuse, NY: Syracuse University Press.

Peet, R. (1998) *Modern Geographical Thought.* Oxford: Blackwell.

Rose, G. (1997) 'Situating knowledges: positionality, reflexivities and other tactics', *Progress in Human Geography*, 21: 305–20.

Sauer, C. (1928) 'The morphology of landscape', in John Leighly (ed.) (1963), *Land and Life: A Selection from the Writings of Carl Otwin Sauer.* Berkeley, CA: University of California Press.

Unwin, T. (1992) *The Place of Geography.* New York: Longman.

Part 1
Philosophies

In this part, leading proponents of different approaches to geographic understanding make 'cases for' different philosophical and theoretical leanings. Some illustrate their arguments with short examples and others argue their case through logic. Although it is far from comprehensive, the part covers a large swathe of philosophical and theoretical perspectives. It is intended not to offer the student a guide to philosophies in geography, but rather to highlight some of the tensions between various ways of knowing.

We do not go out of our way to suggest a linear sequence between these ways of knowing, nor do we attempt to smooth out differences. Although each way of knowing entered the discipline at a particular time and for a particular set of reasons and there are important connections, each philosophical underpinning continues to influence geographical research in different, conflicted ways. Some of the influences maintain a relatively unchanged currency, while others have transformed into different ways of knowing. The part offers practical insight into how philosophies inform work and how research questions are always based on assumptions and choices between different ways of knowing.

The chapters are loosely arranged in chronological order. Using this broad framework as a pedagogic experiment, we introduce the chapters in three ways. The chapters on positivism, humanism, feminism and Marxism (2, 3, 4 and 5) are grouped together because the authors articulate very focused, albeit quite different, intentions. We call these 'Singular Intentions'. Rather than having a specific intention, the chapters on behavioural research, structuration theory and realism (6, 7 and 8) each articulate a basis for understanding our geographical world by suggesting ways that knowledge is structured. We call this 'Constructing Geographical Knowledge in Relation to the World'. Finally, postmodern geographies, poststructuralist theories, actor-network theory and postcolonial theory (Chapters 9, 10, 11 and 12) offer arguments for how knowledge is not easily patterned. We call this 'Beyond Structure'.

Obviously there are important connections and conflicts between all these chapters, and they could have been grouped in a myriad of ways. Students are invited to find other ways that these chapters relate to each other, because there are clearly many.

Singular Intentions

In the first chapter of this part Rob Kitchin (Chapter 2) refers to positivist philosophies in the plural because they elaborate multiple ways of knowing in geography. There is no single positivist way of knowing; rather there are multiple positivistic ways of knowing. As with positivism, the following three chapters each describe a set of philosophical approaches designed with a particular intent. The approaches of positivism are tightly circumscribed around a perceived need to apply scientific principles, rigour and analytic reasoning. The singular intent of positivist geographers, then, is to apply the principles of science to geographic understanding.

The two most common forms of positivism are based on verification (logical positivism) and falsification (critical rationalism). In the former, deductive reasoning is used to formulate theory, and then set and test hypotheses. The latter is based on attempting to undermine theory by identifying exceptions.

Advocates who feel that geography tends towards unsystematic and analytically naive inquiry have propelled a concerted focus on spatial science from the 1950s onwards. With the exception of some behavioural geographers (see Chapter 6), however, there is rarely a meaningful engagement with philosophy by those who embrace science in human geography. This thought is articulated further by Fotheringham in Chapter 22. Kitchin argues that scientific principles applied to quantitative data are seen as factual, objective and universal in nature. He notes that this has led feminist geographers to criticize some spatial science for harbouring alleged hidden, masculinist underpinnings. Few spatial scientists today would claim allegiance to the central tenets of positivism, but they do see their approaches as sensible, robust and, above all, scientific. Its influence on contemporary quantification (Chapter 22) and GIS (Chapter 23) and on the practices of individual geographers (see Chapters 13, 15 and 18) is profound.

In Chapter 3 Nicholas Entrikin and John Tepple consider the emergence of humanistic geography as a loosely structured movement that developed out of a critique of what was regarded as the obsessively narrow focus of positivistic human geography on human decision-makers as rational economic actors whose behaviour could be predicted and modelled. Rather, Entrikin and Tepple show how the singular intent of humanistic geographers has been to demonstrate the importance of individuals' experiences, beliefs and attitudes in shaping the decisions that we make and the ways that we engage with the world. Here, they illustrate how the emphasis within humanistic geography has been on uncovering meanings, values and interpretations in order to incorporate a more complex understanding of human reality into geography. In doing so they highlight the way that humanists drew on a range of philosophies from the humanities (e.g. phenomenology, existentialism) as well as interpretive traditions of fieldwork from disciplines such as anthropology. The implications of this for geographical practice are explored by Paul Rodaway in Chapter 24 where he reflects on the development of people-centred methodologies in geography. The influence of humanistic philosophies on the practices of individual geographers who may not call themselves humanists is reflected in the work of Richa Nagar (Chapter 19), Vera Chouinard (Chapter 17), Larry Knopp (Chapter 20), as well as, more explicitly, in the work of David Ley (Chapter 14) and Robin Kearns (Chapter 16).

The singular focus on the intentional agent in humanistic geography is the subject of conflict with other approaches that have sought to highlight how individuals' choices are constrained by social structures such as patriarchy (Chapter 4) and capitalism (Chapter 5), and with those that have attempted to tease out the complex relationship between agency and structure (Chapter 7). Finally, poststructuralism (Chapter 10) and postcolonialism (Chapter 12) have challenged humanism's very notion of the intentional agent and the universal claims that follow from this.

In Chapter 4, Deborah Dixon and John Paul Jones focus on the ways that feminism has recovered various geographies of gender, first by critiquing and contesting masculinist ways of knowing and then by elaborating its own epistemologies. Most importantly in terms of practice and transformation, they point out that the purpose of feminism is to generate ways of knowing that improve women's lives. Feminist geographers use multiple theories and methods to better understand the sources, dynamics and spatiality of women's oppression and to suggest strategies of resistance. The singular intention of feminists, then, is to better women's lives. To aid this endeavour, their primary focus is on day-to-day social and spatial activities.

Beginning with the myriad of processes through which women geographers and their ways of knowing were marginalized in nineteenth- and twentieth-century Anglo-American geography, Dixon and Jones make a case for how feminism challenged, and continues to challenge, the masculinist bases of the discipline. They show that a large part of the early ostracizing of women's knowledge was based on the patriarchal practices of the discipline – paternalism, misogyny, sexual discrimination, gender-coded language – and the kinds of things geographers studied (e.g. distributions of economic activities). Feminist geography not only critiques these bases of the discipline, but also offers an epistemology that transforms geographic ways of knowing. For example, traditional objects of analysis such as regions, landscapes and places are researched with questions that probe the spatial dimensions of gender divisions of labour. Feminism also introduces new objects of analysis such as homes and bodies. New epistemologies focusing on difference, gender as social relations and gender as a social construction enable feminists to push geography in new, transformative directions that influence methodologies, practices, lived experiences and discursive meanings. In this latter regard, the chapter links in important ways to Kim England's (Chapter 26) discussion of feminist methods and methodologies and how those, in turn, produce feminist geographies. The webs of meaning between ways of knowing and ways of doing are circular and often are mutually reinforcing.

In many ways Marxism parallels positivism as a reaction to less theoretically informed empirical studies in geography. Like feminism and positivism, there are multiple variations in geographic thinking that derive from classical Marxism. Indeed, some claim that a large part of feminism's focus on the exploitation of women comes from Marxist sensibilities. Henderson and Sheppard argue in Chapter 5 that, in the same way that feminism focuses on making women's lives better, the root of Marxism is to bring about 'more just conditions for human flourishing'. Marxism is largely focused on making sense of the geography of capitalism, and its singular intent is to establish spatial justice.

Henderson and Sheppard argue that although established some time ago, Marxist theory is far from anachronistic. That said, many practitioners prefer to call themselves post-Marxists. Henderson and Sheppard argue that many of the tenets of classical Marxism are now subsumed within other ways of knowing such as realism, structuration theory and poststructuralism (Chapters 8, 7 and 10). Further, they suggest that the appellation 'post' refers not to the demise of Marxism but rather to approaches that acknowledge the ongoing influences of a critical way of knowing.

In their attempts to explain the world, Marxists argue that the positivistic bases of science and social science reduce it to a series of stable and well-defined entities connected by causal relations. As an epistemological basis of Marxism, dialectical reasoning traces how the relations between things are constantly changing, altering the entities themselves. Henderson and Sheppard note that at all scales, from humans to the world economy, objects of analysis are internally heterogeneous and always at risk of being torn apart by the very relationships that bring them into being. They go on to note that dialectical reasoning of this kind is not alien to the sciences. The focus of Marxism, however, is primarily on production, consumption, value and exploitation, the accumulation and circulation of capital and class conflict, with attended concerns about political identities and culture industries. Because Marx was concerned with material change and the transformation of people (see here Chapter 25 where Samers looks at the legacy of Marxism for political activism within geography), his emphasis was the social aspects of individual being, Henderson and Sheppard make the claim that he is a contemporary thinker. They are concerned that geographers de-emphasize Marx's focus on the formation of capital and class by looking primarily at gender, race and sexuality. And so this chapter raises conflicts with Chapter 4. Henderson and Sheppard argue that geographers need to understand, first and foremost, the intersections between social processes as articulated through the circulation of capital and the social construction of commodity production and distribution.

Constructing Geographical Knowledge in Relation to the World

Behavioural research is tied in part to positivism but, as Reginald Golledge notes in Chapter 6, it aligns itself to a much larger set of philosophical underpinnings, including phenomenology, symbolic interactionism, postmodernism and transactionalism. As a spatial science, the intent of behavioural research is to seek process explanations for why specific spatial actions are undertaken. The behavioural processes of interest are, for example, perception, learning, forming attitudes, and memorizing. Critical of some of the models and data used by early quantifiers, behavioural geographers today expand what they see as the limited purview of normative science by using qualitative data and exploratory and experimental data analysis techniques, concentrating on individuals and primary data rather than aggregate analysis of secondary data.

Focusing on pithy philosophical questions such as 'What is reality?' and 'What reality is relevant?', Golledge argues that behavioural research in geography challenges some major conceptions of spatial science. For example, by introducing the dimension of time to individuals' spatial activities a more nuanced articulation of behaviour is possible. In practice, this research does not abandon science and positivism because it uses rigorous experimental design to elaborate on people's knowledge, perceptions, and actions. Behavioural research aligns with the basis of positivism articulated in Chapter 2 and with the practice of science articulated in Chapters 22 and 23. Golledge argues that behavioural research tackles age-old questions about what is geographical knowledge and how it is constructed. Although he acknowledges the connections between people and knowledge, he argues that through systematic research we can know something about the processes through which that knowledge is elaborated.

According to Isabel Dyck and Robin Kearns (Chapter 7), structuration theory, as constructed by Anthony Giddens, elaborates a bridging point between humanism and Marxism, but it is also related to the behavioural perspective elaborated through time geography. Giddens' intent is to unravel the complexities between human agency (see, for example, Chapter 24) and structural constraints such as those elaborated by Marxism

(Chapter 5). Time geography points to the possibility of larger societal structures (known as constraints) within which individual behaviours are articulated. Structuration theory constructs society as neither independent of human activities nor their product. Rather, the interdependence of structure and agency is derived through space and time connections. Structure is seen as rules and resources such as those elaborated through institutions; human agency is based on the idea that individuals are perpetrators of events and that there is always choice. Human interaction with institutions is always in space and time and so there is always a spatial and temporal dimension to how social systems change. Dyck and Kearns elaborate structuration theory with relatively straightforward examples involving mothering and health geography. They note that poststructural concerns with difference and identity have tended to divert attention away from the rules and resources through which structure is realized. Argued here is a need to engage with structuration theory as a way of elaborating power relations that stretch over time and space globally.

In Chapter 8, Andrew Sayer takes a slightly different view on how knowledge is constructed. The philosophy of realism, he argues, is not about constructing some basis for reality but rather about assuming that the world is always there and always different from the way philosophers and theorists attempt to get to know it. In short – and this is probably the biggest difference between realism and the behavioural research articulated in Chapter 6 – Sayer argues that the world cannot be got at by systematically researching the processes of knowledge formation. He would suggest further that the views expressed in Chapters 2 and 6 have difficulty entertaining the idea that knowledge can be fallible. Sayer sees social constructions, like those developed in Chapter 5, as wishful thinking unless we are willing to accept that they may not work or make a difference. Realism, then, is about understanding, first and foremost, that the world is mostly independent of our thought processes; and although those thought processes and how they construct the world are important, they are not omnipotent.

Beyond Structure

At some point in the not too distant past, postmodernist geographers argue, the known and structured bases of modernity began to unravel. This might have happened with Auschwitz as suggested by Theodor Adorno or with the destruction of Pruitt–Igoe (the public housing experiment in racial differentiation in St Louis) as suggested by David Harvey. In Chapter 9, David Clarke argues that postmodernism throws into doubt reason as a monolithic and driving concept of western society. Reason, facts and science are not pillars of understanding but simply matters of faith like everything else. Stories are just stories that help us along in the world, and not metanarratives (big stories) with claims to truth and authentic reality (e.g. modern medicine). Metanarratives, argue some postmodernists, are invoked simply as a way for science to legitimate itself because of the belief that universal knowledge is possible and that it grants privileged access to truth. Instead, postmodernists argue, the world is inherently complex, confusing, contradictory, ironic and so forth, and they want to keep it that way.

Geographers have had a lot to say about postmodern spaces; indeed, Clarke avers that we should be quite proud of the confusion we've instilled in the larger academic debates on the relations between space and time. David Harvey, Ed Soja and Mike Dear have been particularly successful in confusing traditional understandings of urban space. Clarke notes further that David Harvey's *The Condition of Postmodernity* (1989) is one of the most famous books on postmodernity. Ironically, a Marxist metanarrative works well for Harvey's analysis and, as is pointed out by Henderson and Sheppard (Chapter 5) and Clarke, this way of knowing seems capable of staving off more contemporary discourses that seek to undermine its credibility.

The problem with postmodernism, notes Clarke, is that it tends to delegitimize social critique and progressive political activism more than neoliberal capitalism and neoconservatism. And so, perhaps, the postmodern moment may be, in actuality, a further engagement with capitalism rather than something radical and transformative. These latter affectations are perhaps better laid at the feet of other 'poststructuralists'.

Paul Harrison in Chapter 10 and John Wylie in Chapter 27 argue that we need to divest ourselves of the constraints of any kind of structural understanding. The problem with postmodernism is that it also divests itself from critical and radical perspectives in favour of whimsy and planned depthlessness. Poststructuralist theories, and particularly those derived from a wider continental tradition (e.g. Foucault and Derrida), move human geography forward because they are primarily critical methods for assessing – in deep, archaeological ways – the insidious power relations embedded in institutions, beliefs and political arrangements. Harrison argues further that poststructuralism is more radical than ways of knowing based on Marxism (such as feminism, realism and structuration theory) because it does not offer a base or a focused intention. And it does not position radicalism as a metanarrative.

Poststructuralism differs from preceding ways of knowing in other, interesting ways. In this introduction, we suggest that Chapters 2, 3, 4 and 5 offer singular intentions and Chapters 6, 7 and 8 offer a singular base. Poststructuralism, according to Harrison, offers neither a particular diagnosis of how the world is organized nor a systematic alternative. Poststructuralism is anti-essentialist and, as such, it offers a particular concern with radical otherness and difference. Moreover, while poststructuralism is often characterized in terms of the need to be endlessly critical and therefore overly negative, Harrison is at pains to emphasize the affirmative nature of poststructuralism and its deeply ethical nature. Here, he highlights the writing of Derrida on deconstruction. (Later in Chapter 27 Wylie demonstrates the importance of Derrida's notion of deconstruction as well as Foucault's articulation of discourse analysis as forms of geographical practice.)

Actor-network theory (ANT) shares the anti-essentialist approach of poststructuralism. Like poststructuralism it aims to understand the complexity of the world. In Chapter 11, Fernando Bosco shows how ANT provides a framework for tracing connections and relations between a variety of actors – both human and non-human (discursive and material objects) – in which geographies are understood to emerge from, or are the effects of, these relations. Here ANT offers a radical reading of agency, seeing it not as the property of intentional human actors (contrast Chapter 3) but rather as the product of things coming together, such that non-human actors (like a fencepost or a pen) might be understood to have agency. Bosco draws attention to the way that ANT uses the term 'actant', rather than 'actor', to distinguish the fact that it attributes no special motivation to individual human actors. Like other relational approaches, ANT is an attempt to escape dualisms such as structure–agency which so preoccupy Dyck and Kearns in Chapter 7, and it is not limited by the Euclidean understandings of space that underpin positivistic approaches (Chapter 2).

Postcolonialism, like poststructuralism (Chapter 10), is concerned with the complexities of identity, difference and representation. In Chapter 12 Barnett identifies postcolonialism's origins in the writings of intellectuals in the mid twentieth century – the time of anti-colonial struggles against European domination. He argues that relations of colonial subordination are embedded in systems of identity and representation. Here, like Harrison (Chapter 10) and Wylie (Chapter 27), he draws on Foucault's notion of discourse to explain the power of cultural representations. From a Foucauldian perspective, postcolonialism does not adhere to any singular way of structuring ways of knowing.

Barnett is at pains to stress the intertwined nature of western and non-western histories and societies. He dwells at length on Said's book *Orientalism* – which demonstrates how western notions of identity, culture and civilization have drawn on imaginings of the non-west or the 'Orient' in a cultural process of 'othering' – to address the relational constitution of representations and identities. He then reflects on some of the broader moral and philosophical concerns raised by postcolonialism in relation to universalism, cultural relativism, and cross-cultural understanding/representation. Like poststructuralism, postcolonialism problematizes textual practices, such as reading, writing and interpreting, and the ways that textual meanings are produced.

In many ways, Chapters 9, 10, 11 and 12 might be thought of as articulating the death of philosophy. At the very least they suggest a deterritorialization that makes a mockery of the ways knowledge is parsed out in each of the chapters in this part. Whether you are comfortable going this far or not, it is clear that no one way of knowing has any more or less legitimacy than any other, that many talk past each other as much as they align themselves, that each competes in academia and in society to say something of worth, and that there is significant tension and conflict that is not just based on the logic of a philosophical argument.

2 POSITIVISTIC GEOGRAPHIES AND SPATIAL SCIENCE

Rob Kitchin

Introducing Positivism

Positivism is a set of philosophical approaches that seeks to apply scientific principles and methods, drawn from the natural and hard sciences, to social phenomena in order to explain them. Auguste Comte (1798–1857) is widely acknowledged as the father of positivism. He argued that social research prior to the nineteenth century was speculative, emotive and romantic and as a result it lacked rigour and analytical reasoning. Unwin (1992) details that Comte used the term 'positive' to prioritize the actual, the certain, the exact, the useful, the organic, and the relative. In other words, he posited that it is more useful to concentrate on facts and truths – real, empirically observable phenomena and their interrelationships – than on the imaginary, the speculative, the undecided, the imprecise. What Comte demanded was the objective collection of data through common methods of observation (that could be replicated) and the formulation of theories which could be tested (rather than as with empiricism where observations are presented as fact). Such testing would be systematic and rigorous and would seek to develop laws that would explain and predict human behaviour. As such, Comte rejected metaphysical (concerned with meanings, beliefs and experiences) and normative (ethical and moral) questions as they could not be answered scientifically. Like with most other 'isms' and 'ologies' there are various different forms of positivism. The two most commonly discussed are *logical positivism* based on verification and *critical rationalism* based on falsification.

Logical positivism was developed by the Vienna Circle (a loose collection of social scientists and philosophers) in the 1920s and 1930s. Like Comte, they posited that the scientific method used in the traditional sciences could be applied directly to social issues – that is, social behaviour could be measured, modelled and explained through the development of scientific laws in the same way that natural phenomena are examined. Such a view is called naturalism and is underpinned by a set of six assumptions as detailed by Johnston (1986: 27–8):

'1 That events which occur within a society, or which involve human decision-making, have a determinate cause that is identifiable and verifiable.
2 That decision-making is the result of the operation of a set of laws, to which individuals conform.
3 That there is an objective world, compromising individual behaviour and that the results of that behaviour which can be observed and recorded in an objective manner, on universally agreed criteria.
4 That scientists are disinterested observers, able to stand outside their subject matter and observe and record its features in a neutral way, without in any respect changing those features by their procedures, and able to reach dispassionate conclusions about it, which can be verified by other observers.

5 That, as in the study of inanimate matter, there is a structure to human society (an organic whole) which changes in determinate ways, according to the observable laws.
6 That the application of laws and theories of positivist social science can be used to alter societies, again in determinate ways, either by changing the laws which operate in particular circumstances or by changing the circumstances in which the laws will operate.'

The Vienna Circle significantly extended Comte's work, however, by formulating rigorous analytical procedures centred on verification. As such, they sought to define precise scientific principles and methods by which social behaviour could be measured and social laws verified (the extent to which scientific theories explained objective reality). The mode of measurement they advocated was one centred on the precise quantitative measurement of facts (e.g. heights, weights, time, distance, wage). These measurements allowed the statistical testing of relationships between variables as a means to test (verify) explanatory laws. Because the method focuses on known facts that are easily collected across large populations (e.g. using the census) it is possible to test and verify laws against very large sample sizes. Here, a deductive approach is employed, wherein a theory is formulated and hypotheses are set and then tested. In cases where the data do not support the hypotheses, the theory can be modified, new hypotheses set, and the data reanalysed. A cumulative process is thus adopted, wherein theories are extended and built up in a structured and systematic manner through the incorporation of new findings and the rejection and resetting of hypotheses. Given that samples are often not perfect, complete verification is understood to be impossible, and logical positivism thus deals with weakly verified statements understood in terms of probabilities (the statistical likelihood of occurrence) that it aims to strengthen (Johnston, 1986). By increasing the strength of the probability that a relationship did not occur by chance and is potentially causal, hypotheses can be tested and theories deductively constructed. In this way, logical positivism provides a method for gaining objective knowledge about the world. Objectivity through the independence of scientists is maintained through conformity to the following five premises (Mulkay, 1975, cited in Johnston, 1986: 17–18):

'1 *Originality* – their aim is to advance knowledge by the discovery of new knowledge.
2 *Communality* – all knowledge is shared, with its provenance fully recognized.
3 *Disinterestedness* – scientists are interested in knowledge for its own sake, and their only reward is the satisfaction that they have advanced understanding.
4 *Universalism* – judgements are on academic grounds only, and incorporate no reflections on the individuals concerned.
5 *Organized scepticism* – knowledge is advanced by constructive criticism.'

In contrast to Comte, the Vienna Circle accepted that some statements could be verified without recourse to experience, making a distinction between analytical statements and synthetic statements. Analytical statements are *a priori* propositions whose truth is guaranteed by their internal definitions (Gregory, 1986a). Such analytical statements are common in the formal sciences and mathematics, where questions are often solved in a purely theoretic form long before they can be empirically tested. Indeed, theoretical physics almost exclusively seeks to provide solutions (based on known laws and properties) to problems that remain impossible to empirically test (see for example Steve Hawking's *A Brief History of Time*). Synthetic statements are propositions whose truth needs to be established through empirical testing because they lack internal definition and are complex. In addition, the Vienna Circle forwarded scientism (that is the claim that the positivist method is the only valid and reliable way of obtaining knowledge, and all other methods are meaningless because they do not produce knowledge that can be verified) and a narrowly defined scientific politics that argued that positivism offers the only means of providing rational solutions to all problems (Johnston, 1986).

Critical rationalism was developed in response to logical positivism and challenges its focus on verification. Forwarded by Karl Popper, it contends that the truth of a law depends not on the number of times it is experimentally observed or verified, but rather on whether it can be falsified (Chalmers, 1982). Here it is argued that rather than trying to provide a weight of confirmatory evidence, scientific validation should proceed by identifying exceptions that undermine a theory. If no exceptions can be found then a theory can be said to have been corroborated. The critique of such an approach is that a theory can never be fully validated as a yet unidentified exception might still be awaiting discovery (Gregory, 1986b). While many geographers would profess to adopting a critical rationalist approach based on falsification, in practice they tend to employ verification, seeking to explain away exceptions or residuals by recourse to statistics based on probability, rather than rejecting a hypothesis outright. A variety of other versions of positivism have been proposed and contemporary positivist philosophy significantly extends the work of the Vienna Circle. That said, debates in geography draw on these older forms of positivism, mainly because positivist geography itself rarely engages in any deep or meaningful dialogue with philosophy and as such its underpinnings have not been advanced with regard to new forms of positivism.

Development and Use of Positivism in Human Geography

> **Positivism is one of the unrecognised, 'hidden' philosophical perspectives which guides the work of many geographers ... [It remains hidden] in the sense that those who adhere to many of its central tenets rarely describe themselves as positivists ... While many boldly carry the banner of their chosen philosophy, the name of positivism is rarely seen or heard in the works of geographers who give assent to its basic principles. (Hill, 1981: 43)**

Until the 1950s, geography as a discipline was essentially descriptive in nature, examining patterns and processes, often on a regional basis, in order to try to understand particular places. From the early 1950s, a number of geographers started to argue that geographical research needed to become more scientific in its method, seeking the underlying laws that explained spatial patterns and processes. For example, Frederick Schaefer, in a paper often cited as the key catalyst for the adoption of scientific method in human geography, argued that 'geography has to be conceived as the science concerned with the formulation of the laws governing the spatial distribution of certain features on the surface of the earth' (1953: 227). In effect Schaefer drew on the arguments of logical positivism to contend that geography should seek to identify laws, challenging the exceptionalist claims of geographers such as Hartshorne (1939) that geography and its method was unique within the social sciences. In other words, geography should shift from an ideographic discipline (fact gathering) focusing on regions and places to a nomothetic (law producing) science focused on spatial arrangement.

The principal concern of the early advocates of geography as a *spatial science* was that geographical enquiry up to that point was largely unsystematic and analytically naive. Geographers were developing empiricist accounts of the world by simply accumulating facts as evidence for generalist theories. The problem with such empiricist endeavours was that they did not distinguish between causal correlations and accidental or spurious (non-causal) associations. For example, environmental determinist accounts suggested that environmental conditions explicitly influenced society in a causal fashion (e.g. high ambient temperatures caused underdevelopment in tropical countries by inducing idleness among local residents) (Hubbard et al., 2002). Moreover, such accounts committed ecological fallacies, that is ascribing aggregate observations to all cases within an area. However, just because two things are observed in the same place at the same time does not mean that one caused the other or that they apply universally. These patterns need to be tested scientifically. Indeed, most people now accept that ambient temperature may influence human behaviour, but it does not determine it, and it has little or no effect on levels of development. For geographers such as Schaefer, geography as a discipline would gain real utility, and by association respectability within the academy, only if it became more scientific. Scientific method would provide validity and credibility to geographic study and it would provide a shared 'language' for uniting human and physical geography.

Quantitative revolution

What followed was the so-called quantitative revolution wherein the underlying principles and practices of geography were transformed (Burton, 1963), with description replaced with explanation, individual understandings with general laws, and interpretation with prediction (Unwin, 1992). In order to employ a scientific method, to transform human geography into a scientific discipline concerned with the identification of geographical laws, a number of geographers started to use statistical techniques (particularly inferential statistics, concerned with measuring the probability of a relationship occurring by chance) to analyse *quantitative* data. Quantitative data were seen as factual, as objectively and systematically measured. They were therefore universal in nature, free of the subjective bias of the measurer and analyst. By statistically analysing and modelling these data, geographers hoped to be able to identify universal laws that would explain spatial patterns and processes, and also provide a basis for predicting future patterns and identifying ways to intervene constructively in the world (e.g. altering policy to engender change). So, just as physics and chemistry tried to determine the general laws of the physical world, geographers adopted a naturalist position (a belief in the equivalence of method between social and natural sciences) to try to determine the spatial laws of human activity.

This transformation in theory and praxis led to a whole variety of different types of laws, most of which did not pretend to be the universal law as portrayed by many critics. For example, Golledge and Amedeo (1968, summarized in Johnston, 1991: 76) detailed four types of law being developed in human geography: '*Cross-sectional laws* describe functional relationships (as between two maps) but show no causal connection, although they may suggest one. *Equilibrium laws* state what will be observed if certain criteria are met. … *Dynamic laws* incorporate notions of change, with the alteration of one variable being followed by (and perhaps causing) an alteration in another … Finally *statistical laws* … are probability statements of B happening, given that A exists' (the first three laws might be deterministic or statistical).

The aim, in short, was to create a scientific geography, with standards of precision, rigour and accuracy equivalent to other sciences (Wilson, 1972). However, as Hill (1981) notes, given that spatial science borrowed the idea of

BOX 2.1 SPATIAL MODELS AND LAWS

Throughout the late 1950s and the 1960s a whole plethora of geographical models and laws, based on scientific analysis of quantitative data and taking the form of mathematical formulae, were developed using a hypodeductive approach. For example, early quantitative geographers tried to find a formula that adequately modelled the interaction of people between places. One of these was Isard et al.'s (1960, detailed in Haggett, 1965: 40) inverse-distance gravity model:

$$M_{ij} = (P_j / d_{ij})\ f(Z_i)$$

where M_{ij} is the interaction between centres i and j, P_j is a measure of the mass of centre j, d_{ij} is a measure of the distance separating i and j, and $f(Z_i)$ is a function of Z_i, where Z_i measures the attractive force of destination i. This advanced earlier models that did not take into account how 'attractive' each location might be in relation to others (for example, in climate or amenities).

scientific method largely without conscious reflection on its philosophical underpinnings it is perhaps better to term it positivistic rather than positivist. Certainly, many positivistic geographers (most of whom would prefer to adopt the label quantitative or statistical geographers) would balk at the scientism and scientific politics of logical positivism, though they would see the scientific method as the most sensible and robust (rather than the only) approach to geographical enquiry (see also Chapter 22).

As with all 'revolutions', certain key sites and people were instrumental in pushing and developing the emerging quantitative geography. In the US, geographers such as William Garrison at Washington State, Harold McCarty at Iowa State and A.H. Robinson at Wisconsin trained a generation of graduate students who became faculty elsewhere, where in turn they propagated their ideas (Johnston, 1991). In the UK, Peter Haggett at Bristol and later Cambridge was a key influence (along with physical geographer Richard Chorley). Indeed Haggett's book *Locational Analysis in Human Geography* (1965) was an important text that helped to strengthen the case for quantitative geography. Such was the pace of adoption that by 1963 Burton had already declared that the revolution was over and quantitative geography was now part of the mainstream. That said, it is important to note that not all geographers were enthusiastic converts to what was increasingly called spatial science, and many continued to practise and teach other forms of geographical enquiry (Johnston, 1991; Hubbard et al., 2002). Nonetheless, the quantitative turn, and its conception of space as a geometrical surface on which human relationships are organized and played out, did change how many of these geographers conceived the notions of space and place.

Harvey's explanation in geography

Despite the rapid growth of quantitative geography throughout the 1960s, as noted, it largely operated in a philosophical vacuum: it focused on methodological form, not the deeper epistemological structure of knowledge production (Gregory, 1978). David Harvey's book *Explanation in Geography*

(1969) was a milestone text for the discipline. Harvey's key observation was that until that point geographers had rarely examined questions of how and why geographical knowledge was produced. And no one had tried to forward a robust and theoretically rigorous methodological (rather than philosophical) base for the discipline. Harvey's text thus sought to provide such a base by explicitly acknowledging the importance of philosophy to geographical enquiry. In particular he drew on the philosophy of science (which can effectively be translated as positivism despite the fact that Harvey never uses the term) to construct a theoretically sound ontology and epistemology – presented as a coherent scientific methodology. As Harvey (1973) himself later acknowledged (see Chapter 9), however, wider philosophical issues were skirted as his aim was to concentrate on formalizing methodology using philosophy rather than philosophy *per se*.

Spatial science as implicit positivism

While Harvey's text was enormously influential, providing an initial, theoretically robust ontological and epistemological base for spatial science, it is fair to say that most geographers employing the scientific method have subsequently paid little attention to its philosophical underpinnings. As Hill (1981) notes, positivism *implicitly* underpins much spatial science work, in that while research seeks to determine casual relationships and spatial laws through statistical analysis and geographical modelling, there is little explicit appreciation or engagement with positivism or other philosophies. As such, while there is the adoption of a scientific method and the use of terms such as law, model, theory and hypothesis, these are often used without an appreciation of what they actually mean or constitute (Hill, 1981; Johnston, 1986). Such research forms a major part of the discipline today, despite criticisms levelled at its positivistic underpinnings. For example, nearly all GIS and geocomputational research is practised as spatial science, although it is fair to say that much of it has actually continued the tradition of empiricism; wherein facts are allowed to 'speak for themselves' and are not subject to the rigours of spatial analysis through statistical testing (for example, most mapping work where the maps are allowed to speak for themselves); it is also increasingly rare to see hypotheses stated and then tested. This is not to say that all *quantitative* geography is implicitly positivist (or empiricist). In fact much is not. Indeed quantitative geography refers to the geographical inquiry that uses quantitative data, and such data can be interrogated from a number of ontological and epistemological positions (it is important never to conflate data type with a philosophical approach).

Criticism and Challenges to Positivist Geography

The period of transformation in geography's method opened the way for a sustained period of reflection on the ontology, epistemology and ideology of geographical inquiry from the late 1960s onwards. This coincided with a period of large social unrest in many western countries when many geographers were questioning the relevance and usefulness of the discipline for engaging with and providing practical and political solutions. Consequently, numerous geographers started to question the use and appropriateness of the scientific method and its new, philosophical base of positivism from a number of perspectives. It is important to note here that many of these critiques were not of using and analysing quantitative data *per se*, but rather of the positivist approach to analysing such data; it was a critique of ontology, epistemology, method and ideology, not data type.

The critiques of positivistic geography came from many quarters. For some, such as

Robert Sack (1980), positivistic geography was a form of spatial fetishism, focusing on the spatial at the expense of everything else. Spatial science represented a spatial separatist position, decoupling space from time and matter, which he argued meant that it had little analytical value: determining spatial patterns would not tell us why such patterns exist or why they might change over time because the approach fails to take account of social and political process.

Marxist and radical critiques developed the latter point. By rejecting issues such as politics and religion and trying to explain the world through observable facts, radical critics noted that spatial science was limited to certain kinds of questions and was further limited in its ability to answer them. It treated people as if they were all rational beings devoid of irrationality, ideology and history, who made sensible and logical decisions. It therefore modelled the world on the basis that people live or locate their factories and so on in places that minimize or maximize certain economic or social benefits. Critics argued that individuals and society are much more complex, with this complexity being impossible to capture in simple models and laws. As a consequence, Harvey, in a noted turnaround, condemned positivistic geography just a few years after writing its 'bible': there is 'a clear disparity between the sophisticated theoretical and methodological framework we are using and our ability to say anything meaningful about events as they unfold around us' (1973: 128). For Harvey, spatial science could say little about issues such as class divisions, Third World debt, geopolitical tensions and ecological problems because it was incapable of asking and answering the questions needed to interrogate them. Moreover, it was noted that positivistic geography lacked a normative function in that it could seek to detail what is and forecast what will be, but could give no insight into what should be (Chisholm, 1971). For Harvey and others, the only way to address such issues was to turn to radical theories such as Marxism which sought to uncover the capitalist structures that underpinned social and economic inequalities and regulated everyday life, and to transform such structures into a more emancipatory system.

Accompanying these radical critiques, from the early 1970s humanist geographers (see also Chapter 3) similarly attacked positivism with regards to its propensity to reduce people to abstract, rational subjects and its rejection of metaphysical questions (Buttimer, 1976; Guelke, 1974; Tuan, 1976). In effect it was argued that spatial science was peopleless in the sense that it did not acknowledge people's beliefs, values, opinions, feelings and so on, and their role in shaping everyday geographies. Clearly, individuals are complex beings that do not necessarily behave in ways that are easy to model. Humanistic geographers thus proposed the adoption of geographical enquiry that was sensitive to capturing the complex lives of people through in-depth, qualitative studies.

In addition, both radical and humanist critics questioned the extent to which spatial scientists are objective and neutral observers of the world, contending that it is impossible (and in the case of radicals undesirable) to occupy such a position. Geographers, it was argued, are participants in the world, with their own personal views and politics, not privileged observers who could shed these values while undertaking their research (Gregory, 1978). At the very least, researchers make decisions over what they study and the questions they wish to ask, and these are not value-free choices.

This argument was supplemented by feminist geographers such as Domosh (1991), Rose (1993) and McDowell (1992) who argued that spatial science was underpinned by a masculinist rationality (see also Chapter 4). That is, positivism was defined by man's quest for a god's-eye view of the world, one which was universal, 'orderly, rational, quantifiable, predictable, abstract, and theoretical' (Stanley and Wise, 1993: 66) and in which the knower 'can separate himself from his body, emotions, values, past and so on, so that he and his thought are autonomous, context-free and objective' (Rose, 1993: 7). They argued that geographical enquiry had to reject such rationality and become much more sensitive to power relations within the research process, and the geographers had to be more self-reflexive of their positionality, supposed expertise, and influence on the production of knowledge. In other words, geographers had to give up the pretence that they could necessarily create a master, universal knowledge of the world and accept that knowledge will always be partial and situated (from a certain perspective). What this meant in practice was that feminist geographers largely dismissed quantitative geography as a viable means of feminist praxis.

In turn, this feminist critique opened the door to a wider debate on the relationship between feminism, epistemology and spatial science in a special forum of *Professional Geographer* (1994: 'Should Women Count?'), that in turn helped (alongside texts such as Pickles, 1995) to fuel the development of critical approaches to GIS in the late 1990s and early 2000s. Critical GIScience draws off feminist, postmodern and poststructuralist theories to rethink the *modus operandi* of spatial science (see Curry, 1998; Kwan, 2002; Harvey, 2003). In many senses it is an attempt to reposition quantitative geography by providing it with a radically different philosophical framework from positivism, one that is more contemporary and robust to traditional criticisms of spatial science and that enables it to address questions that previously it avoided or was unable to tackle.

Positivist Geography Today

Despite the criticism levelled at geographical work underpinned by positivist reasoning, implicit positivism remains strong within human geography. A very large number of geographers argue that they are scientists, employ scientific principles and reasoning, and seek laws or mathematical models that purport to explain the geographical world. However, few seemingly give much thought to the philosophical underpinnings of their scientific method or philosophical debate and critique in general. This leaves much spatial science (and by association, quantitative geography) with relatively weak and unstable philosophical underpinnings (much of it backsliding into empiricism) and vulnerable to theoretic critique and challenge for which it has little response. This is not to say that all spatial science lacks theory; rather it lacks a fundamental and robust ontological, epistemological and ideological base. It also does not mean that spatial science is not useful or valuable within certain limited parameters. The work of spatial scientists clearly does have utility in addressing both fundamental scientific questions and 'real-world' practical problems and therefore has academic merit and worth (and it most definitively has utility in the eyes of policy-makers and businesses). However, by ignoring wider philosophical debate spatial scientists often fail to make a robust case for their approach to fellow geographers. As a consequence, many are seduced by the criticisms levelled at positivism and quantification more broadly, and become suspicious and wary of such research. Rather than tackle these criticisms, spatial science increasingly relies on the commercial and policy cache of GIS to make implicitly positivistic geography sustainable. As the debates in GIScience illustrate, however, the implicit positivism underpinning GIS use is open to challenge, with an acknowledgement that the employment of the scientific method can be practised from more critical perspectives. What might usefully transpire in the long term then is the development of spatial science underpinned by more critical philosophies, with a move away from or reworking of implicit positivism. That said, given the demand for GIS and quantitative geography in the public and private sector, it is likely that unreconstructed, positivistic geography is secure for the foreseeable future.

References

Burton, I. (1963) 'The quantitative revolution and theoretical geography', *The Canadian Geographer*, 7: 151–62.

Buttimer, A. (1976) 'Grasping the dynamism of lifeworld', *Annals of the Association of American Geographers*, 66: 277–92.

Chalmers, A.F. (1982) *What Is This Thing Called Science?* St Lucia: University of Queensland Press.

Chisholm, M. (1971) 'In search of a basis of location theory: micro-economics or welfare economics', *Progress in Geography*, 3: 111–33.

Curry, M. (1998) *Digital Places: Living with Geographic Information Technologies.* New York: Routledge.

Domosh, M. (1991) 'Towards a feminist historiography of geography', *Transactions, Institute of British Geographers*, 16: 95–115.

Golledge, R. and Amedeo, D. (1968) 'On laws in geography', *Annals of the Association of American Geographers*, 58: 760–74.

Gregory, D. (1978) *Ideology, Science and Human Geography.* London: Hutchinson.

Gregory, D. (1986a) 'Positivism', in R.J. Johnston, D. Gregory and D.M. Smith (eds), *The Dictionary of Human Geography*, 2nd edn. Oxford: Blackwell.

Gregory, D. (1986b) 'Critical rationalism', in R.J. Johnston, D. Gregory and D.M. Smith (eds), *The Dictionary of Human Geography*, 2nd edn. Oxford: Blackwell.

Guelke, L. (1974) 'An idealist alternative to human geography', *Annals of the Association of American Geographers*, 64: 193–202.

Haggett, P. (1965) *Locational Analysis in Human Geography.* London: Arnold.

Hartshorne, R. (1939) *The Nature of Geography.* Lancaster, PA: Association of American Geographers.

Harvey, D. (1969) *Explanation in Geography.* Oxford: Blackwell.

Harvey, D. (1973) *Social Justice and the City.* London: Arnold.

Harvey, F. (2003) 'Knowledge and geography's technology: politics, ontologies, representations in the changing ways we know', in K. Anderson, M. Domosh, S. Pile and N. Thrift (eds), *Handbook of Cultural Geography.* London: Sage, pp. 532–43.

Hill, M.R. (1981) 'Positivism: a "hidden" philosophy in geography', in M.E. Harvey and B.P. Holly (eds), *Themes in Geographic Thought.* London: Croom Helm, pp. 38–60.

Hubbard, P., Kitchin, R., Bartley, B. and Fuller, D. (2002) *Thinking Geographically: Space, Theory and Contemporary Human Geography.* London: Continuum.

Isard, E., Bramhall, R.D., Carrothers, G.A.P., Cumberland, J.H., Moses, L.N., Prices, D.O. and Schooler, E.W. (1960) *Methods of Regional Analysis: An Introduction to Regional Science.* Cambridge, MA: MIT Press.

Johnston, R.J. (1986) *Philosophy and Human Geography: An Introduction to Contemporary Approaches*, 2nd edn. London: Arnold.

Johnston, R.J. (1991) *Geography and Geographers: Anglo-American Geography since 1945*, 4th edn. London: Arnold.
Kwan, M.-P. (2002) 'Feminist visualization: re-envisioning GIS as a method in feminist geographic research', *Annals of the Association of American Geographers*, 92: 645–61.
McDowell, L. (1992) 'Doing gender: feminism, feminists and research methods in human geography', *Transactions, Institute of British Geographers*, 17: 399–416.
Mulkay, M.J. (1975) 'Three models of scientific development', *Sociological Review*, 23: 509–26.
Pickles, J. (ed.) (1995) *Ground Truth: The Social Implications of Geographic Information Systems*. New York: Guilford.
Rose, G. (1993) *Feminism and Geography*. Cambridge: Cambridge University Press.
Sack, R. (1980) *Conceptions of Space in Social Thought*. London: Macmillan.
Schaefer, F.K. (1953) 'Exceptionalism in geography: a methodological examination', *Annals of the Association of American Geographers*, 43: 226–49.
Stanley, L. and Wise, S. (1993) *Breaking Out Again: Feminist Ontology and Epistemology*. London: Routledge.
Tuan, Y.-F. (1976) 'Humanistic geography', *Annals of the Association of American Geographers*, 66: 266–76.
Unwin, T. (1992) *The Place of Geography*. Harlow: Longman.
Wilson, A.G. (1972) 'Theoretical geography', *Transactions, Institute of British Geographers*, 57: 31–44.

3 HUMANISM AND DEMOCRATIC PLACE-MAKING

J. Nicholas Entrikin and John H. Tepple

Introduction

Humanistic geographers study topics such as the cultural construction of place and landscape, the cartography of everyday life, the power of language and meaning to create and transform environments, place and identity, religious symbolism and landscape, and geographical myths and narratives (Ley and Samuels, 1978; Adams et al., 2001a). Common to all of these research interests is a concern with understanding meaningful, humanly authored worlds (Tuan, 1976). Beginning students in geography may recognize these topics as part of their course work and readings and wonder why they may never have been taught about humanistic geography. How could a geographical orientation that has been associated with so many themes of current interest be relegated primarily to discussions of the recent history of the field? How, for example, could it be so little noted in the many announcements of the so-called 'cultural turn' in contemporary geography? An intellectually satisfying answer would require a historical analysis that would be inconsistent with the forward-looking goals of this volume. However, discussion surrounding one issue, the nature of the geographical subject or agent, helps both to give insight into the recent disciplinary amnesia about humanistic geography and at the same time to recognize its currency. Before turning to this theme, it is first useful to provide some background.

Origins

The term 'humanistic geography' creates some confusion in that it has both a general and a specific meaning. In its current usage humanistic geography is typically associated with a specific intellectual orientation in geography in the 1970s and 1980s. Broadly understood, however, the same term may be applied to the relatively undernourished roots of modern human geography in the study of the humanities. Such a link has arguably existed since at least the time of Alexander von Humboldt in the early nineteenth century (Tuan, 2003; Bunkse, 1981). The primary sources of its recent reappearance were a series of influential articles by Yi-Fu Tuan (1976), Anne Buttimer (1974; 1976), and Edward Relph (1970; 1977) and a book, *Humanistic Geography: Prospects and Problems*, edited by David Ley and Marwyn Samuels (1978). Taken together these works offer a glimpse at what might be described as a highly varied and loosely structured movement in late-twentieth-century geography. It was held together more by a strong sense of what was being opposed than by what was being advocated. Its proponents shared a common vision of the narrowness of the positivistic human geography of the period. The spatial analytic tradition had emerged as a powerful force in the 1960s after the so-called 'quantitative revolution' and its only opposition was the diminished and intellectually aging legacy of a descriptive landscape and regional geography. Both of these research orientations generally lacked a robust concept of individual agency or an interest in how geography might contribute to an understanding of 'being-in-the-world' (Entrikin, 1976).

In the mathematical models of spatial analysis, rational actors made decisions based on perfect information in the non-environments of isotropic planes, homogeneous and featureless spaces in which the only variable was the relative distance between and among points or locations on the plane. Regional and landscape studies tended not to focus on individual actors, but rather to look for the sources of geographical variation in the ways of life of social and cultural groups. Framed in the scholarly and social context of the 1970s, humanistic geographers saw their work as opening the discipline to more realistic conceptions of humans as geographical agents. In this belief, humanists saw their efforts as intellectually liberating. They argued that agents draw on their experiences, attitudes, and beliefs, as well as their moral and aesthetic judgment, in making decisions that shape their environments. Geographical agents were not only economic actors seeking to enhance their material wellbeing, but also moral and cultural beings. Stated another way, one could say that humanistic geographers resisted the reductionist tendencies in human geography. They challenged what they viewed as an overemphasis on analytic simplicity that seemed to distance human geography from the creative and chaotic flux of everyday life. This laudatory goal, however, brought with it the difficult challenge of working with a complex conception of geographic agency. As David Ley wrote:

> **An aspiration of humanistic perspectives is to speak the language of human experience, to animate the city and its people, to present popular values as they intersect with the making, remaking and appropriation of place. (1989: 227)**

To achieve such a goal required an emphasis on the human subject as the creator and interpreter of meaning. This emphasis was evident in the semantic depth that humanist geographers gave to traditional concepts such as place, region, space, landscape, and nature and the extended reach of these concepts into studies of literature and art. Where once these concepts referred to an underlying world of natural and cultural elements, the humanists made visible their relation to human projects and the subjects who created them. Meaning is not something to be found in objects, but instead must be understood in relation to subjects. Thus place, region, and landscape are not simply spatial categories for organizing objects and events in the world, but rather processes in the ongoing dynamic of humans making the earth their home and creating worlds out of nature (Tuan, 1991b). It is for this reason that meaning, imagination, and human agency are so closely associated with humanistic geography. It is also for this reason that humanistic research and writing are often seen as being relatively subjective. The humanistic project in geography may be understood as part of what Denis Cosgrove describes in another context as an 'alternative geographical tradition', one associated with 'a more self-reflective *moral* project the primary goal of which is the wisdom … that comes from self-knowledge as an embodied being in the world, and wherein action is governed by the examined life, rather than a calculus of power' (2003: 867).

Humanistic geography was one of two major 1970s intellectual movements that grew out of a discontent with spatial analysis. The other was a concern with social relevance and politics that took its most coherent form in a structural neo-Marxism (Harvey, 1973; Duncan and Ley, 1982). Both of these challenges to the reigning orthodoxy in human geography sought to incorporate a more complex human reality than was evident in the models of spatial analysis.

Humanistic geography has been seen as the less theoretical and political of the two movements, a view that is not incorrect but that has often been overstated to the point that humanistic geography has been inaccurately labeled as being atheoretical and apolitical. Each of these alternative positions was attacked by positivistic geographers for undermining human geography's scientific credentials in their apparent move away from objective research methodologies toward more subjective and seemingly value-laden studies (see Chapter 2). The rigor of data collection, measurement, hypothesis testing, and model building seemed to be giving way to philosophical pronouncements and 'soft' studies of interpretation and meaning or to ideologically driven analyses.

Some of these criticisms were valid, but others reflected an underlying scientism, in which science is manipulated and transformed from a form of inquiry into a rhetorical weapon. Such criticisms rarely found their target, however, as it became clear that humanistic geography for the most part did not directly challenge spatial analysis, but rather encouraged an expansion of the scope of human geography (Bunkse, 1990). Humanists posed different types of questions that emphasized meaning, values, and interpretations. For example, humanists would argue that individuals choose pathways within the city not just in terms of distance, time, or cost minimization, but also because of the perception of danger, a sense of belonging, or aesthetic sensibilities. Regions are not only agglomerations of economic activities and the functional dependencies of employment opportunities and residential spaces, or even spatial categories of human activities, but also part of the identity of individuals and groups, part of how they see themselves in relation to others and how they give meaningful order to their experience of near and distant places. Geographical studies of migration are more than simply origin and destination studies of push–pull factors; they also include the experiences of attachment, dislocation, alienation, and exile that constitute the experiential reality associated with leaving one's home. Thus movements of people through space could also be seen in terms of the experience of place.

Critics found humanistic geography to be a diffuse target. For example, the humanistic geography described by Tuan (1976; 1986; 1989) emphasized the relatively neglected tie of geography to the humanities and the ideals of liberal education (Entrikin, 2001). His version of this movement consisted of a broadening of scholarly horizons of geography rather than a direct challenge to the intellectual legitimacy of spatial models. Tuan sought to expose geographers to different modes of experience and different ideas of 'the good life' and its ideal environments, believing that such exposure contributes to a deeper understanding of our own experiences and beliefs. His approach to geography was thus not an imperialist one of colonizing substantive areas of neighboring disciplines but rather one of adding dimensionality and depth to the traditional themes and concepts of geography. To understand fully the ways in which humans transform their environments, the geographer could not leave unexplored the meanings that cultures have given to nature, the values and goals that shape their actions in building places and landscapes, or their imaginative explorations of other possible environments.

Other geographers, such as David Ley (1977) and Edward Relph (1977), presented a somewhat different argument in support of a humanistic geography as an alternative to the prevailing naturalist model of social science, a model that emphasized the methodological unity of the natural and social sciences. In presenting a vision of geography as an interpretive social science, these humanists sought to open the discipline to the unique qualities of human beings as intentional agents who act in the world in relation to their projects. Proponents considered various philosophies that emphasized intersubjectivity, including phenomenology, existentialism, structuralism, and hermeneutics.

It is interesting to note in this context that geography has a humanist orientation while related human sciences such as sociology and anthropology have no such broadly applied label. This is no doubt related to the strong hold of materialist orientations in geography that retarded the development of interpretive traditions in the field (Lowenthal, 1961). Such traditions in other disciplines, for example, ethnomethodology, phenomenological sociology, symbolic interactionism, ethnography, and so on, all preceded their equivalents in geography. Geography had a late start in the study of meaning, and thus it is not surprising that all of the various forms of interpretive study were lumped under the broad conceptual umbrella of humanistic geography. The term 'humanistic geography' became the signature in geography for the study of meaning and experience and the move beyond the traditional concern with linking concepts to their referents toward an interest in relating meaning to subjects.

A less frequently cited third direction for this orientation emphasized the close connections of humanism to the historical perspective in geography. This linkage was made especially clear in Cole Harris' (1978) arguments concerning the historical mind and later supported by Denis Cosgrove (1989) and Stephen Daniels (1985). Daniels wrote that:

> If humanistic geography is to be more than a criticism of positivist geography it needs a more thoroughly reasoned philosophical base, a closer understanding of the conventions through which human meanings are expressed, a more adequate account of what humanistic methods are or might be, and above all a greater historical understanding. (1985: 155)

For Cosgrove,

> Any humanist endeavour is inevitably historical and, to a degree, reflexive for it concerns the nature and purposes of conscious human subjects together with their individual and collective biographies. (1989: 189)

These different directions indicate that humanistic geography was less a coherent strategy than a shared spirit of opening the field to alternative forms of analysis. Its initial formulations had a liberationist quality to them. In all of this variety certain common themes stood out, primarily the importance of the individual as an intentional agent, whose actions are shaped not only by material needs but also by a geographical imagination that included moral and aesthetic ideals. This fully dimensional geographical agent is the primary legacy of the humanistic movement of the 1970s. It would soon turn into the most visible target of those seeking to transcend this orientation.

Transcendence

The social scientific path of humanistic geography led to early conflicts with those who questioned the individualist and voluntaristic character of such studies. Surely, critics would argue, human beings are not free agents in the world but rather are constrained through economic, political, and cultural structures that limit the possibilities for choice and action. The life choices of an American academic living in Los Angeles are clearly very different from those of a newly arrived immigrant from Mexico working in the garment industry in downtown Los Angeles. Some sort of balance between constraint and choice is necessary, and humanist geographers and their critics contested this middle ground (Gregory, 1981; Duncan and Ley, 1982). This argument merged into the so-called 'structure–agency' debates of the 1980s, when alternatives such as structuration theory (see Chapter 7) and critical realism (see Chapter 8) claimed to offer positions that could incorporate the humanist's concern with agency with a theorized set of structural constraints.

Structuration theory emphasized the ways in which actions and practices interacted with structural constraints to both transform and reproduce social structures. Everyday habitual routines – such as going to work or to school, choosing where to live to be with others who share similar qualities and tastes, staying home to raise children – reproduce and reinforce structures of existing social relations. The actions that structurationists studied, however, tended to be the routine and the customary with very little attention paid to questions of meaningful or intentional action. Thus the seeming absorption of the humanist concern with agency into the more broadly conceived social theory of structuration in fact largely neglected this core humanist theme. Critical realism, a view about what is real (ontology) that is based upon an analytically prior commitment about how the world works (theory), also emphasized structures and largely ignored agency.

The structure–agency debates took on the appearance of both ends struggling to gain the middle ground but gradually became more and more one-sided as meaningful agency became an increasingly 'thin' theoretical concept. Thus, in spite of its name, the so-called structure–agency debate diminished rather than strengthened the role of human agency in geographical studies. The seamless transition made by many of the participants of this debate into support for theories of practice, whether derived from the work of Anthony Giddens, Pierre Bourdieu, or, more recently, the actor-network theory of Bruno Latour, contributed further to the demise of the intentional agent and the affective in geography. The intentional agent, so central to the humanist perspective in geography, is less transcended by the move toward theories of practice as it is submerged and eventually forgotten.

Humanism, Pre, Post, and Anti

Actor-network theory is unique among theories of practice in its denial of the intellectual separation between the human and the non-human (see Chapter 11). Latour challenges the legacy of humanism, which he associates with 'the free agent, the citizen builder of the Leviathan, the distressing visage of the human person, the other of a relationship, the consciousness, the cognito, the hermeneut, the inner self, the thee and thou of dialogue, presence to oneself, intersubjectivity' (1993: 136). All such agency-related concepts remain for Latour hopelessly asymmetrical and one-sided in giving value, importance, and causal power exclusively to humans and relegating the non-human to the inert world of things and objects. He claims that this division needs to be overcome in order to see more clearly the world as it presents itself to us in the form of networks, relations, and hybrids that cross the artificial boundaries drawn between culture and nature, the worlds of people and of things. Global warming, genetic engineering, and deforestation offer examples of such networks that cannot be understood properly within the confines of humanist orientations. His is a position that one might label as pre- or posthumanism, since for Latour (1993) humanism is an element of modernism, and his argument against both modernism and postmodernism is that we have never been modern.

From poststructuralism (see Chapter 10) came a more fundamental attack on humanism and the outlines of an anti-humanism. In an era in which geographers engage poststructuralism, postcolonialism, ecocentrism, and radical feminism, the intentional agent or subject of the humanist movement has been attacked as being a fiction, a construction used by those presenting false universal claims about humanity. These universal claims about the human condition were seen as the partial views of privileged cultural elites, defined variously by sex, gender, ethnicity, nationality and other differentiating qualities (Bondi, 1993; Rose, 1993). The alternative positions, many of which are discussed in detail in other parts of this volume, are large and diverse but they share an important topical concern with the construction of the subject. Their views of the subject are often opposed to a humanist subject which is described variously as Eurocentric, masculinist, racist, and so on, and associated with a hegemony that actively hides difference and silences the voices of the culturally less powerful and oppressed. The reflexive, centered subject and the individual intentional agent have been replaced by the decentered subject or subject position, given form and shape by social forces that underlie and produce the competing discourses of modern life. What is gained is a strong program for the socially constructed self, and what is lost is the autonomous, intentional agent.

This transformation is especially evident in discussions of place and space. For example, Robert Sack, in his book *Homo Geographicus* (1997), provides a theoretical scaffolding for a humanist conception of self and place. In his argument place and self are mutually constitutive. Each may be seen as influenced by forces of nature, society, and culture, but the self as autonomous agent is the core mechanism of place-making and in turn place facilitates and constrains the agent. The place-making actions of individuals and groups transform environments, from the simple act of consumption to the potentially more consequential and larger-scale acts of collective agencies, such as communities, corporations, and governments. The humanist geographer recognizes the socially or humanly constructed nature of these places but does not characterize this world solely in terms of impersonal forces of nature and society and the power of some groups to dominate all groups. Human agents are the primary place-makers, both as individuals and as members of collectives. To the extent that individuals are aware of their roles and responsibilities as place-makers, they are autonomous agents able to make moral decisions about the value of places in relation to the goals of human projects. Such agency neither presupposes a completely unconstrained, autonomous actor nor privileges consensus and cooperation over resistance and conflict.

Poststructuralism in geography shifted the concept of the individual subject from centered to decentered (Pile and Thrift, 1995; Pile, 1996). Indeed the active subject and subjectivity are transformed into subject positions, created by the contingent intersection of often conflicting and multiple discourses. Place is similarly the contingent intersection of multiple and sometimes conflicting social processes (Massey, 1994). Such views were often presented from those seeking to introduce a more explicit ethnic, gendered, or sexual vocabulary to geographical discourse and who saw humanistic geography as too prone to ignoring difference. This same anti-universalistic position was adopted through postcolonialism as a reflection of a Eurocentric position. The autonomous moral agent of the humanist was transformed into the socially constructed subject, adrift and rudderless in the discursive currents of power relations. Critics argued that humanism and its subject had been transcended by politically progressive alternatives. However, political progressiveness without moral agency and a sense of moral growth, so central to the humanistic project, is difficult to imagine let alone to measure or prove.

For the current critics of humanism, all is power, and humanistic geographers are frequently judged as being naive about this issue. However, power has not been ignored in humanist geography (Tuan, 1984). In making his case for an appreciation of Renaissance humanism, Cosgrove notes the fundamental ambiguity in humanism (and by implication, humanistic geography) between its liberationist and dominating character:

> It is of direct contemporary relevance because humanism has always been very closely aligned in practice with the exercise of power both over other human beings and over the natural world. In large measure geography has developed as one of the instruments of that power. At the same time, and herein lies the subtlety of the dialectic, both humanism and geography offer the opportunity to subvert existing power relations (1989: 190)

This opportunity depends in part upon resuscitating moral and political agency.

Possibilities

What is the current relevance of humanism in geography? As noted earlier, one understanding of humanism in geography refers to a very specific moment in the history of the discipline, a moment that has now passed. At that time, humanistic geography offered a critical voice against the narrowing of the field and an alternative vision of humans as complex intentional agents. As developments in other fields transformed geography, this once revolutionary insight became a taken-for-granted aspect of geographic research. The critical, liberationist voice of the humanist was muted by other voices that claimed the title of 'critical geography' (Adams et al., 2001b). For this group, humanists were insufficiently radical or were even conservative. The relatively atheoretical and apolitical quality of much of the work presented under the title of humanistic geography gave support to this view (Barnes and Gregory, 1997).

At the same time that humanistic geography was being challenged by its various critics, neo-Marxist geography was undergoing a transformation from a structural form, deeply antagonistic to the perceived idealism and subjectivism of humanistic geography, to a cultural Marxism. Although still hostile to idealism and subjectivism, many of its practitioners appeared to have incorporated the humanist concern with place and identity, symbolic landscapes, and the geography of everyday life into their own research (Harvey, 1996; 2000). Raymond Williams and Henri Lefebvre were cited as the intellectual ancestors of such a move away from earlier, more economistic and structural forms of analysis. One would expect to find a role for humanistic geography in this so-called cultural turn, but it is largely missing from this literature.

The current geographical research most closely and explicitly tied to the legacy of humanism is the work in moral geography and ethics (Sack, 2003; Proctor and Smith, 1999; Smith, 2000). In fact, humanists have long expressed interest in moral issues. The recent interest in ethics has been evident in articles, books, special sections of journals, and relatively new journals, such as *Ethics, Place and Environment* and *Philosophy and Geography*. It is within this emerging subfield that the humanist's concerns with the autonomous intentional agent and humans as the creators and interpreters of meaning are most evident. Like humanistic geography, moral geography is a broad field organized loosely around several themes, such as social and environmental justice, contextual ethics, professional ethics, and geographical understanding as a source of moral judgment. Two notable contributors, David Smith and Robert Sack, have sought to make geography fundamental to an understanding of modern ethics.

Smith (2000) presents a contextual ethics, which examines the influence of proximity, distance, and boundaries in an ethics of care. Sack (2003) looks at place-making as human projects guided by instrumental and intrinsic judgments of goodness. Each of their positions on moral geography requires a strong conception of the autonomous agent, one linked in varying degrees to community and culture, but with the capacity to make independent moral judgments. As Sack states, 'moral theory cannot force us to behave well' (2003: 31); it can only offer the tools for helping humans choose well, and one of those tools is geographic understanding. The work of Smith, Sack, and others in this area seemingly belies the claim of critics that humanistic geography is somehow antagonistic to theory.

Indeed, it is through this connection to moral theory that one can also reclaim the overt political aspects of humanistic geography. More specifically, humanism, and by implication humanistic geography, is closely tied to conceptions of liberal democracy. Tzvetan Todorov makes this point forcefully in his study of the origins of French humanism, *Imperfect Garden: The Legacy of Humanism,* where he writes that:

> Liberal democracy as it has been progressively constituted for two hundred years is the concrete political regime that corresponds most closely to the principles of humanism, because it adopts the ideas of collective autonomy (the sovereignty of the people), individual autonomy (the liberty of the individual), and universality (the equality of rights for all citizens). (2002: 31)

He cautions, however, that:

> Nonetheless, humanism and democracy do not coincide: first, because real democracies are far from perfect embodiments of humanist principles (one can endlessly criticize democratic reality in the name of its own ideal), then because the affinity between humanism and democracy is not a relationship of mutual implication exclusive of any other. (2002: 31)

The autonomous individual is central to humanism, but it is not an autonomy that ignores others as critics often assert or imply. This point is evident in Todorov's three non-reducible 'pillars' of humanism:

> the recognition of the equal dignity of all members of the species; the elevation of the particular human being other than me as the ultimate goal of my action; finally, the preference for the act freely chosen over one performed under constraint. (2002: 232)

When understood as a form of life, as opposed to a set of institutions, liberal democracy may be described as an ongoing project that, like all human projects, takes place. The basic political tension within this form of life is between the individual and the community, between the autonomy of the individual and the good of the community. Democratic communities face the continuous challenge of maintaining a healthy balance between the private and public spheres, of creating a viable civil society. Rules of place are used to maintain this distinction and to achieve this balance (Sack, 2003; Cresswell, 1996; 2004). The collective is constantly challenged to avoid the breakdown of the group caused by the retreat of individuals into the private sphere. Equally, a healthy democracy requires vigilance against the loss of the private sphere to the public and against the model of the people-as-one that characterizes totalitarian regimes. Thus moral theory in support of the democratic concerns the other-directedness of the individual whose identity is determined not only in the private sphere of family and personality formation but also in the engagement of others in the public realm. Humanistic geography as a form of moral education contributes to the goals of democratic community building by exposing individuals to other ways of life, different experiences, and different interpretations of experience. The moral horizons of individuals are expanded and their moral geographies are fused with the moral geographies of others.

Such fusion links the moral to the geographical imagination. Places are often imagined before they are realized (Tuan, 1990). Giving material form to imaginative worlds leads to a continuous process of transforming environments. The world may be characterized as a dynamic composite of places of differing scales that are being continuously made, unmade, and remade in relation to human projects. Places may be made unintentionally, through habit and custom, or intentionally through planning and forethought. They may be made undemocratically through fiat and the imposition of absolute power or they may be made democratically through collective discussion, planning, action, and participation of all those affected by communal decisions. They may be created through the official language of the state or through the alternative language of the dispossessed – a language that does not appear on maps but that constructs a geography of everyday life in marginalized communities. The most powerful tools that individuals have in creating places are language and imagination and these are part of the core of the humanist perspective in geography (Lowenthal, 1961; Tuan, 1991a; Wright, 1947).

The current fascination with the decentered subject emphasizes subject positions constructed through the intersection of social forces and the interplay of discourses. Power relations are their collective starting point, and themes of domination, oppression, transgression, and difference are woven into the traditional geographic concerns of place, space, and landscape. The politics is described as democratic but in an agonistic form, where the democratic is viewed as a continuous and unending process of conflict, in which consensus and community are viewed with mistrust as hidden forms of domination, of the tyranny of the powerful over the powerless (Entrikin, 2002a).

The humanist engages the same question of constructed geographies with a centered subject interpreting the world, a subject who acts as a relatively autonomous agent whose power rests primarily in language, the power to create meaning, and the ability to communicate and cooperate with others. Places are socially produced, not as the contingent intersection of social processes, but rather through the cooperative actions of individuals in communities. It is the geography created by meaningful action that is of greatest concern to the humanist and the one most evidently missing from many contemporary geographic practices.

In their inherent contingency, places as tools for human projects are a means of giving order and stability to the flux of experience. Places are necessary conditions of social existence, and the making and unmaking of places are essential elements in the struggles to give this existence specific form. They are the sites for recognizing difference as well as for forging the common bonds of humanity. They are the sites of goodness as well as evil. The mysteries that in part motivate humanistic geographers are not those related to individuals and groups competing over projects or that some groups seek to impose their will on others. Instead, they involve the recognition that place-making may also result from meaningful cooperation and that some projects may even be based on shared commitment to achieving some form of moral progress (Entrikin, 2002b). Todorov's description of the humanist project describes as well the geographic project of democratic place-making, which

> **could never bring itself to a halt. It rejects the dream of a paradise on earth, which would establish a definitive order. It envisages men in their current imperfection and does not imagine that this state of things can change; it accepts, with Montaigne, the idea that their garden remains forever imperfect. (2002: 236)**

Humanistic geography offers no specific design for an ideal cosmopolitan place, but it does offer a means of seeing the possibility of individual and collective moral progress toward the creation of more humane and democratic environments.

References

Adams, P.C., Hoelscher, S. and Till, K. (eds) (2001a) *Textures of Place: Exploring Humanist Geographies.* Minneapolis: University of Minnesota Press.

Adams, P.C., Hoelscher, S. and Till, K. (2001b) 'Place in context: rethinking humanist geographies', in P.C. Adams, S. Hoelscher and K. Till (eds), *Textures of Place: Exploring Humanist Geographies.* Minneapolis: University of Minnesota Press, pp. xii–xxxiii.

Barnes, T. and Gregory, D. (1997) 'Agents, subjects and human geography', in *Readings in Human Geography: The Poetics and Politics of Inquiry.* London: Arnold, pp. 356–63.

Bondi, L. (1993) 'Locating identity politics', in M. Keith and S. Pile (eds), *Place and the Politics of Identity.* London: Routledge, pp. 84–101.

Bunkse, E. (1981) 'Humboldt and an aesthetic tradition in geography', *Geographical Review*, 71: 127–46.

Bunkse, E. (1990) 'Saint-Exupéry's geography lesson: art and science in the cultivation of landscape values', *Annals of the Association of American Geographers*, 80: 96–108.

Buttimer, A. (1974) *Values in Human Geography.* Commission on College Geography 24. Washington: Association of American Geographers.

Buttimer, A. (1976) 'Grasping the dynamism of lifeworld', *Annals of the Association of American Geographers*, 66: 277–92.

Cosgrove, D. (1989) 'Historical considerations on humanism, historical materialism, and geography', in A. Kobayashi and S. Mackenzie (eds), *Remaking Human Geography.* Boston: Unwin, pp. 189–205.

Cosgrove, D. (2003) 'Globalism and tolerance in early modern geography', *Annals of the Association of American Geographers*, 93: 852–70.

Cresswell, T. (1996) *In Place/Out of Place: Geography, Ideology, and Transgression.* Minneapolis: University of Minnesota Press.

Cresswell, T. (2004) *Place: A Short Introduction.* Oxford: Blackwell.

Daniels, S. (1985) 'Arguments for a humanistic geography', in R. Johnston (ed.), *The Future of Geography.* London: Methuen, pp. 143–58.

Duncan, J. and Ley, D. (1982) 'Structural Marxism and human geography: a critical assessment', *Annals of the Association of American Geographers*, 72: 30–59.

Entrikin, J.N. (1976) 'Contemporary humanism in geography', *Annals of the Association of American Geographers*, 66: 615–32.

Entrikin, J.N. (2001) 'Geographer as humanist', in P.C. Adams, S. Hoelscher and K. Till (eds), *Textures of Place: Exploring Humanist Geographies.* Minneapolis: University of Minnesota Press, pp. 426–40.

Entrikin, J.N. (2002a) 'Democratic place-making and multiculturalism', *Geografiska Annaler*, 84B: 19–25.

Entrikin, J.N. (2002b) 'Perfectibility and democratic place-making', in R. Sack (ed.), *Progress: Geographical Essays.* Baltimore: Johns Hopkins University Press, pp. 97–112.

Gregory, D. (1981) 'Human agency and human geography', *Transactions of the Institute of British Geographers*, n.s. 6: 1–18.

Harris, C. (1978) 'The historical mind and the practice of geography', in D. Ley and M. Samuels (eds), *Humanistic Geography: Progress and Problems.* Chicago: Maroufa, pp. 123–37.

Harvey, D. (1973) *Social Justice and the City.* London: Arnold.

Harvey, D. (1996) *Justice, Nature, and the Geography of Difference.* Oxford: Blackwell.

Harvey, D. (2000) *Spaces of Hope.* Berkeley, CA: University of California Press.

Latour, B. (1993) *We Have Never Been Modern*, trans. C. Porter. Cambridge: Harvard University Press.

Ley, D. (1977) 'Social geography and the taken-for-granted world', *Transactions of the Institute of British Geographers*, n.s. 2: 498–512.

Ley, D. (1989) 'Fragmentation, coherence, and limits to theory in human geography', in A. Kobayashi and S. Mackenzie (eds), *Remaking Human Geography.* Boston: Unwin and Hyman, pp. 227–44.

Ley, D. and Samuels, M. (eds) (1978) *Humanistic Geography: Prospects and Problems.* Chicago: Maroufa.

Lowenthal, D. (1961) 'Geography, experience, and imagination: towards a geographical epistemology', *Annals of the Association of American Geographers*, 51: 241–60.

Massey, D. (1994) *Space, Place, and Gender.* Minneapolis: University of Minnesota Press.

Pile, S. (1996) *The Body and the City: Psychoanalysis, Space, and Subjectivity.* London: Routledge.

Pile, S. and Thrift, N. (1995) 'Introduction', in S. Pile and N. Thrift (eds), *Mapping the Subject: Geographies of Cultural Transformation.* London: Routledge, pp. 1–12.

Proctor, J. and Smith, D. (eds) (1999) *Geography and Ethics: Journeys in a Moral Terrain.* London: Routledge.

Relph, E. (1970) 'An inquiry into the relations between phenomenology and geography', *Canadian Geographer*, 14: 193–201.

Relph, E. (1977) 'Humanism, phenomenology, and geography', *Annals of the Association of American Geographers*, 67: 177–83.

Rose, G. (1993) *Feminism and Geography: The Limits of Geographical Knowledge.* Minneapolis: University of Minnesota Press.

Sack, R. (1997) *Homo Geographicus: A Framework for Action, Awareness, and Moral Concern.* Baltimore: Johns Hopkins University Press.

Sack, R. (2003) *A Geographical Guide to the Real and the Good.* London: Routledge.

Smith, D. (2000) *Moral Geographies: Ethics in a World of Difference.* Edinburgh: Edinburgh University Press.

Todorov, T. (2002) *Imperfect Garden: The Legacy of Humanism.* Princeton, NJ: Princeton University Press.

Tuan, Y.-F. (1976) 'Humanistic geography', *Annals of the Association of American Geographers*, 66: 266–76.

Tuan, Y.-F. (1984) *Dominance and Affection: The Making of Pets.* New Haven, CT: Yale University Press.

Tuan, Y.-F. (1986) *The Good Life.* Madison, WI: University of Wisconsin Press.

Tuan, Y.-F. (1989) *Morality and Imagination: Paradoxes of Progress.* Madison, WI: University of Wisconsin Press.

Tuan, Y.-F. (1990) 'Realism and fantasy in art, history, and geography', *Annals of the Association of American Geographers*, 80: 435–46.

Tuan, Y.-F. (1991a) 'Language and the making of place: a narrative-descriptive approach', *Annals of the Association of American Geographers*, 81: 684–96.

Tuan, Y.-F. (1991b) 'A view of geography', *Geographical Review,* 81: 99–107.

Tuan, Y.-F. (2003) 'On human geography', *Daedalus*, 132: 134–7.

Wright, J. (1947) 'Terrae incognitae: the place of imagination in geography', *Annals of the Association of American Geographers*, 37: 1–15.

4 FEMINIST GEOGRAPHIES OF DIFFERENCE, RELATION, AND CONSTRUCTION

Deborah P. Dixon and John Paul Jones III

Introduction

Feminist geography is concerned first and foremost with improving women's lives by understanding the sources, dynamics, and spatiality of women's oppression, and with documenting strategies of resistance. In accomplishing this objective, feminist geography has proven itself time and again as a source for innovative thought and practice across all of human geography. The work of feminist geographers has transformed research into everyday social activities such as wage earning, commuting, maintaining a family (however defined), and recreation, as well as major life events, such as migration, procreation, and illness. It has propelled changes in debates over which basic human needs such as shelter, education, food, and health care are discussed, and it has fostered new insights into global and regional economic transformations, government policies, and settlement patterns. It has also had fundamental theoretical impacts upon how geographers: undertake research into both social and physical processes; approach the division between theory and practice; and think about the purpose of creating geographic knowledge and the role of researchers in that process. Finally, feminist geography has helped to revolutionize the research methods used in geographic research.

Feminist geography, however, cannot neatly be summed up according to a uniform set of substantive areas, theoretical frameworks, and their associated methodologies: hence the plural 'geographies' in the title of this chapter. To facilitate our survey of feminist geography, we draw out three main lines of research. Each of these holds the concept of gender to be central to the analysis, but they differ in their understanding of the term. Under the heading of *gender as difference*, we first consider those forms of feminist geographic analysis that address the spatial dimensions of the different life experiences of men and women across a host of cultural, economic, political, and environmental arenas. Second, we point to those analyses that understand *gender as a social relation*. Here, the emphasis shifts from studying men and women *per se* to understanding the social relations that *link* men and women in complex ways. In its most hierarchical form, these relations are realized as patriarchy – a spatially and historically specific social structure that works to dominate women and children. A third line of inquiry examines the ways in which *gender as a social construction* has been imbued with particular meanings, both positive and negative. Not only are individuals 'gendered' as masculine or feminine as a form of identification, but also a wealth of phenomena, from landscapes to nation-states, are similarly framed. In practice, there is quite a bit of overlap among each of these lines of research. Yet it is still useful to make a distinction in so far as each body of work lends itself to a particular set of research questions and associated data and analyses.

Recovering the Geographies of Gender

Before we get started, it would be helpful to set a context for our survey of these three theoretical perspectives. This involves thinking about the discipline's traditional male-centeredness, which we can categorize in three ways: institutional discrimination, substantive oversights, and 'masculinist' ways of thinking and writing. We begin by noting that geography, in both the US and Europe, was formed out of a late-nineteenth- to early-twentieth-century academic setting that was highly exclusionary in terms of class, race/ethnicity, and gender. Early universities were dominated in the main by upper-class white men. Within Anglo contexts, a small number of women academics were primarily concentrated in the teaching and helping professions (e.g. nursing). Few were to be found in the disciplines from which modern geography was established, such as geology and cartography. During the nineteenth and well into the mid twentieth centuries, a crude form of biological stereotyping underlay not only conceptions of what women were able to achieve intellectually, but also their physical capacity. This was despite the fact that many women in the early years of the discipline – Mary Kingsley is a celebrated example – were engaged in intellectually stimulating and physically rigorous explorations of their own. Moreover, women also played a central role in educating geographers within teacher training institutions.

It was out of this broader, academic climate that 'expert' societies arose so as to establish geography as a specialized, intellectual endeavor. The goal of these societies was to define the discipline as a science (as opposed to lore) by debating theory, the kinds of phenomena to be investigated, and appropriate methodologies, and to work within universities to establish programs at both the undergraduate and graduate levels. The two most influential of these, the Association of American Geographers in the US and the Royal Geographical Society in the UK, were not open, as they are today, to anyone interested in promoting geographical knowledge. Instead, their members first had to be nominated and then elected. These and other rules and practices had a filtering effect on membership by specifying who was considered a legitimate scientist. For example, the early constitution of the Association of American Geographers reserved full membership for those who had previously published original research. Yet with few women included in graduate training, most women writers published their research in a style and in venues not deemed scholarly. Not surprising, then, is the fact that of the 48 original members of the Association of American Geographers, established in 1904, only two were women: Ellen Churchill Semple and Martha Krug Genthe (who, among the entire original membership, held the only PhD in geography, obtained in Germany).

All told, the male-oriented culture of these academic societies and university departments had a significant negative impact on the number and status of women in the discipline. Many women reported a range of obstacles and difficulties in negotiating the field, from a benign paternalism to outright misogyny, and from tokenism to blatant sexual discrimination. The resistance of male geographers to women conducting independent fieldwork lasted well into the 1950s: geography's expeditionary legacy continued to lead some to a nostalgic belief that only 'stout hearted men' were capable of such research (sometimes referred to as 'muddy boots' geography). Overall, geography's culture offered few opportunities for constructive engagement to the vast majority of college educated women, as evidenced by the much larger proportion of women found in the humanities and some cognate social sciences, such as anthropology and sociology.

Bearing this institutional discrimination in mind, it is not surprising to find substantive oversights in which male-dominated activities constituted the norm of geographic research. This presumption is strikingly revealed in the gender-coded language geographers have used in their research. In reading the literature produced by geographers up to and including the 1970s, what appears to be a mere semantic peccadillo – as in the 'Man–Land' tradition or the assumption of 'economic man' – actually reveals an underlying assumption about what constitutes primary human activities and who constitute economic, political, cultural, and environmental agents (who, for example, makes history and geography in the book, *Man's Role in Changing the Face of the Earth*?). So blatantly sexist is some of this writing that geographers' citing their predecessors today often liberally pepper their quotations with '[*sic*]' – a term used to indicate that what has just preceded is reproduced exactly as written. Though some may regard this practice as pedantic, it does allow contemporary writers the opportunity to explicitly distance themselves from sexist (or racist, etc.) language.

More significant than the stylistic substitution of 'man' for 'human' are the ways in which putatively male activities have been the primary focus of analysis across each of geography's traditional objects of inquiry, be these landscapes, regions, spatial variations, or the environment. As many feminist geographers have pointed out, the discipline's prioritization of traditionally male, productive activities has in one way or another worked to marginalize the study of women's lives. It has meant, for example, that geographers have spent more time examining steel manufacturing than, say, day care. We can see this bias reproduced in a wide range of substantive research areas. Traditional cultural geography, for example, was concerned to evaluate how different cultures made use of the earth and its resources in the process of making a living and constructing built environments in accord with these demands. Traditional regional geography focused in turn on a complex of interrelations that gave a specific and interactive character to areal divisions, but here the categories to be integrated mirror the list of productive activities listed above – the only significant additions being the physical environment, the distribution of population (typically distinguished by ethnicity only), and the (largely male) arena of formal politics. And, in the period of spatial science prior to the development of a social relevance perspective, location theory took this abstraction of the productive activities of society to its furthest extent, deploying the assumption of economic man in an idealized space and tracing its implications for the distribution of economic activities (as in assessments of the models of von Thünen, Weber, Lösch, Alonso, and Christaller, as well as various versions of the gravity model).

As a result, those geographers interested in working on activities such as childraising, education, neighborhood organization, and social welfare (i.e. activities known as 'social reproduction' as opposed to productive economic ones) did so in a vacuum. Thus, though there existed specialized study groups within Anglo-American academic societies devoted to transportation, industry, economic development, and land use, specialty groups devoted to gender, children, education, and sexuality are more recent phenomena. And at the interdisciplinary level, the focus on production relative to reproduction within geography meant that spatially minded social scientists who wanted to examine, say, the family, health care, or social welfare, would have to look to other disciplines more sympathetic to their study (e.g. sociology, social work) for their graduate training, thereby diminishing the scope and ultimately the academic significance of the field.

Completing our discussion on the silence of gender is the claim that, prior to the arrival of feminist geography, the discipline operated with what is termed a masculinist epistemology. This epistemology is based on a way of knowing the world (through universalism), framing the world (through compartmentalization), and representing the world (through objectivity). Universalism is the belief that there exists a 'god's-eye' position from which the world can be surveyed in its totality. Such a position lifts one out of the messy, complicating facts of class, race/ethnicity, national origin, political persuasion, and, of course, gender and sexuality, which would otherwise 'bias' the investigative process. Yet, as feminist researchers have pointed out, the attempt to transcend such facts of life is ultimately driven by the belief that they *could* and *should* be transcended. Such goals carry an air of omniscience and infallibility, which cements the role of the scientist and 'his' position of authority.

Compartmentalization relies on the use of rigidly fixed boundaries to comprehend the qualities and characteristics of phenomena, such as nature and culture, male and female, plant and animal. The rationale for this obsession, wherein everything has its place, is rigor, a stance that guards against any ambiguity that might undermine scientific analysis of cause and effect. Feminist critics of excessive compartmentalization point to the homogenizing effects of taxonomies, which result in a tendency to overlook difference within and across research objects. In highlighting difference, feminists focused less on the objects contained within categories than on how these categories were formed in the first instance. This led feminists to develop relational as opposed to discrete understandings of phenomena, in which they argued that objects were defined not by their supposedly intrinsic characteristics (e.g. biology) but by interrelations within the social world (e.g. gender divisions of labor).

Related to both universalism and compartmentalization is a masculinist strategy of representing the world as an objective observer. To achieve this, the researcher purposively excludes any trace of their own thoughts and feelings by adopting a third person, passive tense style that is stripped of self-referencing, hesitation, emotive phrasing, or rhetorical flourishes. Such writing attempts to use clear prose that can be commonly understood, even while invoking the necessary technical terminology. Marked adherence to this mode of communication assures other scholars that the research reported has not been biased by personal or societal influences. Moreover, it is assumed to enable researchers to systematically compare findings in a manner untainted by individual presentation styles, thereby bolstering the belief that objectivity contributes to a growing stock of scientific knowledge. Underlying this assumption, however, is a belief in a 'common' frame of reference wherein everyone does indeed clearly understand what is being said. Such a style also serves to distance the author from any responsibility for the reception of her or his work: even though they may recognize that some research could be used toward socially undesirable ends, authors adopting this stance ultimately affirm that science is inherently apolitical.

These dimensions of masculinist epistemology are not the subject of feminist debate and critique alone, for scholars have long debated the advisability and possibility of conducting distanced research. For example, the field of hermeneutics, the origin of which lies in the exegesis of the Bible, explicitly deals with the complicated role of researchers in relation to their research contexts. The contribution of feminist thought has been to recognize that universalism, compartmentalization, and objectivity have traditionally been associated with male faculties of sense and reason, whilst their oppositions – particularism, relationality, and subjectivity – have been constituted as the domain of unreasoning, female faculties driven by mere sensibility. A major area of feminist research, therefore, has involved charting the ways and means by which this gendering of epistemology took place, and an assessment of its repercussions in terms of the marginalization of women within and beyond academia.

In an ironic turn, therefore, the discipline's traditional disregard for gender has provided research material, as well as a professional challenge, for feminist geographers. Moving beyond mere critique, feminist geographers have produced an alternative, feminist epistemology that not only transforms how geography's accepted objects of analysis – regions, landscapes, places, etc. – are to be researched, but has brought to light a range of other objects of analysis, such as the body. Today, feminist approaches intersect with all of human geography's domains. Feminist geography spans traditional geographic foci in development, landscapes, and the environment, contributes to what has been labeled the 'new' regional, cultural, and economic geographies, and has realized numerous connections to other fields, especially philosophy, English, cultural studies, anthropology, postcolonial studies, economics, and sociology, among others. As a result of these developments, feminist geography now constitutes a fully fledged subdisciplinary and interdisciplinary endeavor, complete with a specialized journal, *Gender, Place and Culture*. In the following sections, we draw out three theoretical approaches toward gender that have emerged over the past 30 years, and discuss how each has made, and continues to make, its own contribution to geographic research.

Gender as Difference

The geographic concepts of location, distance, connectivity, spatial variation, place, context and scale have all been enriched through the lens of feminist theory, which focuses on the difference that gender makes to a host of social processes. Feminist geographers transform the question, 'Where does work take place?', for instance, by the more targeted one, 'Who works where?' This more specific question can help researchers better understand the spatial dimensions of gender divisions of labor and their effects on women's economic wellbeing. Likewise, studies that look at connectivity have been enriched by an examination of the gendered character of the subjects undertaking the activity, whether in migration, commuting, or communication.

As part of a project's research design, researchers often separately measure for men and women variables such as unemployment rates, income, and educational levels, typically collected across geographic units. The differential spatial experiences of men and women can then be analyzed. Comparing the spatial variation of women's and men's unemployment rates, for example, can yield insights into the particular processes that contribute to the economic marginalization of women as opposed to men. With the understanding that these processes may not operate equally for all women across space, moreover, researchers can raise questions of place context – a term used to refer to the combination of cultural, economic, political, and environmental dimensions that give character to a particular setting. A focus on place has researchers address how a particular context influences women's lives, and can be the basis for cross-context comparisons among women for any number of research problems. For example, one might find that the degree and type of women's political involvement in different places are influenced by contextual factors such as the gender division of labor in local economies, the quality of education in the localities, or the severity of local environmental problems.

An emphasis on gender as difference also enhances studies employing different scales of analysis. Key here is the fact that processes influencing spatial patterns of women's lives work across different scales, with some operating at relatively local levels and others more extensively. In examining women's economic viability, for example, researchers would find useful an investigation of the presence of local social networks that partly influence their job searches; at the same time, they cannot neglect how the place context within which women are seeking employment is itself positioned relative to global capital flows, which will affect the type and availability of employment, as well as the local gender division of labor.

Still another avenue for feminist geographic research on difference can be found under the heading of sense of place, which refers to the perceptions people have of particular places and natural and built environments more generally. Studies of sense of place emphasize psychosocial influences upon one's interpretations, evaluations, and preferences regarding places or their representation in one medium or another. Such studies are often based upon the collection of primary data, and are therefore particularly well suited for asking questions of difference, since the researcher can purposively include both men and women respondents. One might, for example, compare men's and women's mental maps of a local neighborhood, using the detail therein to help answer questions about their perception of dangerous and safe zones across the study area. The range of places for which men and women respondents might be expected to differ is large, from classrooms, wilderness areas, and spaces of the home, to sports venues, drinking establishments, and shopping malls. This knowledge has practical relevancy in that it identifies places that are enabling for women, and might offer guidelines for the construction of environments that are non-threatening.

By introducing gender difference into all manner of geographic concepts, feminist geography has initiated new lines of inquiry in geography, thereby redressing the research imbalances noted earlier in this chapter. Recall that the theorization of these spheres has traditionally marked a separation between presumptively male-oriented productive economic activities and female-oriented reproductive activities. Feminist geographers of difference have made two significant contributions with respect to this framework. First, they have brought to light the role of women in the economy by noting, for example, the contribution of First World women who work in suburban back-offices devoted to processing credit applications, and that of Third World women whose labor in branch manufacturing plants makes possible the production of low-wage consumer items, such as electronics. Second, feminist geographers have expanded substantive domains, including new research on women's roles in neighborhood associations, household survival strategies in Third World countries, inequalities in the provision of day care facilities, and efforts to eliminate environmental pollution and toxic waste hazards through grassroots organizing. To uncover these geographies, feminist geographers have become leaders in the collection of primary, field-based data, precisely because such data are necessary to reveal women's everyday spatial experiences. Though such methodologies as interviews, focus groups, ethnographies, participant observation, and surveys are more time consuming than simply downloading data from secondary sources, such as the census, they are necessary to bring to light the complexities of those experiences.

Gender as a Social Relation

In turning their attention to gender relations, feminist geographers shift their focus from men and women as discrete objects of inquiry, which, as we noted above, is itself a masculinist formulation, to the structured interconnections that routinely intertwine their life experiences. Patriarchy is one of the key structures studied by feminist geographers. The term 'patriarchy' describes the systematized exploitation, domination, and subordination – in short, oppression – of women and children through gender relations. It is held together through language, as when men speak loudest, longest, and last, and is given form through rules of behavior and legal statutes that stamp in gendered terms what types of activities are desirable, appropriate, expected, or sanctioned, and that specify, formally or informally, who maintains access to material resources. Thus, patriarchy is legitimated and perpetuated by a host of social norms and moral rules – as when, for example, women 'naturally' assume the major burden of raising children. The relations that link the lives of men and women take place within and between a variety of specific sites, such as the family, school, and church, each of which is infused with patriarchy, the effects of which range from a patronizing paternalism to outright violence.

Research into specific examples of patriarchal relations is complicated by the recognition that this structure is always socially, historically, and geographically specific. In other words, there is no single patriarchy, but a multitude of variations. This variation ensues in large part from the way patriarchy intersects with other kinds of social structures, and one important line of inquiry in feminist geography has been to study the intersection of patriarchal and capitalist social relations. For Marxist or socialist feminists, capitalism is the key structure in modern life: through it one can apprehend the ways in which labor is expropriated from the working class by those capitalists who control the means of production. These feminists study the way capitalist social relations are formed in conjunction with patriarchal gender relations, the result of which are variations in women's economic position. In addition, they note that capitalism has extended its power into the home, resulting in a class of unpaid women whose household labor is expropriated by the male wage earner and, by extension, his employer. For these feminists, capitalism determines the specific form that patriarchy takes. It is the complex and differentiated intersection of these relations that gives us varieties of women's exploitation across the globe.

In radical feminism, by contrast, patriarchy is prioritized. These scholars note that, historically speaking, patriarchy predates capitalism, and facilitated its emergence within specific sociohistorical contexts. These feminists ground their prioritization not in the control over labor, but in men's control over women's bodies – a control exercised in sexual relations and childrearing, and maintained through patriarchal ideology and violence. Still other theorists have attempted to create a *rapprochement* between these positions, arguing that the two structures, while analytically distinct, are co-present in everyday practice. As such, they can be studied as mutually enabling structures in a wide variety of contexts.

While socialist and radical feminists were debating the theoretical primacy of patriarchy and capitalism, black, Latina, and Third World feminists developed extensive critiques of the Eurocentricity of these debates, drawing attention to the extent to which women's lives are also indelibly racialized and colonized. These feminists have pointed to the global diversity of patriarchal and class relations and their intersection with other global-to-local structures. Still another structural relation contextualizing patriarchy is heteronormativity, a concept developed by queer theorists. This term describes the widespread assumption that heterosexuality is the natural form of sexual relations, while homosexuality is an aberration. Like patriarchy and other structures, heteronormativity is a social relation with its own language, norms, and practices. As a consequence of these arguments, contemporary research undertaken to illuminate the structural dynamics and locational specificity of patriarchy needs to contend not only with class relations under capitalism, but minimally also those of race, colonial history, and sexuality. Like some feminists researching gender as difference, those who study gender as a social relation often rely upon research strategies that involve 'talking to women' through interviews, focus groups, and the like. But in addition, they are equally likely to pay attention to the subtle ways that patriarchy, class, race, nation, and sexuality are formed and perpetuated through everyday forms of representation, including political rhetoric (e.g. speeches, policy documents), media imagery (e.g. film and video, magazines, the Internet), and bodily adornment and comportment (e.g. dress, mannerisms, habits).

Gender as a Social Construction

Social constructivists are interested in the ways in which 'discourses' establish distinctions – or differences – between individuals and groups, made and natural objects, types of experience, and aspects of meanings. They argue that none of these are naively given to us as unmediated parts of reality; instead, all are framed through categorizations that enable us to comprehend them. In this view, people, objects, experiences, and meanings have no intrinsic meaning until their qualities and boundaries have been framed in discourse. We use the term 'discourse' to refer to particular framings, most of which rely upon one or another binary opposition, such as nature/culture, male/female, individual/society, objective/subjective, and orderly/chaotic. Discursive construction refers to the social process by which these categories are produced and filled with objects and meanings. Though discourses are enabled and reproduced through language, to constructivists 'discourse' is a term more complicated than its everyday use as 'mere words', for it refers not only to the processes of categorization (see above) but also to everyday social practices – from raising children to dancing – that, like language, are also imbued with meaning and hence also signify something about the world. Through discourse we come to understand where things fit in the world, literally and figuratively. We also come to comprehend the relationships among categories that have been established. And, discourses tell us a great deal about what is appropriate and what is inappropriate, what is valued and what is devalued, and what is possible and what is impossible.

Applied to feminism, discursive construction points to gender codings as key elements in establishing difference and policing categories. Feminist geographers working with theories of social construction of gender, for example, are interested in the ways in which discursive categories, particularly male/female and masculine/feminine, are brought into play at specific times and in specific places in order to establish spaces of exclusion and inclusion. Drawing on all feminists' concern for difference, feminist social constructivists also examine how these explicitly gendered categories seep into other socially constructed ones, such as 'race' and 'sexuality', 'production' and 'reproduction', 'nature' and 'culture', and so on.

Take for example the concept of 'maleness'. To constructivists, this is not a term that refers to a naturally given object with essential characteristics. It instead describes a social construct, formed out of ideas concerning what it is to be male, as opposed to what it is to be female. This binary construct is determined and maintained by a gender-specific language about people's beliefs, actions, and qualities. Thus, words such as 'caring', 'tenderness', and 'empathetic' have different gendered connotations than words such as 'stoic', 'noble', and 'boisterous'. Importantly, the meaning, significance, and social value of these terms are not fixed, but vary from one context to another: hence, tenderness could conceivably take on a masculine quality. At the same time, however, the connotations among these terms are socially determined, and hence linked to dominant forms of power (which can be defined as the ability to construct and maintain difference through language and practices). Finally, once maleness has been granted the status of 'normal', the social relations that ensue – such as patriarchy – may also be regarded as natural and, hence, enduring.

Perhaps nowhere is discursive analysis more illuminating and yet controversial than in the analysis of sexual difference. Some constructivists see such difference as another example of discourse, one rooted in biological categorizations of physicality, from shape and form to genes and voice. They argue that biological and other discourses continually impact the body, through ideas and practices surrounding medical protocols, labor practices, legal statutes, and reproductive capabilities. The discourses that circulate in these domains are so encoded on bodies that we seldom take time to think about the everyday reinforcements that buttress male/female difference. (Think, for example, about the discursive work silently undertaken in public buildings, with their separate bathrooms for men and women.) These insights have led some feminists to question the very foundation of the terms 'male' and 'female': they see biological discourses as creating the conditions by which we 'perform' our sex and gender. Others accept biological differences as 'prediscursive', but are equally attuned to the ways in which all aspects of the body, including the mind, are gendered, raced, and sexualized through their embeddedness in discourse. Thus, while we cannot lift ourselves outside socially derived significations that structure our understandings of male and female, there is at base a materiality upon which these significations are attached (even if we cannot really know or experience that materiality outside discourse).

Given either emphasis on the construction of gender, how do geographically informed constructivist analyses proceed? The primary goal of most such analyses is to understand how sexed and gendered meanings are at work in all aspects of everyday spatial life, policing what is thought and delimiting places and identities. To undertake such work, feminist geographers look first to those sites from which knowledge concerning gender is articulated, such as schools, churches, media outlets, the home and government agencies, and consider how these sites collect information, rework it as knowledge, and then proceed to disseminate that knowledge through particular networks. How, for example, do the 'real' life stories in teen magazines configure and reproduce a socially and spatially specific audience (e.g. 'white mallrats')? How does housing design both reflect and reproduce ideas about what kinds of (gendered) activities occur where, and by whom? How has the teaching of geography within schools helped to construct it as a primarily male discipline? Second, feminist geographers look to the geography of discourses through which people are gendered, as well as to the other discourses they intersect, such as race, nationality, ethnicity, sexuality, nature, and so on. What complex gendered codings, for example, lie at the intersection of the term 'Mother Nature', and what undercurrents ensue for how the environment is 'managed'? How is it that a day care center is regarded as a traditional worksite for domesticated women, while a garden allotment is considered an escapist landscape for married (but not gay) men? What complex gendered and raced meanings accompany partitions of space such as 'ghettos', 'working-class areas', and 'farmsteads'? Even entire countries, such as Australia and France, tend to be gendered differently in popular media (e.g. as 'laddish' vs 'sensualist'). Third and finally, one can consider how the everyday operation of these discourses can affect a form of 'discursive violence', foisting onto people an identity they may not wish to adhere to, and rendering other forms of identity that do not fit into the accepted categories as aberrant or unnatural. This is especially true when bodies or identities are marked as 'queer' and made to feel uncomfortable in what is largely a heterosexually coded built environment. In all of these sorts of analyses, feminist social constructivists turn to qualitative methodologies to trace the subtle plays of discursive constructions in all sorts of representations, including not only those of the media and everyday speech, but those in the built environment itself.

Indeed, the traditional concept of the 'field' itself – whether the home, the workplace, the urban neighborhood, or the remote village – has been opened up considerably by feminist geographers. Classically, in geographical research, the researcher remains mysterious, distant and silent while the field subject discloses more and more information: in this case, the visibility of the researched obscures the presence of the researcher. In contrast, feminists emphasize that, like her or his objects of analysis, the researching subject is likewise constituted – or positioned – by gender relations of social power. Gender relations form part of a broader, social context within which research takes place – from the individual biographies and social structures influencing both the geographer and her research subjects, to the subdiscipline of geography within which she works, and on to the funding agencies, the universities, and the place contexts, both global and local, that inform and bracket the work. This, then, is the new 'feminist field': a fluid, complex, and spatially stretched set of relations that bear little resemblance to older notions of expert geographers researching people in particular places.

By Way of Conclusion: Suggested Reading

In this concluding section, we offer a brief roadmap through some texts and articles that have been important in the development of feminist geography. We also point to a few classic debates and emerging lines of inquiry. We begin with the note that feminist geography exerted a compelling critique of geography as a male-oriented discipline in the mid 1970s. Mildred Berman's (1974) article on sexual discrimination within the academy was matched by Alison Hayford's (1974) assessment of the wider, historical 'place' of women. Later work by Linda McDowell (1979) and Janice Monk and Susan Hanson (1982) expanded on the substantive oversights and masculinist presumptions of geographic research. Both Hanson and Monk were later elected President of the Association of American Geographers (two of only five women elected). Susan Hanson's (1992) address challenged geographers to consider the commonalities between feminism and geography and to transform both disciplines. Janice Monk's (2004) presidential address to the Association takes a historical approach to recover the work lives of women geographers who taught and practiced during long periods of professional exclusion. Two years after Monk and Hanson's (1982) essay, the Institute of British Geographers' Women and Geography Study Group published *Geography and Gender: An Introduction to Feminist Geography* (1984). A ground-breaking text in many ways, *Geography and Gender* focused attention on the specificities of women's experiences, within and beyond the academy. Students interested in a feminist examination of the history of the discipline should also read the article by Mona Domosh (1991), as well as Alison Blunt's (1994) analysis of nineteenth-century explorer and writer Mary Kingsley.

The 1980s and early 1990s saw the emergence of a large body of work on the intersection between gender, work, and space. A key early text in this regard is by Linda McDowell and Doreen Massey (1984); they historicize the geographies produced by the intersection of gender and class relations. A couple of years later, the relative role of patriarchy vs capitalism in explaining women's exploitation became the topic of a lively theoretical exchange in the journal *Antipode*. The radical vs socialist feminist division is clearest in the essays by Jo Foord and Nicky Gregson (1986) and Linda McDowell (1986), and in the response by Gregson and Foord (1987). Readers might also want to consult Sylvia Walby's *Theorizing Patriarchy* (1990), which offers a good sociological account of patriarchy in capitalist societies. Doreen Massey's book *Space, Place and Gender* (1994) collects her works through the mid 1990s and gives insights into one of feminist economic geography's most original thinkers. Another good choice for those interested in women and work is Susan Hanson and Gerry Pratt's *Gender, Work and Space* (1995), Nicky Gregson and Michelle Lowe's *Servicing the Middle Classes* (1994), and Kim England's *Who Will Mind the Baby? Geographies of Childcare and Working Mothers* (1996).

In the 1980s and 1990s, geographers began to engage academic debates surrounding postmodernism (see Chapter 9), one key vector of which was the relationship between this then-new area of thought and feminism. Interested readers might follow debates in Liz Bondi (1991), Gerry Pratt (1993), and J.-K. Gibson-Graham (1994). Another key debate circulating through postmodernism and feminism was sparked by David Harvey's *The Condition of Postmodernity* (1989). His political economic analysis of culture under late capitalism was roundly criticized by Doreen Massey (1991) and Rosalyn Deutsche (1991). Reading this along with Harvey's (1992) rejoinder is helpful, but a better sense of his thinking on the intersection between class and gender can be found in Chapter 12 of *Justice, Nature, and the Geography of Difference* (1997).

Especially since the mid 1990s, feminist geographers have produced a substantial amount of work at the intersection of bodies, identities, and space/place. An early edited collection of important works is found in Nancy Duncan's *BodySpace* (1996). Other feminist readings of bodies can be found in Heidi Nast and Steve Pile's edited volume *Places through the Body* (1998), in Ruth Butler and Hester Parr's *Mind and Body Spaces: Geographies of Illness, Impairment and Disability* (1999), in Elizabeth Teather's *Embodied Geographies: Space, Bodies and Rites of Passage* (1999), and in selected chapters of Linda McDowell's *Capital Culture: Gender at Work in the City* (1997a). Also see Robyn Longhurst's *Bodies: Exploring Fluid Boundaries* (2000) and Pile's *The Body and the City: Psychoanalysis, Space and Subjectivity* (1996). Some of the essays in the above collections were harbingers of a shift toward queer theory in geography. An early book in this area was the edited collection by David Bell and Gill Valentine, *Mapping Desire* (1995). Michael Brown unpacks the geographies of the 'closet' in *Closet Space: Geographies of Metaphor from the Body to the Globe* (2000). Finally, geographic approaches to masculinity have appeared in works by Peter Jackson (1991), Steve Pile (1994), and Richard Phillips (1997).

There is a wealth of feminist research on the interplay of gender, nature, and development (including 'post' or 'anti' development theory). Readers might consult the collection edited by Janet Momsen and Vivian Kinnaird (1993), as well as work by Cathy Nesmith and Sarah Radcliffe (1993) and Radcliffe (1994). A good collection of work in feminist political ecology is by Dianne Rocheleau, Barbara Thomas-Slayter, and Esther Wangari (1996). Feminist geographers have also drawn on postcolonial theorizations to better understand the global construction of gender, race, nation, and class. Students should consult Alison Blunt and Gillian Rose's edited volume *Writing Women and Space: Colonial and Postcolonial Geographies* (1994) and Alison Blunt and Cheryl McEwan's *Postcolonial Geographies* (2003).

There are a number of good sources to turn to for feminist research methods in geography. A 1993 collection in *The Canadian Geographer* traced the contours of feminist epistemology alongside in-depth qualitative research methods. Heidi Nast's edited collection, 'Women in the "Field"', appeared in *The Professional Geographer* (1994). Many of the articles offer interesting and introspective examinations of feminist methods as they played out in the work of the assembled geographers. Other good assessments of the 'field' are by Cindi Katz (1992) and Ann Oberhauser (1997). As discussed above, feminist research methods are typically qualitative (e.g. Nash, 1996), but there has been a lively debate on the role of quantitative methods in feminist research (Kwan, 2002). See the collection in *The Professional Geographer* (1995), which appeared under the heading, 'Should Women Count?' Pamela Moss's edited collection *Feminist Geography in Practice* (2002), as well as the volume by Melanie Limb and Claire Dwyer (2001), offer students a wealth of direction in the pursuit of feminist research. The 2003 special issue on 'Practices in Feminist Research' in *ACME: An International E-Journal for Critical Geographies* considers what holds feminist methodology together as a distinct approach given the spread of critical methodologies within geography more generally. Students interested in praxis should consult the collection in *Antipode* (1995), as well as essays by Vicky Lawson and Lynn Staeheli (1995) and Susan Smith (2002).

Thorough overviews of feminist geography can be found in Linda McDowell and Jo Sharp's *A Feminist Glossary of Human Geography* (2000), as well as in collections by Linda McDowell (1997b; 1999) and John Paul Jones III, Heidi Nast, and Susan Roberts (1997). Gillian Rose's *Feminism and Geography* (1993) provides an extended, close reading of geography's masculinist bias, largely as read through the field's twentieth-century history. The conclusion attempts to rethink space by reconfiguring a number of key binaries that have influenced thinking in geography. Students would be well advised to peruse the current and back issues of *Gender, Place and Culture*, while the online bibliography of feminist geographic research found at http://www.emporia.edu/socsci/fembib/is an excellent source of material.

References

ACME: An International E-Journal for Critical Geographies (2003) 'Practices in Feminist Research', collection, 2(1): 57–111.

Antipode (1995) 'Symposium on Feminist Participatory Research', collection, 27: 71–101.

Bell, David and Valentine, Gill (eds) (1995) *Mapping Desire: Geographies of Sexuality*. London: Routledge.

Berman, Mildred (1974) 'Sex discrimination in geography: the case of Ellen Churchill Semple', *The Professional Geographer*, 26: 8–11.

Blunt, Alison (1994) *Travel, Gender and Imperialism: Mary Kingsley in West Africa*. New York: Guilford.

Blunt, Alison and McEwan, Cheryl (2003) *Postcolonial Geographies*. London: Continuum.

Blunt, Alison and Rose, Gillian (1994). *Writing Women and Space: Colonial and Postcolonial Geographies*. New York: Guilford.

Bondi, Liz (1991) 'Feminism, postmodernism, and geography: space for women?', *Antipode*, 22: 156–67.

Brown, Michael (2000) *Closet Space: Geographies of Metaphor from the Body to the Globe*. New York: Routledge.

Butler, Ruth and Parr, Hester (eds) (1999). *Mind and Body Spaces: Geographies of Illness, Impairment and Disability*. London: Routledge.

Canadian Geographer (1993) 'Feminism as method', collection, 37: 48–61.
Deutsche, Rosalyn (1991) 'Boy's town', *Environment and Planning D: Society and Space*, 9: 5–30.
Domosh, Mona (1991) 'Towards a feminist historiography of geography', *Transactions of the Institute of British Geographers,* 16: 95–104.
Duncan, Nancy (ed.) (1996) *BodySpace: Destabilizing Geographies of Gender and Sexuality.* London: Routledge.
England, Kim (ed.) (1996) *Who Will Mind the Baby? Geographies of Childcare and Working Mothers.* London: Routledge.
Foord, Jo and Gregson, Nicky (1986) 'Patriarchy: towards a reconceptualization', *Antipode*, 18: 186–211.
Gibson-Graham, J.-K. (1994) 'Stuffed if I know! Reflections on post-modern feminist research', *Gender, Place and Culture*, 1: 205–24.
Gregson, Nicky and Foord, Jo (1987) 'Patriarchy: comments on critics', *Antipode*, 19: 371–5.
Gregson, Nicky and Lowe, Michelle (1994) *Servicing the Middle Classes: Class, Gender, and Waged Domestic Labor in Contemporary Britain.* New York: Routledge.
Hanson, Susan (1992) 'Geography and feminism: worlds in collision?', *Annals of the Association of American Geographers*, 82: 569–86.
Hanson, Susan and Pratt, Gerry (1995) *Gender, Work and Space.* London: Routledge.
Harvey, David (1989) *The Condition of Postmodernity.* London: Blackwell.
Harvey, David (1992) 'Postmodern morality plays', *Antipode*, 24: 300–26.
Harvey, David (1997) *Justice, Nature, and the Geography of Difference.* London: Blackwell.
Hayford, Alison (1974) 'The geography of women: an historical introduction', *Antipode*, 6: 1–19.
Jackson, Peter (1991) 'The cultural politics of masculinity: towards a social geography', *Transactions of the Institute of British Geographers*, 16: 199–213.
Jones, John Paul III, Nast, Heidi and Roberts, Susan (eds) (1997) *Thresholds in Feminist Geography: Difference, Methodology, Representation.* Lanham, MD: Rowman and Littlefield.
Katz, Cindi (1992) 'All the world is staged: intellectuals and the projects of ethnography', *Environment and Planning D: Society and Space*, 10: 495–510.
Kwan, Mei-Po (2002) 'Feminist visualization: re-envisioning GIS as a method in feminist geographic research', *Annals of the Association of American Geographers*, 92: 645–61.
Lawson, Vicky and Staeheli, Lynn (1995) 'Feminism, praxis and human geography', *Geographical Analysis*, 27: 321–38.
Limb, Melanie and Dwyer, Claire (2001) *Qualitative Methodologies for Geographers: Issues and Debates.* New York: Arnold.
Longhurst, Robyn (2000) *Bodies: Exploring Fluid Boundaries.* London: Routledge.
Massey, Doreen (1991) 'Flexible sexism', *Environment and Planning D: Society and Space*, 9: 31–58.
Massey, Doreen (1994) *Space, Place and Gender.* Minneapolis: University of Minnesota Press.
McDowell, Linda (1979) 'Women in British geography', *Area*, 11: 151–5.

McDowell, Linda (1986) 'Beyond patriarchy: a class-based explanation of women's subordination', *Antipode*, 18: 311–21.

McDowell, Linda (1997a) *Capital Culture: Gender at work in the City.* Oxford: Blackwell.

McDowell, Linda (ed.) (1997b) *Space, Gender, and Knowledge: Feminist Readings.* Oxford: Oxford University Press.

McDowell, Linda (1999) *Gender, Identity, and Place: Understanding Feminist Geographies.* London: Blackwell.

McDowell, Linda and Massey, Doreen (1984) 'A woman's place?', in Doreen Massey and John Allen (eds), *Geography Matters!* Cambridge: Cambridge University Press, pp. 128–47.

McDowell, Linda and Sharp, Jo (eds) (2000) *A Feminist Glossary of Human Geography.* London: Arnold.

Momsen, Janet and Kinnaird, Vivian (eds) (1993) *Different Places, Different Voices: Gender and Development in Africa, Asia, and Latin America.* New York: Routledge.

Monk, Janice (2004) 'Women, gender, and the histories of American geography', *Annals of the Association of American Geographers*, 94(1): 1–22.

Monk, Janice and Hanson, Susan (1982) 'On not excluding half of the human in human geography', *The Professional Geographer*, 34: 11–23.

Moss, Pamela (ed.) (2002) *Feminist Geography in Practice.* Malden, MA: Blackwell.

Nash, Catherine (1996) 'Reclaiming vision: looking at the landscape and the body', *Gender, Place and Culture*, 3: 149–69.

Nast, Heidi and Pile, Steve (eds) (1998) *Places through the Body.* London: Routledge.

Nesmith, C. and Radcliffe, Sarah A. (1993) '(Re)mapping Mother Earth: a geographical perspective on environmental feminisms', *Environment and Planning D: Society and Space*, 11: 379–94.

Oberhauser, Ann M. (1997) 'The home as "field": households and homework in rural Appalachia', in John Paul Jones III, Heidi Nast and Susan Roberts (eds), *Thresholds in Feminist Geography: Difference, Methodology, Representation.* Lanham, MD: Rowman and Littlefield, pp. 165–83.

Phillips, Richard (1997) *Mapping Men and Empire: A Geography of Adventure.* London: Routledge.

Pile, Steve (1994) 'Masculinism, the use of dualistic epistemologies and third spaces', *Antipode*, 26: 255–77.

Pile, Steve (1996) *The Body and the City: Psychoanalysis, Space and Subjectivity.* London: Routledge.

Pratt, Gerry (1993) 'Reflections on poststructuralism and feminist empirics, theory and practice', *Antipode*, 25: 51–63.

Professional Geographer (1994) 'Women in the "Field": Critical Feminist Methodologies and Theoretical Perspectives', collection, 46: 426–66.

Professional Geographer (1995) 'Should Women Count? The Role of Quantitative Methodology in Feminist Geographic Research', collection, 47: 427–58.

Radcliffe, Sarah (1994) '(Representing) postcolonial women: difference, authority and feminisms', *Area*, 26(1): 25–32.

Rocheleau, Dianne, Thomas-Slayter, Barbara and Wangari, Esther (eds) (1996) *Feminist Political Ecology: Global Issues and Local Experiences.* New York: Routledge.

Rose, Gillian (1993) *Feminism and Geography: The Limits of Geographical Knowledge.* Minneapolis: University of Minnesota Press.

Smith, Susan (2002) 'Doing qualitative research: from interpretation to action', in Melanie Limb and Claire Dwyer (eds), *Qualitative Methodologies for Geographers: Issues and Debates.* New York: Arnold.

Teather, Elizabeth (ed.) (1999) *Embodied Geographies: Spaces, Bodies and Rites of Passage.* London: Routledge.

Walby, Sylvia (1990) *Theorizing Patriarchy.* Oxford: Blackwell.

Women and Geography Study Group of the IBG (1984) *Geography and Gender: An Introduction to Feminist Geography.* London: Hutchinson.

5 MARX AND THE SPIRIT OF MARX

George Henderson and Eric Sheppard

Marxism is an extraordinarily rich tradition, with numerous offshoots, variants, and adherents. It offers both commentary and substantive thought on economics, development, urbanization, agriculture and industry, politics and governance, international relations, social classes and divisions, science, literature, painting, film, history, the environment, ethics, and more. It is interested in what the world is like and who makes it that way; in what knowledge and feelings people have about their situations and how those perceptions arise from those very situations. It is at root an inquiry into human endeavors of all sorts, with the aim of bringing about more just conditions for human flourishing. To cover Marxism in a single chapter, even those Marxist approaches undertaken in geography, is a tall order. We can only cut it down to size and offer a few starting points and guide posts, with the hope of raising more questions than can be answered here.

We begin by discussing why Marxism remains relevant, notwithstanding recent material and intellectual developments. Then, we discuss the geographical resonance of its philosophical foundations. Turning our attention, like Marx, to capitalism, we discuss Marx's political economic analysis and how it has been developed to make sense of the geography of capitalism. Finally, we show how Marxist geographers embrace a much broader set of questions than the economy, illustrating the richness and ongoing vitality of this approach. Students of human geography should be wary, therefore, of jettisoning Marx along with some objectionable bathwater.

The deployment of Marx's ideas in geography has an interesting history. At one level, interest in Marxism has declined markedly in popularity since 1990. At that time, the most influential and widely cited geographers, following David Harvey, drew heavily on his ideas. Since, Harvey's scholarship has undergone steady criticism, and many of the most influential geographers now position themselves as post-Marxist. At another level, however, the apparent decline of Marxism marginalizes the fact that a number of its most important insights have been internalized within human geography. We argue here that any 'death of Marx' is premature, by demonstrating what makes his ideas of ongoing relevance to geography.

One reason advanced for the death of Marxism is historical: the dissolution of those regimes in the Soviet Union and Eastern Europe that self-identified as Marxist. The confinement of Marxist-influenced political systems to North Korea, Cuba and China, with democracy supposedly on the rise everywhere, is widely alleged to prove the bankruptcy of Marx's theories (although China invites pause for reflection). This argument is problematic. First, the Marxism practiced in such countries has had little in common with the ideas of Marx. Marx focused much of his attention on understanding capitalism rather than developing schemes for socialist and communist societies, and would have found little to praise in those autocratic and repressive regimes. Second, global political and economic developments since 1990 have resulted in a world that looks more like the capitalism that Marx envisioned than at any time since he wrote. We now live in a world dominated by neoliberal capitalist ideologies and practices, with non-capitalist economic practices at a historically low ebb. While there are many similarities between the rapidly globalizing British-dominated and free-trade-oriented world economy studied by Marx and today's US-dominated neoliberal globalization, the prevalence of capitalist practices and discourses is far greater. Powerful nation-states take few actions without reflecting on how they will be assessed by 'the market'. In the global south, which for decades sought a 'Third World' between capitalism and state socialism, elites now adopt capitalism as their chosen development vehicle. Everyday discourse is dominated by stock market returns, interest rates, markets and consumption. Unions and progressive political parties have either lost influence or reinvented themselves to adapt to neoliberalism. Seeking someone to help make sense of these developments, Cassidy (1997) described Karl Marx in *The New Yorker* as 'the next great thinker'.

A second argument advanced against Marx, influential in geography, is philosophical. French postmodern and poststructural thinkers, expressing their own frustration with the French left's ongoing dalliance with Soviet Stalinism, began to argue in the 1960s against Marx, and other grand theorists. It is from this position that critical Anglo-American geographers have expressed increasing skepticism about the relevance of Marx's ideas, notwithstanding shared political sentiments. These debates have become quite personal, as Marxist and post-Marxist geographers each advance their position by caricaturing and denigrating (othering) their opponents. We argue, however, that there is far more heat than light in these disputes. The appellation 'post' refers not to a distinctly new time period with radically different ideas and practices, but to an approach that acknowledges the ongoing influence of pre-existing ideas and practices even as it seeks to deconstruct them. Marx remains a key starting point for the grand philosophers of the 'post' variety (Lacan, Foucault, Derrida, Deleuze, Spivak), and similarly cannot be set aside by critical geographers. The often unacknowledged continuities between Marxism and the 'posts', we argue, also point to the ongoing relevance of Marx's thought.

Philosophical Foundations: Materialism and Dialectics

One of geography's distinguishing features is a concern for nature–society relations. This is also at the center of Marx's desire to develop a materialist account of society. *Materialism* is the view that whatever exists depends on matter. As Marx developed the idea, he pointed to how human activities are grounded in the biophysical environment from which sustenance, clothing and shelter are drawn. This pristine 'first nature' formed a materially necessary foundation for human society. At the same time, he noted that society has progressively transformed, reshaped and commodified nature, creating a 'second nature'. Our contemporary world, where climate, ecosystems, organisms and the landscape are profoundly reshaped by human activities, certainly is one where almost all aspects of nature have been transformed. Yet nature is also shaped by biophysical processes that continually break out of the boxes into which humans seek to cram nature (think of global warming, or mad cow disease), biting back in ways that show how second nature remains crucial to social life, and always partially beyond the control of capitalism. This ongoing, complex and interdependent relationship between societal and biophysical processes is well captured by applying Marx's favored dialectical analysis (see Smith, 1990; Castree, 1995, for a more detailed account).

A second distinguishing feature of a geographical way of knowing, therefore, is to pay attention to the relational nature of the world. Waldo Tobler famously quipped that the first law of geography begins with the presupposition that everything is related to everything else: nature with society; economic, political, cultural processes with one another; local places with one another and with global processes; and so on. Such relational thinking was also central to Marx, who developed a form of relational dialectical thinking in opposition to the Aristotelian (i.e. essentialist) logic that dominates such mainstream approaches to science as logical empiricism. *Dialectical reasoning* (dating back to Greek philosophy) focuses on analyzing the relations between things, rather than the things themselves. Harvey describes this well:

> **Dialectical thinking emphasizes the understanding of processes, flows, fluxes, and relations over the analysis of elements, things, structures and organized systems all ... There is a deep ontological principle involved here ... that elements, things, structures and systems do not exist outside of or prior to the processes that create, sustain or undermine them ... Epistemologically, the process of enquiry usually inverts this emphasis: we get to understand processes by looking either at the attributes of what appear to us ... to be self-evident things or at relations between them ... On this basis we may infer something about the processes that have generated a change in state but the idea that the entities are unchanging in themselves quickly leads us to a causal and mechanistic thinking. Dialectical reasoning ... transforms the self-evident world of things ... into a much more confusing world of relations and flows that are manifest as things. (1996: 49)**

Whereas mainstream science and social science seek to explain the world by reducing it to a series of stable and well-defined entities (quarks, organisms, human agents) connected by stable causal relations, dialectical reasoning traces how these interrelations are constantly changing, altering the entities themselves. As long as they are reproduced by the relations constituting them, entities seem to be stable and well defined. These 'permanances' are illusory, however. Entities of all scales, from humans to the world economy, are internally heterogeneous and always at risk of being torn apart by the very relationships that brought them into being.[1]

Marx insisted that materialism must be both dialectical and historical. Dialectical materialism focuses on relations between the material world and our ideas about it, arguing that each shapes the other, but that the mind always remains dependent on material processes supporting human life. A logical extension of this way of thinking, in other words, is that commonly held dualisms, such as society and nature or culture and nature, must be carefully scrutinized. Indeed, this is a project that quite a few Marxist-inspired geographers have undertaken in lively fashion in recent years (cf. Castree, 2003).

Historical materialism holds that any mode of production is beset with contradictions that can undermine its viability and tear it apart – even capitalism. Thus, instead of accepting global capitalism as a utopian end-state in which markets optimally allocate society's wealth amongst its members, Marx sought to identify its contradictions and potential limits, and (to a much lesser extent) to speculate on how those limits might be reached and what might succeed capitalism. Marx's historical materialism is sometimes misleadingly represented as a teleological sequence of stages of societal development: slavery → feudalism → capitalism → socialism → communism. As a nineteenth-century white male thinker, there is little doubt that Marx had a pretty

Eurocentric view of the world, and at times did resort to a stagist shorthand whereby the historical sequence experienced in Europe was taken to be universal. Yet the idea of historical materialism simply holds that any such mode of production has internal contradictions that can undermine it, without specifying fixed sequences, or even a predetermined set of possible modes of production. Such a prior identification of entities would be inconsistent, of course, with dialectical reasoning.

Marx's Political Economy

While Marx did not offer a canned historical sequence of social change, he did argue that each place and time is characterized by a predominant mode of production: a socially organized way in which humans provide for the material basis of their existence, by coordinating production with the social relations necessary to support it. For example, under slavery owners of the means of production also owned labor, whereas under capitalism individuals own their own labor, which they bring to the labor market. These different systems are governed by different norms and legal principles about who counts as a free person. Moreover, Marx argued that the social relations deemed 'necessary' for the material basis of human life were outcomes of antagonistic group relations, struggles over ideas, and often the exertion of brute force by a ruling class. Social relations are not *naturally* necessary but are historically wrought through struggles over material and political interests. An important corollary is that all historically significant modes of production must be capable of producing a *surplus* (i.e. more at the end of the year than at the beginning), or else they would have no reserves to tide them through a period of crisis. Marx recognized that diverse modes of production typically coexist (e.g. slavery has not been eliminated in today's capitalist world economy), but analysis should begin with the dominant one.

Capitalism

Marx spent much of his lifetime applying this philosophical approach to unravel and demystify the capitalist mode of production, which he saw as a definite improvement over slavery and feudalism but still riven with hidden contradictions and inequalities. Under capitalism, he argued, the production of goods to support human life takes the form of commodity production. Commodity production occurs when goods are produced to make a profit, rather than to meet a particular need. Historically, some possess the means of production to make these goods (capitalists), or the land and other natural resources used for commodity production (landowners), whereas others provide the labor power necessary for production (workers). Workers are free to choose where to work (in theory, if not in practice), but do not have the means to undertake commodity production themselves.[2] There are thus distinct class positions that different individuals occupy in a capitalist mode of production, shaping their interests and identities, which Marx was careful to distinguish as 'class-in-itself' (the interests shared by those associated with a particular class position) and 'class-for-itself' (the creation of a shared identity, necessary for members of a class to pursue their interests through collective action). In addition, capitalism entails legal institutions, particularly those creating and guarding private property rights; political institutions, a state seeking to mitigate the contradictions of capitalism without disaffecting the different classes; and a set of discourses legitimizing and normalizing capitalism, such as those of neoliberalism (cf. Jessop, 2002).

Marx provided distinctive insights into the inequities and contradictions of capitalism, quite different from the classical political economists who preceded him or the marginal economists who came after. On the face of it, under capitalism it seems that the marginal value of goods to society is given by their market price; that capitalists and workers earn what they deserve. It seems that capitalists make their profit as fair payment for the capital they invest, reflecting the value of that capital to society. By the same token, wages are interpreted as a measure of the social utility of labor. According to Adam Smith's principle of the invisible hand, the free operation of markets means that the self-interested actions of individuals produce a fair redistribution of society's wealth amongst its members. Yet Marx's analysis showed that things are not as they seem, whether we consider production or consumption. Marx's notion of capitalism is a far cry from those who equate capitalism with markets. Rather, capitalism is a distinctive way of socially organizing *production,* in which the market is necessary for capitalism's existence[3] while also obscuring the nature of capitalist production.

Production, value and exploitation

Marx was terribly concerned with what is real and tangible and with what the seemingly real and tangible obscures. The very first words of *Capital* display this focused zeal: 'The wealth of those societies in which the capitalist mode of production prevails, presents itself as "an immense accumulation of commodities" ... Our investigation must therefore begin with the analysis of a commodity' (Marx, 1967 [1867]: 35). One of Marx's first questions was a basic one: what gives these commodities their value? He argued that market prices are only one way in which their value can be assessed. He stressed that commodities have use value, a measure of their utility to those purchasing and selling them. They also have labor value, measured by the labor time socially necessary to make those commodities (including the labor invested in the materials and machinery used up during production). 'Socially necessary' refers to the labor time associated with prevailing production methods. Marx showed that capitalism can make a surplus only if the labor value contributed by workers to the production process is greater than the labor value of the real wage they receive (see Box 5.1).

Without exploitation in labor value terms, monetary profits are impossible. Marx also argued that in the long term, goods will exchange in the market in proportion to their labor value: supply and demand fluctuations can drive market prices away from these proportions, but only in the short term. These propositions have been controversial, but they have turned out to be robust. Money profits indeed cannot be made without exploitation in labor value terms, and empirical calculations have confirmed that long-run prices correlate closely with labor values.

Labor power is the only commodity where these propositions do not apply, because the free labor market does not extend inside the factory gate:

> **We now know how the value paid by the purchaser to the possessor of this peculiar commodity, labor-power, is determined. The use-value which the former gets in exchange, manifests itself only in the actual usufruct, in the consumption of the labor-power. The money-owner buys everything necessary for this purpose, such as raw material, in the market, and pays for it at its full value. The consumption of labor-power is at one and the same time the production of commodities and of surplus-value. The consumption of labor-power is completed, as in the case of every other commodity, outside the limits of the market or of the sphere of circulation. Accompanied by Mr Moneybags and by the possessor of labor-power, we therefore take leave for a time of this noisy sphere, where everything takes place on the**

BOX 5.1 MARX'S THEORY OF EXPLOITATION

Whereas in money value terms it may seem that workers earn a fair day's pay for a fair day's work by participating in the labor market, in labor value terms they are *exploited*. A component of the labor value they contribute is captured as capitalists' profits:

Value of one hour of labor (1)
= labor value L received by workers as their wage + surplus value S retained by capitalists

$1 = L + S$; rate of exploitation $(E) = S/(L + S)$

> surface and in view of all men, and follow them both into the hidden abode of production, on whose threshold there stares us in the face 'No admittance except on business.' Here we shall see, not only how capital produces, but how capital is produced. We shall at last force the secret of profit making ... On leaving this sphere of simple circulation or of exchange of commodities, which furnishes the 'Free-trader Vulgaris' with his views and ideas, and with the standard by which he judges a society based on capital and wages, we think we can perceive a change in the physiognomy of our dramatis personae. He, who before was the money-owner, now strides in front as capitalist; the possessor of labor-power follows as his laborer. The one with an air of importance, smirking, intent on business; the other, timid and holding back, like one who is bringing his own hide to market and has nothing to expect but – a hiding. (Marx, 1967 [1867]: 175–6)

Under the capitalist mode of production the full value of what working people make cannot return to them. If it did, capitalism would not exist. In so far as market exchange is perpetuated, so are the social relationships that define capitalism. Not only are the economic and social realms inseparable, but Marx redefines each in terms of the other.

Consumption and commodity fetishism

In all of this there is a pervasive though peculiar irony. Marx wrote that even while commodity production is central to capitalism, any given commodity tells us nothing about the social conditions and relationships that went into its making. Even though social relationships are a necessary, definitive aspect of commodity exchange under capitalism, 'Value,' Marx quipped, 'does not stalk about with a label describing what it is. It is value, rather, that converts every product into a social hieroglyphic' (1967 [1867]: 74). Commodities, once in people's possession, are understood by them merely as use values – by the comfort, utility or pleasure they bring, their distinctiveness, and the individual style they confer on their owners. The conditions under which they are produced are utterly lost to our eyes, except for the occasional exposé about the sweatshop labor behind Nike shoes or mistreatment of the cattle in our hamburgers. Under capitalism, it becomes easier to desire these appealing *things* than it does to question the social manner of their production. Marx called this strange state of affairs *commodity fetishism*: once labor power takes the form of commodities, labor power as a social phenomenon, binding together a whole class of individuals with a common interest, becomes imperceptible. More ironically, commodity fetishism has itself become a commodity, the raw material for the advertising industry that has exploded since Marx's time.

In putting commodity production front and center, Marx is saying that several issues are at stake. First, whereas all people have always relied upon the non-human world ('nature') for their survival, under capitalism commodity production mediates our relations with the non-human world. In particular, attitudes toward nature and material appropriations of it tend to be highly instrumental, very often wasteful, and generally subservient to the interests of private property, resource extraction, and the like. (Neil Smith, 1990 and Denis Cosgrove, 1998 [1984] stress, however, that such alienation from nature has also fueled strong movements romanticizing and aestheticizing nature, like the World Wildlife Fund, or the growing interest in eco-tourism.)

Second, commodity production structures relationships between people, whose capacity for labor becomes a commodity itself, lying beyond their direct control. Those things and human relationships that fit the requirements of commodity production will dominate in capitalist society. Furthermore, these phenomena come to dominate people themselves, as we bring our own interests into greater and greater alignment with those of commodity production, and measure our worth by our wealth. The primary way for us to meet our needs is to purchase the goods produced under capitalism, provided we have the wherewithal and opportunity to sell our labor power. But needs are a trickier matter than it seems at first. Capitalism, Marx says, can indeed provide many people's needs but these accord with a restrictive sense of what it might mean to live a fulfilling life.

Third, the idea of commodity fetishism hints that capitalism entails a struggle over and for knowledge. In so far as the social relations of capitalism become, in a sense, fixed by the requirements of commodity production, they also become opaque to us. This is because the things that capitalism makes, and which surround us in our everyday existence, do not reveal to us the social dimensions of their own origins and existence. But it is precisely an understanding of these social dimensions Marx saw as crucial for emancipatory transformation.

Finally, because there are real interests at stake on all sides of the capital–labor relation, we would expect that the making of commodities is not a one-time affair. It is and must be repeated but is at the same time the stuff of historical modification and geographical modification from place to place.

Time and Space: the Accumulation and Circulation of Capital

Marx argues that capitalism is inherently dynamic: accumulation of capital must ever expand. He predicted, for example, that more and more kinds of things would become commodities, more and more 'needs' would be invented, more and more parts of a country's population would become wage workers ('alienated' from their own means of production) who were not wage workers before, and more and more non-capitalist societies around the world would become drawn into capitalism's orbit. At the same time, he predicted that even as the size of the working classes grew, a countermovement would also develop whereby in many economic sectors machines would replace human labor. To a very large degree, developments during the twentieth century bore him out.

Capital accumulation is a continuous though expanding cycle or circuit centering, as has been noted, on the production of

M...C (LP, MP)...P...C′...M′...C (LP, MP)...P...C′...

Figure 5.1 The circulation of capital

commodities. In Figure 5.1 (adapted from Harvey, 1982: 70), money M is used to purchase two kinds of initial commodities C – labor power LP (typically *advanced* by workers) and the means of production MP – for purposes of making new and different commodities. During production P, new commodities C′ are made by consuming LP and (a portion of) MP, producing a *surplus* value at the end. C thus grows to C′ because a labor process has imparted surplus value to the commodity. These new commodities are taken to the market where a successful sale realizes more money M′, so that the whole process can begin again.

Figure 5.1 implies that value, once created by labor, must be preserved or translated throughout this whole process. Only labor creates value during production; commodity exchanges and sales after the fact serve to preserve and circulate that value, at least if all 'goes well'. Figure 5.1 also illustrates that the circulation of capital comprises three *individual* circuits: the circulation of money capital, productive capital, and commodity capital (M–M′, P–P, and C–C′, respectively). Capital thus takes different forms whose needs continually contradict each other. Let us take a closer, if very brief, look through each of these windows.

In the circuit of money capital, M is advanced to pay for production, and M′ is that plus the profits made as a result of a successful sale. Thus capitalists must be concerned with preserving (or enhancing) the value of money itself, retaining enough of it to invest in successive rounds of production. Money must circulate as efficiently and rapidly as possible; time is money and delays are costly. Yet, money capital is ultimately used for production, which takes time and material resources.

During the circulation of productive capital, some portion of money must be at rest, in order for production to occur, contrary to the interest of capital in its raw money form. It takes time to assemble inputs, make commodities and then get them to market and sell them. The longer this takes, the slower money capital can move, cutting into profit rates. This can be counteracted by accelerating the circulation of money capital. In addition, capitalists have to manage processes that take far longer than the time necessary to produce and sell a commodity: machinery and buildings, but also new product development, require investments that can only be paid back after many production cycles. While firms seek to set aside profits to pay for such fixed costs, the credit market is essential to enable capitalists to borrow for these longer-term investments. The second circuit also has its own internal contradictions. Capitalists want the value of productive capital to last for its lifetime. Yet capitalist competition is based on developing new products and technologies, which have the effect of undermining the value of already existing products and technologies. This undermines the value of older productive capital, which may become obsolete before it has paid for itself.

The circuit of commodity capital begins when C′ becomes M′, i.e. when productive capital generates the money necessary to pay for a new round of production. But much can go wrong between shipping a product to market and realizing profits. Production and sales are separated over time and space. Capitalists may fail to predict demand correctly, particularly in distant markets; products may be delivered too late; and productive and money capital can get lost *en route*. Also, there is no guarantee that commodities will remain 'socially useful'. Intense competition among capitalists places an extraordinary premium on stylistic change and innovation.

Particular roles 'crystallize' out of these circuits: finance capitalists have much to say about where money capital is invested; industrial capitalists organize and supervise production; and wholesalers and retailers manage the sale and purchase of commodities. As finance, industrial, and merchant capitalists have sought greater efficiencies, a vast array of new strategies has been devised: sophisticated and risky forms of financial securities, just-in-time methods of production, relocation of assembly plants and processing facilities, and stunning concentration in retail markets, to name only a few. These have their corollaries in waves of plant closures, local and regional deindustrialization, rollbacks of labor and environmental standards, and fraud and bankruptcies on Wall Street.

The seeming functionality of Figure 5.1 hides the impossibility of its smooth and seamless practice over time and space. It demands coordination of a great many things, not least formal arrangements for human exploitation, only to lead down a path of regular economic and social crisis.

Yet we should not forget a fourth circuit: what Marx called 'variable capital'. Variable capital refers to the contributions of labor to the value of commodities. This circuit entails the making and perpetuation of working classes and the suite of daily practices, social differences, supportive networks, and cultural logics that attend our daily return to work (e.g. K. Mitchell et al., 2003). It is as complex as any of the other three: it has to do not simply with how the capability for work is established but with the circumstances whereby particular groups of people, demarcated by gender, race, nationality, sexuality and ethnicity, are targeted for particular kinds of jobs. It depends upon forms of unpaid, often hidden and women's labor, raising generations of future workers and returning already existing paid workers, clothed and fed, to work. It also involves maintaining a surplus army of workers, populations of unemployed, who must be sustained through periods of economic crisis to be drawn upon during periods of recovery. Finally, it entails struggles over people's identities as workers versus a fuller conception of human being. It is through the circuit of variable capital and the struggles that shape it that living standards and an acceptable quality of life are hammered out.

Time is also important in wider senses than circuits of capital. The pressure to turn M into M′ has been expressed in a politics of the working day: specifically, how long it should be, how fast work should proceed, how many days should comprise the work week, and how much time should be allowed for vacation, family, and sick leave. Such innovations as the eight-hour day, the weekend, and paid (or unpaid) leave are all outcomes of the struggle over time and to whom it rightly belongs. Depending upon how history is periodized and narrated, the concern for profit rates can become palpable in other ways. It is therefore common to speak of the Great Depression of the 1930s or the Downturn of the 1970s. While it is hotly debated whether profit rates must *necessarily* fall in some coordinated fashion throughout the whole of the capitalist world, one thing is clear. The sense of time itself is often measured in ways that articulate to how capitalism structures everyday life.

Marx had much less to say about space, but the work of a generation of economic geographers has come a long way in remedying this shortcoming. First, a dialectical approach means that space (and time) cannot be treated as externally given entities – coordinates and dates:

> **Space and time are neither absolute nor external to processes but are contingent and contained within them. There are multiple spaces and times (and space–times)**

> **implicated in different physical, biological and social processes. The latter all *produce* – to use Lefebvre's (1974 [1991]) terminology – their own forms of space and time. Processes do not operate *in* but *actively construct* space and time and in so doing define distinctive scales for their development. (Harvey, 1996: 53)**

The space–times of capitalism are created as a result of capitalist political economic processes seeking to facilitate, or in the case of class struggle, sometimes to disrupt or tame, capital accumulation. Under capitalism, time took the form of clock time, so that hours worked and profit rates could be calculated, whereas space was precisely measured, to define property ownership, the location of resources, labor, inputs and markets, and their accessibility. The spatial organization of activities is important to the functioning of capitalism. The greater the socio-economic distance between inputs, production, and markets, the more time it takes for the circuit of capital to be completed. Greater distances are also a source of greater uncertainty and disruption of circuits. Enormous effort has been put into reducing these spatial barriers, coming up with geographic technologies that enhance ease of travel and communication, ranging from sextants and sailing clippers, to GPS and the Internet (Hugill, 1993; 1999). Marx termed this the tendential 'annihilation of space by time' (1973; Harvey, 1985b). In brief, temporal barriers to the turnover of capital can be reduced by reducing socio-economic distance. The complement to this ever-shrinking world, though, is that it is selectively shrinking, leaving whole areas and peoples abandoned and delinked from this dominant economy of time–space compression. This delinking has been dubbed 'time–space expansion' (Katz, 2001; Crump, 2003).

Second, production of the built environment is also a vital aspect of the spatiality of capitalism. Urban morphologies can enhance capital accumulation by providing appropriate land use plans, buildings and spatial arrangements. Company towns located workers in company-built housing close to factories, enhancing control over labor. The operation of the land market also sorts urban areas out into land use patterns in which those with more money can demand favored locations, whereas others (workers) take what is left over. In early industrial cities, this meant that workers crowded into relatively expensive but undesirable central city locations near factories.

Communications infrastructures and built environments that once best served capital accumulation subsequently can become a barrier to capital accumulation, however. As production methods change, and previously desirable inner city locations become polluted, congested and run down, there is strong pressure to pull investments out of these landscapes and to restructure the landscape in order to satisfy the distinct locational preferences of a new era of capitalism. For example, suburbanization was a way of avoiding congested inner cities, and also catalyzed inner city decline in the USA. Not only wealthy and white residents, but also retailing, manufacturing, and real estate capital left inner cities for the greener pastures of the metropolitan fringe with the space to build big box stores and single-floor factories, catalyzed by state subsidization of highway systems and housing purchases, and by the emergence of independent municipalities on the urban fringe with the power to exclude undesirable land uses and offer residents lower taxes and a cleaner environment (Harvey, 1972; Walker, 1981). Over time this would produce a 'rent gap' in the inner city that recently has been attracting investments back downtown as gentrification, enterprise zones, and assorted public–private partnerships (Smith, 1996). Investment is thus a see-saw process whereby capital is pushed into and pulled out of places over time (Smith, 1990).

Such periodic spatial restructuring, typically occurring during periods of economic turndown, can be observed at spatial scales ranging from the intraurban to the international. Indeed, recent Marxist research has devoted considerable attention to how even these scales are produced and transformed as capitalism evolves, arguing that the recent acceleration of globalization is associated with a declining importance of the nation-state scale relative to global scales and subnational scales – glocalization. National economies both are increasingly shaped by supranational forces and devolve responsibilities for economic prosperity and competitiveness from the nation to cities and industrial districts (Swyngedouw, 1997; Brenner, 1999).

Third, political processes governing the capitalist economy are generally organized territorially, as national, regional and local states. Territorial governance structures experiment with different schemes to coordinate capitalist production and demand, since the market is not self-regulating (Tickell and Peck, 1992; Painter, 1997). For example, between 1945 and 1973 First World nation-states catalyzed a Fordist regime of mass production, complemented by political and legal structures that permitted higher wages to pay for those products, and a welfare safety net for those left behind. After Fordism ran into difficulties in the 1970s, new regulatory systems sought, with mixed success, to shift to a more flexible regime of accumulation, complemented by what Bob Jessop (1994) has dubbed the Schumpeterian workfare state – a mode of social regulation replacing safety nets with welfare-to-work programs. Different territories also compete with one another, seeking to attract geographically mobile capital to employ their relatively immobile workforce. This local entrepreneurialism has intensified with time–space compression and glocalization, pitting all residents in one place in competition with their counterparts elsewhere, irrespective of class. Harvey argues that time–space compression has made different entrepreneurial conditions in different places the critical factor in such competition, although others argue that geographical location as such still matters (Sheppard, 2000).

After a review of all of these processes and conditions, one would be excused for thinking it all a bit chaotic. The 'circulation' of capital, while an absolute necessity, is indeed a vast, inconstant, and risky attempt to coordinate the transfer of value from one incarnation of capital to the next and from one place to another. Geographical regimes of capital accumulation are unsteady over the long haul, making uneven geographical development (i.e. persistent spatial inequalities in wealth and economic growth) an inherent feature of capitalism.

In historical and geographical reality there is an extraordinary variety of ways for accumulation to proceed. Because capitalism has internal contradictions, a variety of accumulation strategies necessarily results. Because capitalism is peopled – i.e. it *always* confronts and intersects with ongoing traditions, practices, and power relations that are geographically and historically variable – it is also necessarily different from place to place and time to time. This raises an important question. If, as Marx says, capital accumulation is an expansive process over time and space, what is lost and what is gained by interpreting historical and geographical differences through the 'timeless' lens of Marxism's concepts (Chakrabarty, 2000)?

Class conflict

Marx saw that the value of labor power would become a major object of struggle between classes, as we suggested above. Such struggles have become a regular feature of capitalist societies. People know these struggles both in specific terms and more generally as the desire to preserve or enhance their standards of living and quality of life. The value of labor power may rise or fall, but when it rises high enough to threaten capital accumulation, capitalism necessarily enters a period of crisis.

Yet competition between places exemplifies how geography vastly complicates Marx's analysis of class conflict. In Marx's analysis, class location shapes interests and identities, pitting classes against one another. Whereas neoclassical economists claim that wages and profits are set by some stable competitive equilibrium at which each is paid according to its social utility, Marxist economists have shown that this makes sense only in an economy with no class divisions. The historical reality of classes belies this claim. Whatever the profits and wages at any point in time, both classes seek to shift this ratio in their favor, claiming a bigger piece of the pie for their collectivity and destabilizing the economy. The state seeks to manage such conflicts, typically by favoring one side over the other. The economic policies of the Bush administration in the early twenty-first century are an excellent case in point, offering tax breaks to the wealthy while cutting welfare and health benefits for the poor.

Class is a notoriously slippery issue, however. It is not at all clear how or whether shared interests result in a shared identity and collective action. The real economy is made up of all kinds of economic actors who do not fall neatly into one class or another, such as children employed in the family business, workers with pension funds in the stock market, and middle managers (Wright, 1985). Geography further complicates this. When localities compete with one another, the interests of workers in one location are pitted against those of workers elsewhere, in ways that call into question the optimism of Marx and Engels, in *The Communist Manifesto*, about workers' social power (Harvey, 2000: Chapter 3). When capital accumulation favors some places over others, workers in those places systematically benefit at the expense of workers elsewhere, making it in their immediate interest to align with capitalists to protect communities. For example, the AFL-CIO, seeking to sustain the elite status of the US working class, has often supported American international economic policy, itself aimed at maintaining American economic hegemony at a global scale.

Many opportunities exist for workers to influence and even direct capital investments. The new field of 'labor geography' seeks to understand not so much how workers respond to capital's lead regarding plant closings, regional growth or decline, or other indications of uneven development, but how workers themselves play a role in shaping the geography of capital in the first place (Herod, 1998; 2001).

Consciousness, Culture and Representation

The foregoing implies the importance of not losing sight of people's intentionality, agency, and self- or group understanding, even if these are not exercised freely. The issue is a broad one and multiplies rapidly. What *is* the nature of people's consciousness under capitalism? How much does capitalism determine what we think, feel, believe, and communicate? How does it shape society more broadly by entering the realm of symbol and meanings? These questions may be put in the context of Marx's belief that capitalism is voracious in transforming everything it comes into contact with.

Capitalism is mediated by powerful cultural and social processes that have proven uniquely useful to its perpetuation. Capitalism is insinuated in the most basic of institutions, where meanings are given material force. For example, the legal system of the United States gives meaning to the force of law (as does that of any nation-state). It fundamentally protects and legitimates the institution of private property and often gives vent to the class struggle between employer and employee interests.

Legislatures have codified into law the privatization of public assets and basic social services, during certain historical periods, so as to minimize the socialization of the economy and reinforce the disciplining of the labor force (Peck, 1996). It remains the case also that certain quarters of knowledge production (e.g. natural resource management, telecommunications policy, pharmaceutical patents, book and music copyrights) reflect the basic tensions in a society where almost everything becomes a commodity.

The exact path of many cultural and social transformations is by no means clear or predictable, however. Most contemporary work in Marxist geography accords with the notion that people are not mere dupes. There are many issues, then, to be raised about our manifold ways of making meanings and putting meanings into practice. Paraphrasing Marx, however, we do so through forms not of our own making. These forms include class but also citizenship, religion, race, sexuality, gender, ethnicity, and nationality: what goes on in these realms, including their emergence and how they are mutually composed, is not simply reducible to capitalism. Yet Marx himself was not exactly clear on these matters and it has been left to his interpreters to try to work things out.

Some Marxists sought to relegate the sphere of cultural belief and tradition, religion, politics, law, science, literature and the arts, and so forth to a so-called superstructure. This was said to rest upon a 'base' (i.e. capitalism, narrowly defined). In this view, the needs of capital and the material interests of capitalists dominate in capitalist society; legal and political systems, religion, schooling and cultural works would be arranged to support, foster, and ultimately reflect those needs and interests. Little more needed to be said. While popular for a time, this 'base and superstructure' model left far too much unaccounted for, and contradicted much of what is known about social and cultural history.

This model did not take root in Marxist geography. Instead, while asserting that the requirements of capital do indeed make it a voracious force, geographers such as David Harvey were interested in how culture, state, the law, the arts, etc. were inextricably bound with the circulation of capital. Harvey (1982) argued that capitalism in different times and places is actually made differently precisely because it intersects and even relies upon local cultural and social variation. Capitalism need neither flatten out geographical differences in ways of life and modes of power nor put a definitive, totalizing stamp on all those things. This was a restricted claim about culture: capital could not get along without it, but there was more to culture than capital. Harvey elaborated on these views in a classic essay on mid-nineteenth-century Paris (1985a) and again in his popular book *The Condition of Postmodernity* (1989). Denis Cosgrove took up similar questions in his book *Social Formation and Symbolic Landscape* (1998[1984]), looking at how the rise of landscape painting and aesthetics in the 1700s and 1800s reflected ways in which land came to be defined as property to be privately owned and *visually* possessed during the transition from feudalism to capitalism in Western Europe. He did not claim landscape painting's link to capitalism was the only and sufficient link to be made, stressing also quite separate traditions in the art and technology of painting, and the unique biographies of artists and patrons.

The role of culture in capitalism has been given a different spin by Don Mitchell, who is interested in how culture becomes reified in capitalist settings. He means this in two senses: first that 'culture' is a concept, a label we give to those kinds of phenomena that we deem 'the cultural'; and second that 'culture', once deployed in that fashion, becomes static, objectified, and commodified. For a Marxist like Mitchell these are interesting observations to make for what they tell us about the contradictions of the sort of society we live in.

Positing culture in static, proprietary terms leads to what Mitchell and others have called culture wars. Culture wars refer, among other things, to struggles over who has control over the manner in which a nation's history, literature, and peoples will be represented, and over how certain kinds of heritages and legacies should be defined and reproduced. Such culture wars are evident in the US (and in different regions and municipalities within the US) over such issues as public school textbook adoption, public education on sexualities, funding for 'controversial' artists, the role of free speech in public space, and indeed who has the right to occupy public space. Mitchell is especially interested in how capitalist urban restructuring has become expressed through cultural (and legal) battles over the occupation of public space. As many central cities have felt the stresses and strains of disinvestment and interurban competition, strategies for attracting mobile capital have included privatization of and increased surveillance over the use of public space. In this manner 'culture' is said to be put to 'work' in capitalism (D. Mitchell, 2000; 2003).

Capital circulation fuels another static and proprietary view of 'culture': its embeddedness in commodities. Drawing upon Adorno and Horkheimer's notion of a 'culture industry', Mitchell (2003) notes how consumers are encouraged to identify themselves with the goods they buy or desire. In this sense, we are what we eat, wear or drive, a process that differentiates consumers. Very specific consumer identities are promulgated along the lines of, for example, gender, race, and sexuality (Dwyer and Jackson, 2003; Rushbrook, 2000). What happens to people when their identities become shaped by what they purchase? For Adorno and Horkheimer, when culture is a 'consumable', this wreaks havoc with the possibility that culture can help us imagine more transcendent modes of social being. Mitchell and others ask us to examine the consumer-culture logic of shopping spaces, such as the enclosed themed malls outside Denver, Colorado and Minneapolis, Minnesota (Goss, 1993). In order to market the vast range of commodities that capitalism is so adept at producing, these spaces indicate that capitalists steadily have eroded the divide between commodities as such and the appealing environments that surround them. The more consumption includes the human environment as such, the more natural it feels to be 'consumers' and the more 'thingified' culture becomes. As the French 'situationist' Guy Debord (1970) once sardonically observed about capitalism: that which appears is good and all that is good appears. While these theorists of consumer capitalism throw down the gauntlet and demand we really think about what is asked of us as consumers, it should be noted that other scholars put more emphasis on the indeterminacy of meaning in the realm of commodity production and consumption (Dwyer and Jackson, 2003; Sayer, 2003).

The embeddedness of cultural meanings in commodities, and of commodities in cultural meanings, has proved to be central to what some have dubbed the 'cultural turn' in human geography. To some degree this cultural turn indicates the influence (if not wholesale acceptance) of semiotics, particularly as inspired by French semioticians Roland Barthes and Jean Baudrillard. These thinkers fashioned an important set of arguments concerning the 'sign' value of commodities that mobilizes desire and thus accompanies, augments, and potentially shapes the traditional Marxist categories of exchange and use value. The cultural turn in Marxist or Marxist-inspired geography does not stop at the symbolic value of commodities, however. Other work in this tradition turns to examine the sphere of production and labor. This branch of Marxist-inspired research maintains a deep engagement with ethnographic approaches and asks how cultural meanings are intrinsic to the labor process itself, such that one cannot be known without the other. Some of the most interesting research in this area looks at how ostensibly capitalist and non-capitalist economic processes articulate to produce forms of labor and accumulation that defy familiar categories of understanding. For example, Vinay Gidwani (2000; 2001) argues that labor processes can be shaped by the cultural logics of working people, for whom labor has purposes other than a paycheck. He sheds a fascinating light on what the capacity to labor means in different places. This capacity may be used on the labor market, or as a conscious withdrawal from the labor market to signify the status and distinction that comes with leisure. Thus labor is not a purely economic category; it is laden with meaning, and how those meanings are acted upon has a direct effect on actual work arrangements and the production process. Therefore, in order to explain what happens in production, labor must be understood in a *meaning-full* sense.

To return to the opening thought of this section, the issue is not only that social and cultural institutions and beliefs are useful (or not) to capitalism. It is that cultural practices and meanings make capitalism different from place to place. How much so, and whether difference presides to the extent of making capitalism as such unknowable to us, is an interesting question for people of any political persuasion. But in debating what is lost and/or gained by interpreting historical and geographical differences through the 'timeless' lens of Marxism's concepts, we should remember what we are also trying to get at: how shall we account for and what shall be done about the causes of extraordinary inequality, cataclysmic boom and bust, and profound uneven development?

Conclusion: Social Construction, Social Relations, and Social Difference

Marx held that any mode of production is an essential shaper of the society where that mode prevails. People make their worlds (albeit not under conditions of their own choosing) and those worlds make them: 'By thus acting on the external world and changing it, he [*sic*] at the same time changes his own nature' (1967[1867]: 177). There is in that short passage great consonance between Marx's ideas about society and the considerable current interest in social constructionism. On this basis we wish to make a case for Marx as a contemporary thinker, necessary though not wholly sufficient for our times. It is a truism that societies are not fixed entities. Paraphrasing Marx, nature does not make capitalists on the one side and laborers on the other' (1967[1867]: 169). Marx's emphasis on the social aspects of individual being, and the mutability of those social aspects, we believe, make him a powerful intellectual and political resource. This is not an unproblematic resource, however, for what are also socially constructed are the processes whereby agents decide what dimensions of social life to politicize, and what scales of political practice to engage in. History shows that it is nigh impossible to predict how politics arise and over what concerns. These are matters for social agents in their historical and geographical contexts to decide.

Some human geographers have recently de-emphasized or even forgone analyses of how capital and class are formed, seeking to train their vision on gender, race, sexuality, nationality, and/or ethnicity. It is vital to understand how these modes of difference are wrought, through what mutual intersections, and with what repercussions; it is just as vital to understand how they intertwine with alienation and exploitation. Marxism accepts prima facie that capitalism and the social positions in and around commodity production and distribution are socially constructed, precisely through their intersection with myriad social processes. Through their attention to such intersections, geographers working in or alongside the Marxist tradition have illuminated a wide variety of social processes. These include such diverse examples as the making of an African American industrial working class in Birmingham, Alabama (Wilson, 2000a); the adjustment of civil rights struggles to rounds of capitalist restructuring (Wilson, 2000b); the devaluing of 'women's' labor in the *maquiladoras* of Ciudad Juárez, Mexico (Wright, 2001); the making of gendered labor in the banks of the City of London (McDowell, 1997); the spatial divisions of gendered labor in Worcester, Massachusetts (Hanson and Pratt, 1995); and geographies of money, nature, and race in rural California (Henderson, 1999).

Marxist geographers have even sought to direct attention away from capitalism, as when Gibson-Graham's (1996) poststructural, anti-essentialist Marxism seeks to make visible alternative modes of production embedded in capitalism, in order to imagine non-capitalist alternatives. While some Marxists and non-Marxists, alike, question whether such studies can be labeled 'Marxist', the influence of Marx's ideas, and the vitality of Marxism, is clear in each case. There are indeed many arenas into which Marxist work can extend: how specifics of time and place determine the manner in which capitalism fosters already ongoing forms and processes of racialization, sexuality, gender, and nationality (or some other geographical allegiance); how resistance to capital sometimes gains expression through non-capitalist social relations; and how political coalitions and alliances may form around multiple axes of oppression and domination. Such lists could be extended indefinitely, but geographers interested in asking whether a capitalist earth can provide just conditions for human flourishing cannot afford to neglect the Marxist approach.

NOTES

1 Dialectical reasoning is also applicable to the natural sciences (cf. Levins and Lewontin, 1985; Smolin, 1997).
2 In nineteenth-century Europe and the contemporary global south, reduced access to rural land for food production has pushed people to join the urban industrial proletariat.
3 What is sufficient is the existence of a wage labor market in particular. Slave societies may be market societies, but they are not capitalist societies in Marx's sense of the term, because slave labor is not wage labor.

References

Brenner, N. (1999) 'Beyond state-centrism? Space, territoriality, and geographical scale in globalization studies', *Theory and Society*, 28: 39–78.

Cassidy, N. (1997) 'The return of Karl Marx', *The New Yorker*, 20 and 27 October, 248–59.

Castree, N. (1995) 'The nature of produced nature', *Antipode*, 27: 12–48.

Castree, N. (2003) 'Geographers of nature in the making', in K. Anderson, M. Domosh, S. Pile and N. Thrift (eds), *Handbook of Cultural Geography*. London: Sage, pp. 168–83.

Chakrabarty, D. (2000) *Provincializing Europe*. Princeton, NJ: Princeton University Press.

Cosgrove, D. (1998) [1984] *Social Formation and Symbolic Landscape*. Madison, WI: University of Wisconsin Press.

Crump, J. (2003) 'Housing labor's unrest: economic restructuring and the social production of scale', in W. Falk, M. Schulman and A. Tickamyer (eds), *Communities of Work*. Athens, OH: Ohio University Press, pp. 194–224.

Debord, G. (1970) *The Society of the Spectacle*. Detroit: Black and Red.

Dwyer, C. and Jackson, P. (2003) 'Commodifying difference: selling eastern fashion', *Environment and Planning D: Society and Space*, 21: 269–91.

Gibson-Graham, J.K. (1996) *The End of Capitalism (As We Know It)*. Oxford: Blackwell.
Gidwani, V. (2000) 'The quest for distinction: a re-appraisal of the rural labor process in Kheda District (Gujarat), India', *Economic Geography*, 76: 145–68.
Gidwani, V. (2001) 'The cultural logic of work: explaining labor deployment and piece-rate contracts in Matar Taluka (Gujarat), India, Parts I and II', *Journal of Development Studies*, 38: 57–108.
Goss, J. (1993) 'The "magic of the mall": an analysis of form, function, and meaning in the contemporary retail built environment', *Annals of the Association of American Geographers*, 83: 18–47.
Hanson, S. and Pratt, G. (1995) *Gender, Work, and Space*. New York: Routledge.
Harvey, D. (1972) *Society, the City and the Space-Economy of Urbanism*. Resource Monograph 18. Washington, DC: Association of American Geographers.
Harvey, D. (1982) *The Limits to Capital*. Oxford: Blackwell.
Harvey, D. (1985a) *Consciousness and the Urban Experience*. Oxford: Blackwell.
Harvey, D. (1985b) 'Money, space, time and the city', in D. Harvey (ed.), *Consciousness and the Urban* Experience. Oxford: Blackwell, pp. 1–35.
Harvey, D. (1989) *The Condition of Postmodernity*. Oxford: Blackwell.
Harvey, D. (1996) *Justice, Nature and the Geography of Difference*. Oxford: Blackwell.
Harvey, D. (2000) *Spaces of Hope*. Berkeley, CA: University of California Press.
Henderson, G. (1999) *California and the Fictions of Capital*. New York: Oxford University Press.
Herod, A. (ed.) (1998) *Organizing the Landscape*. Minneapolis: University of Minnesota Press.
Herod, A. (2001) *Labor Geographies: Workers and the Landscapes of Capitalism*. New York: Guilford.
Hugill, P. (1993) *World Trade since 1431*. Baltimore, MD: Johns Hopkins University Press.
Hugill, P. (1999) *Global Communications since 1844: Geopolitics and Technology*. Baltimore, MD: Johns Hopkins University Press.
Jessop, B. (1994) 'Post-Fordism and the state', in A. Amin (ed.), *Post-Fordism: A Reader*. Oxford: Blackwell, pp. 251–79.
Jessop, B. (2002) 'Liberalism, neoliberalism, and urban governance: a state-theoretical perspective', in N. Brenner and N. Theodore (eds), *Spaces of Neoliberalism: Urban Restructuring in North America and Western Europe*. Oxford: Blackwell, pp. 105–25.
Katz, C. (2001) 'On the grounds of globalization: a topography for feminist political engagement', *Signs: Journal of Women in Culture and Society*, 26: 1213–34.
Lefebvre, H. (1974) [1991] *The Production of Space*. Oxford: Blackwell.
Levins, R. and Lewontin, R. (1985) *The Dialectial Biologist*. Cambridge, MA: Harvard University Press.
Marx, K. (1967) [1867] *Capital: A Critique of Political Economy, Vol. 1*. New York: International.
Marx, K. (1973) *Grundrisse: Foundations of the Critique of Political Economy*. Harmondsworth: Penguin.
McDowell, L. (1997) *Capital Culture: Money, Sex and Power at Work*. Oxford: Blackwell.
Mitchell, D. (2000) *Cultural Geography*. Oxford: Blackwell.
Mitchell, D. (2003) *The Right to the City: Social Justice and the Fight for Public Space*. New York: Guilford.

Mitchell, K., Marston, S. and Katz, C. (2003) 'Introduction: Life's Work: An introduction, review and critique', *Antipode*, 35: 415–42.

Painter, J. (1997) 'Regulation, regime, and practice in urban politics', in M. Lauria (ed.), *Reconstructing Urban Regime Theory: Regulating Urban Politics in a Global Economy*. Thousand Oaks, CA: Sage, pp. 122–44.

Peck, J. (1996) *Work-Place: The Social Regulation of Labor Markets*. New York: Guilford.

Rushbrook, D. (2000) 'Cities, queer space and the cosmopolitan tourist', *Gay Liberation Quarterly*, 8.

Sayer, A. (2003) '(De)commodification, consumer culture, and moral economy', *Environment and Planning D: Society and Space*, 21: 341–57.

Sheppard, E. (2000) 'Competition in space and between places', in E. Sheppard and T. J. Barnes (eds), *Companion to Economic Geography*. Oxford: Blackwell, pp. 169–86.

Smith, N. (1990) *Uneven Development: Nature, Capital and the Production of Space*. Oxford: Blackwell.

Smith, N. (ed.) (1996) *The New Urban Frontier*. New York: Routledge.

Smolin, L. (1997) *The Life of the Cosmos*. Oxford: Oxford University Press.

Swyngedouw, E. (1997) 'Neither global nor local: 'glocalization' and the politics of scale', in K. Cox (ed.), *Spaces of Globalization: Reasserting the Power of the Local*. New York: Guilford, pp. 137–66.

Tickell, A. and Peck, J. (1992) 'Accumulation, regulation and the geographies of post-Fordism: missing links in regulationist research', *Progress in Human Geography*, 16: 190–218.

Walker, R. (1981) 'A theory of suburbanization', in M. Dear and A.J. Scott (eds), *Urbanization and Urban Planning in Market Societies*. London: Methuen, pp. 383–430.

Wilson, B. (2000a) *America's Johannesburg: Industrialization and Racial Transformation in Birmingham*. Lanham, MD: Rowman and Littlefield.

Wilson, B. (2000b) *Race and Place in Birmingham: The Civil Rights and Neighborhood Movements*. Lanham, MD: Rowman and Littlefield.

Wright, E.O. (1985) *Classes*. London: Verso.

Wright, M. (2001) 'A manifesto against femicide', *Antipode*, 33: 550–66.

6 PHILOSOPHICAL BASES OF BEHAVIORAL RESEARCH IN GEOGRAPHY

Reginald G. Golledge

Perhaps much of the confusion that lies at the heart of geography today results from an awareness that there are simply many geographies and many possible worlds. Uncertainty arises because we know not which geography to choose, nor which possible reality we should aim for. We run the risk of becoming dogmatic by trying to force all worlds into one very limited format, and in doing so we ignore, belittle, or forget the others. (Golledge, 1981: 21)

Introduction

Behavioral research in geography (I have never liked the term 'behavioralism', for it leads to unthinking categorization) has been one of the least understood and most misrepresented fields of the discipline. The most common misrepresentation (produced in part by a frenetic desire to label this work with yet another 'ism') has been to confuse the extreme 'behaviorism' of Skinner, Pavlov, and company with 'behavioral' research based on works by Tolman (1948), Lewin (1951), Piaget (1950), Piaget and Inhelder (1967), and others emphasizing perceptual- and cognitive-based research.

Distinctive Features of Behavioral Research

As opposed to geographic interest in the movement traces and physical characteristics (e.g. distance, frequency, and volume of movement) of people as they performed activities required for everyday living (such as shopping, recreating, journeying to work, changing residence, and migrating), behavioral geographers evinced an interest in seeking process explanations for why specific spatial actions were undertaken. A first distinctive feature of behavioral research, then, was that it emphasized process rather than form (Gale and Olsson, 1979). Rather than examining the form or pattern of spatial behavior, an emphasis was transferred to examining the spatial characteristics of behavioral processes such as perception, learning, forming attitudes, memorizing, recalling, and using spatial thinking and reasoning to explain variations in human actions and activities in different environmental settings. But there were no data sets for reference and use in these endeavors. Rather, primary data had to be collected by survey, interview, observation, or recall. Such data could be either quantitative or qualitative (see Part 3), thus requiring behavioral geographers to move beyond the initial offerings of the quantitative revolution and its normative models and parametric, population-based statistical analysis, to also include unidimensional and multidimensional scaling, hierarchical clustering, analysis of variance, and exploratory analysis generally. This required borrowing techniques from other disciplines or modifying existing techniques for use in the spatial domain. And, as a necessary adjunct, the need to evaluate results obtained from the new approaches stimulated a search for new theories and models that did not require economic and spatial rationality or assumptions that removed environmental and personal variation from the problem space. Lastly, behavioral researchers focused more on individuals than on groups or populations, and this disaggregated research departed markedly from the typical aggregated analysis of secondary data (e.g. censuses) that typified much of the other research in human geography (particularly economic and urban research).

Epistemologies such as environmental determinism, naturism, empiricism/positivism, rationalism, transactionalism, interactional constructivism, phenomenology, and post-modernism have all influenced behavioral research at different times. In particular, Gary Moore and I (1976) described how the form-oriented ideas of Kant (1724–1804) influenced the theoretical and quantitative 'revolution' in the 1960s and consequently impacted some emerging areas of behavioral research. We pointed out that, based on the Kantian view, it was difficult to separate the process of knowing from the knowledge itself. This confusion influenced a gradual change among human geographers generally who began emphasizing process approaches of neo-Kantian, cognitive behavioral, and phenomenological perspectives.

It is apparent that behavioral geography emerged in response to a need to know more about human–environment relations than could be obtained from the analysis of secondary data on population attributes and *post hoc* examination of human movements. In the balance of this chapter, I first examine how behavioral geographers dealt with expanding their universe of discourse to include both objective and subjective reality. Thereafter, I follow a timeline, illustrating how this area of interest has changed as the influence of theories and concepts from spatial perception and spatial cognition became more marked, and as the tie between behavioral geography and cognitive psychology became stronger.

The Behavioral Geographer Looks at the Nature of Reality

Given a philosophy that emphasized human thinking and behavior, then one would assume that, intervening between an external world and a mass of sensate beings trying to impose structure on the chaotic mass of sensory messages that bombard them on a day-by-day basis, are our internalized reflections of this external flux. What challenged behavioral researchers was to find out the nature of any isomorphisms between the flux of external reality and the realities constructed in the minds of sensate beings. It had to be recognized that our own constructions are not independently invented, as on a *tabula rasa*, but derive from the concepts handed down to us in language, literature, image, gesture, and behavior.

Thus, relevant theory at that time suggested that, when we talk about a child learning an environment, we recognize that the learning process is constrained by language perhaps as much as by experience, and that, probably, more or less traditional interpretations of experience and easily identifiable objects dominate what is coded and stored in the early learning processes. The adult, having learned the language and other modes of information processing and communication, lives and behaves in a world of concepts that relate both to real objects (which can be directly perceived) and to imagined or hypothetical constructs that can be identified and comprehended (e.g. the concept of a cognitive map). Perceptions provide intuitive data that facilitate interpretation. Reality to an adult, then, consists of experienced, perceived, and remembered features, objects, events, and behaviors to which a person has been exposed or which she/he has experienced. Behavioral researchers realized that, to deal with an environment that consists of objective reality, one assumes (1) that each individual places themselves and others in a common external world,

(2) that elements of this external world exist and will continue to exist even after a person's interactions with them cease, and (3) that objects in external reality will continue to exist as part of a total external environment quite independent of particular human awareness. Behavioral researchers also assumed (4) that knowledge of the existence of places and features can be retained in memory even after interaction ceases and a person is physically removed from the experienced place, and (5) that sensate beings function in objective environments by disengaging the objects of such environments from their own actions (Golledge, 1979: 110). The adoption of such premises marks the beginnings of an approach that goes beyond objectification and accounts, at least in part, for our joint or common ability to specify the *nature of objective reality as it appears to humans*. It also underlies the premises that we can collect information about such realities and represent them in ways that allow many different people to comprehend the representation, to reach a common level of understanding of what an environment consists of, and to communicate about places, features, objects, and events. This foray into the human senses thus assumed that objective worlds become 'real' to us when we add people into their setting.

Lee (1973) proposed that reality is anything that can be apprehended in perception and grasped in thought and understanding. In this view, experiences are real, concepts are real, facts are real, and reality is essentially the ongoing flux of process. The task of overcoming this 'nature of reality' barrier presented a great conceptual problem for many behavioral researchers, for it involved accepting a world of both objective and subjective reality. Being unable to justify such a reality – because of the difficulty of recovering and representing it – some researchers became disenchanted with the potential utility of the approach and, during the 1970s, moved away from behavioral research. What was needed at that time was a way of overcoming (or at least bypassing) the barrier to thinking about and working with the world in the mind and the human–environment relations that helped make up that world. The need to provide a richer base for continued research meant excursions into new theory and methods. The result was the development of what I have called a 'process view of reality' which transcends into a 'process philosophy of everyday life'. In essence, this view is based on the assumptions that we develop human–environment relational knowledge via an interactional or experiential process, and that our behaviors reveal how we have bridged the gap between information encoded and stored in long-term memory, our sensing of the world around us, and the hard facts of objective reality. This process philosophy of everyday life has justified many new directions for research and has provided a rationale for exploring subjective as well as objective worlds. For example, Susan and Perry Hanson (1993) have written a compelling work on 'The geography of everyday life' which derives from a view that both naive and taught spatial knowledge derives from the processes of living and interacting. It has provided the basis for rationalizing an interest in non-traditional problems (such as the spatial concerns of variously disabled populations). And it has provided a satisfactory way of understanding the rationale behind daily, weekly, monthly, yearly, or longer episodic patterns of human activity. This process philosophy allowed researchers to assume that everything is real in some sense or another, and the problem facing one is finding the right sense in which something becomes real. Essentially, it suggests that only by understanding the processes that guide thinking, reasoning, and acting can we fully comprehend the geospatial patterns found in human–environment relations.

This basis carries with it another set of assumptions, this time about humans. For example, it assumes that each being has a means of constructing a reality in a way that facilitates comprehension. This may involve recognizing systems of spatial relations among objects, between people, and among people and objects. So, each being needs an understanding of itself in relation to external objects and settings in which the objects exist (i.e. must pay attention to *the way* our mind relates to or reflects the flux outside its boundary). People in general must have the ability to form simultaneous spatiotemporal understandings that incorporate both physical relations and the human interpretation of those relations.

In the decade following this search for identity, psychologist Sholl (1987) suggested that spatial knowledge consists of self-to-object and object-to-object relations, both of which are needed for environmental knowing. Self-to-object (egocentric) understandings allow an individual to comprehend where he/she is in an environment, regardless of scale, by relating environmental objects and features to self in a vector-like manner. Self-to-object relations are dynamic and must be spatially updated after every spatial movement or turning motion. Object-to-object relations, on the other hand, remain stable, even after human behavior has changed the self-to-object relations. This perspective has dominated geographic thinking and reasoning for decades. Here people are treated as objects, just like other environmental features, and are grouped and classified by objective factors such as location, age, sex, occupation, and income. For example, the object-to-object locational arrangement of cities in the US remains the same, regardless of from where it is observed. Object-to-object relations form the bases of communal understanding and allow us to ask and answer questions such as 'Where am I?', 'Where are you?', 'Where am I in relation to you?', 'Where are we in relation to other things in the world?', and 'Where are things located in the world?' Traditionally, the study of object-to-object relations has been a significant part of both human and physical geography. But it became apparent to behavioral researchers that not only do we have to assume that it is possible for a human to develop this type of objective spatial understanding, but we also have to assume that internalized reflections of the external flux might have a structure and a set of commonalities that differ from objective reality. This, in turn, implies that internalized spatial knowledge structures must have constancy in their imaging, sensing, and perceptions, and a reasonable degree of uniformity in terms of concepts that are constructed by mind so that elements of the external flux can be recognized and elaborated within a personalized communicable/information system. Behavioral researchers thus began following a path which assumed that conscious perception involved an act of selecting from unconscious experience those elements in which there is repetition and similarity. Thus, percepts become clearer and more precise as the involved concepts become clearer and more precise.

Formalizing Behavioral Approaches over Time

By the early 1970s, behavioral researchers had begun asking perplexing questions – such as 'What is reality?' and 'What reality is relevant?' For some experts, it seemed that reality was simply what most people recognized it to be. To philosophers such as Bertrand Russell, it was the sum total of atomistic facts in the universe. Reality as a collection of atomic facts seems to be behind what Wittgenstein discussed as 'what is the case'. As they acquired more knowledge about human–environment relations, behavioral geographers began rejecting this atomistic view of reality as the sole basis for environmental knowing. They also began suspecting the relevance of economic and spatial rationality to describe real human states. Thus began their shift from normative theory to subjective reality. This move was facilitated by Simon's treatise on *Models of Man* (1957), introduced to geographers by Wolpert (1964). Simon offered realistic assumptions about human decision-making and choice behavior that quickly proved to be more realistic than the simple-minded assumption of economic and spatial rationality that helped bring powerful normative models into the discipline.

As early as 1967, Lowenthal had recognized three approaches that typified what was soon termed 'environmental perception and behavior research' in the discipline. These were (1) a research theme based on human attitudes to extreme environmental events (e.g. natural and technological hazards), (2) examination of spatial characteristics of landscape aesthetics and perceived environments (e.g. studies of landscape tastes and emotional states brought on by experiencing different places), and (3) research on spatial decision-making and choice behaviors. The first of these divisions (hazard research) grew rapidly by focusing on the nature of hazards and human reactions to them (e.g. in terms of attitudes and risk evaluation). The second linked to cultural and historical geography, emphasizing 'in temporal context' as a critical concept of environmental descriptions. The third began focusing on the cognitive processes involved in spatial decision-making and choice behavior. After the publication of the first collection of behaviorally oriented papers by Kevin Cox and myself (1969) in which David Stea introduced the concept of cognitive maps to geography, interest rapidly expanded beyond that which had previously been due to the works of Lynch (1960) on city images and Gould (1966) on space preference surfaces. In particular, cognitive behavioral research was influenced by Roger Downs' (1970) paper on cognitive images of shopping centers which showed that subjective reality had recoverable dimensions that connected subjective and objective worlds and could be used to help explain spatial actions.

There was, however, some holdover from the early 1960s in the philosophy behind this research area. To achieve acceptable explanations, things being perceived and reactions to them had to be placed in 'the right category', which would help introduce clarity, order, validity, and reliability into subjective experiences. On the surface, this appears to be a retreat into objectification. But, by understanding that categorization is a cognitive process that helps to impose order on what may sometimes be considered chaos, this perceived retreat was turned into a vigorous positive force. Adopting this perspective led further to accepting notions that information must be transformed into knowledge before it is useful, and that knowledge must be communicable and have some common base. In other words, we must pay attention to the way our mind relates or reacts to the flux outside its boundary, and we must have our senses select and send relevant messages to be stored in memory and manipulated by mind. This point of view thus recognized that knowledge is created in the mind and results from the reaction between mind and messages that emanate from the external universe and which are filtered through the human senses. Our knowledge of a particular environment or setting must therefore be composed of selections from the mass of messages that bombard a person's senses. Only those recognized as personally relevant among the mass of 'to whom it may concern' messages bypass the sensory filtering, are encoded, and take residence in long-term memory.

Transactionalism, Interactional Constructivism, and Cognitive Behavioral Research

As a result of articulating these sentiments, one research stream (the cognitive behavioral) began to be identified more with behavioral research than the other two segments. Hazard research branched into its own setting, eventually separating somewhat into hazard and risk segments. Landscape geography gradually became enmeshed in a new cultural geography that became more concerned with human values and beliefs. While both areas retain a substantial tie to behavioral theory, methods, and concepts, they also stand today as separate specialties.

Cognitive behavioral researchers assumed that mind orders its content and builds it into a yet unknown structure. They assumed that mind manipulates sensory experience and stores information via manipulation activities such as 'spatial thinking' and 'spatial reasoning'. Particular examples of how research was being practiced appeared in the early 1970s in a highly influential book *Image and Environment*, edited by Roger Downs and David Stea (1973). This quickly became the bible for cognitive behavioral research and its contents were the inspiration for two decades of research on spatial behavior.

The results of internal manipulations of sensed and encoded data are inputs to decision-making and choice behavior and, as such, influence human actions and activities. Accepting that knowledge is not immediate or innate assumes that it is constructed by the mind, often as the result of experience during a learning process.

An interactional-constructivist view first began to seriously influence ways of thinking by behavioral geographers after Gary Moore and I published an edited collection, *Environmental Knowing* (1976). In our introduction to this volume, we suggested that behavioral researchers should adopt an interactional-constructivist perspective to guide their thinking and reasoning activities. An interactional-constructivist view acknowledges that, while everyone may well live in the same external world, people may perceive it as a set of different realities, and, consequently, because of different transactional patterns between humans and an environment, no two persons may construct their reality in precisely the same way. This does not mean we all exist only in unique realities, for we still have to communicate and interact to survive. This requires a base of some sort of common structure.

Finding out what is common and what is idiosyncratic in personal realities has always been at the heart of the cognitive behavioral approach in geography. Researchers examined the differently constructed realities to find what is common, constant, and communicable and to identify what is idiosyncratic, individual reality dependent, and communicable to others only after detailed explanation of the construct base. Having advanced this thesis in a joint introduction to *Environmental Knowing*, Gary Moore and I argued that it must be obvious that such a position is inconsistent with pure objectification, extreme scientism, and other bogeymen and straw men that have been used to critique behavioral research. People still confuse the extreme Skinnerian-type behaviorism of stimulus–response, operant conditioning, and behavioral modification with a cognitive or perceptual behavioral approach to creating spatial knowledge. Helen Couclelis and I (1983) tried to rectify these misconceptions by detailing how behavioral geography differed from behavior modification theories, but to this day there is still some residual confusion in texts examining the history of geographic thought. Part of our thesis was that a 'positivist' (or scientific-experimentalist) view that stresses a need for vigorous knowledge discovery based on use of scientific method to produce reliable and valid statements can also be a humane approach and, indeed, was an important part of the human science in which cognitive behavioral research was embedded.

In Chapter 2, Kitchin provided an overview of positivism. In essence, this is a scientific and empirical philosophy that stresses experimentation and empirical analysis as a way of producing valid and reliable outcomes. Positivists engage in physical, natural, biotic, and human science. The positivist philosophy that underlies much behavioral geography is *not* the stimulus–response, approach–avoidance, conditioned response or behavior modification type of experimental research that is common in some areas of psychological research (particularly on animals). This is a classic misinterpretation by writers on the nature of geography who do not explore thoroughly the different forms of experimental research. Those conducting research in behavioral geography have emphasized cognitive and perceptual processes and various learning theories as the basis of their science and experimentation, not the stimulus–response models so often identified with it. Experiments with humans are not undertaken lightly or carelessly. This type of research has always been subject to scrutiny by national and local human subject committees who evaluate whether or not a particular project will put a participant at risk in any personal, social, medical, psychological, or even economic way.

Many filters, evaluations, and constraints are placed on the behavioral researcher before any interaction involving data collection with human participants can be undertaken. This ensures that participants are protected and treated humanely and courteously, with great concern for their wellbeing. Factors such as boredom, mental and/or physical fatigue, and safety are matters of supervisory importance. Few of these things are realized or acknowledged, especially by geography's critical thinkers who have never designed or carried out such experimentation.

Because of this type of monitoring, there has been a constant evolution of cognitive behavioral research from a 'hard' base dominated by empirical research via experimental data collection and quantitative processing, to a base that involves collecting both 'hard' and 'soft' data and using quantitative or qualitative experimental procedures and methods of analysis. As we will see later, the result has been that behavioral research has expanded into many new areas.

How Philosophy has Enabled the Practice of Behavioral Research

The emphasis on personal data about knowledge, perceptions, and actions required the development of experimental designs for laboratory and field data collection. It should be no surprise that scientific methods and positive philosophy were prominent. While still fundamental in much cognitive behavioral research, the need to observe and examine people's feelings and beliefs, as well as their reactions to structured task situations, necessitated an expansion of the epistemological bases and the design of data collection processes. Two such bases were transactionalism and interactional constructivism. Transactionalism acknowledged the dynamic nature of human–environment relations, that most behaviors are in a state of flux and may change as environments change or as time passes. It accepted that experience was the key to spatial knowledge acquisition and directed attention away from static, cross-sectional experience to behavior that existed in space–time. Interactional constructivism likewise focused on the dynamic nature of human–environment relations, but emphasized that people live and interact in constructed realities. The roles of perception and cognition were paramount in understanding why people performed actions and selected activities.

The transactional and interactional-constructivist base for thinking about spatial behavior raised questions as to how spatial knowledge accumulates, how spatial learning takes place, and how age, sex, ethnicity, race, culture, or membership in social and economic groups influence human–environment transactions and impact the development and comprehension of spatial concepts and constructs. Pursuing answers to these questions has taken behavioral geographers on a wild ride. Forays have been made into the realms of consumer behavior, housing markets, leisure and tourist activities, disaggregate behavioral modeling of human activities, children's environments, hazards, emotional reactions to environments, environmental personalities, human wayfinding, movies, risk, and many other areas. With each excursion into a novel area, methodological changes were necessary to suit the type of transactions being examined. As these areas were explored, the methodological armory of the behavioral researcher expanded. On the quantitative analytic side, new representational and analytical procedures included things such as computational process models for computer simulation of household movement; logistic models of mode choice for urban travel; and laboratory experiments to clarify spatial abilities. On the qualitative analytic side, new approaches included the use of focus groups to clarify knowledge, attitudes, and opinions; non-probability surveys such as 'person-in-the-street' interviewing, quota sampling, case studies, and experimental observation; toy play that revealed children's latent spatial awareness; and photographing play and activity environments to reveal customs and practices that might otherwise not be discovered.

Looking at Alternative Realities

Research sensitive to the existence and needs of special populations has portrayed a very different reality from that experienced by the general population. For example, persons with mental problems may construct realities that are personally unique due to their limited ability to comprehend two- and three-dimensional geospatial concepts and constructs. Autistic children may construct realities that are very different from those experienced by others around them, such that communication becomes difficult if not impossible. People with limited intellectual development appear to create highly linear realities. The realities of some groups of disabled people (such as those who are vision impaired) likewise can be very linear, and, consequently, appear to differ from the realities of many able-bodied people in terms of far fewer known environmental features, less awareness of distant places and landmarks, and poorer knowledge of the number of known routes. For a homeless person, the world may shrink to a habitable neighborhood. For disabled people living in group homes, however, a hostile neighborhood filled with NIMBY sentiments may surround them and restrict local movement. For a wheelchair person, a curb without a curb cut may prove to be an impassable barrier that constrains interactions. For an able-bodied person, such an environmental feature may be a negligible one and not even be represented in memory, or be considered only as part of a 'street' construct. A blind person walking down a sidewalk filled with street furniture and written signage constructs a reality that may be very incomplete, fragmented, and linear compared to the able-bodied person's reality which, using readily available information from visual sources, may bear a much closer resemblance to the objective physical world. Given such a perspective, it is reasonable to assume that some disabled groups can only experience parts of the environment because of the existence of what to them constitute barriers to travel or to experience (e.g. people restricted in travel may never know what is in the next block to their right). Consequently, the information to which they are exposed is but a fraction of the information to which an able-bodied person is exposed. The resulting memory structures available for internal manipulation by the mind are consequently lean and limited compared to the information available for manipulation by the mind of an able-bodied person, even though equivalent cognitive processes of thinking and reasoning are available to both groups. In the past, some critics have failed to comprehend this difference (Gleeson, 1996; Imrie, 1996). Understanding that some groups, because of disability, lack of education, lack of experience, or lack of language, may have impoverished memory sets has led behavioral geographers to pursue burning questions related to the nature of the realities constructed by members of such groups, the differences between them, and the impacts that a relatively impoverished memory bank has on negotiating a setting.

The Barrier of Methodology

The development of concept knowledge relevant to spatial behavioral research at times stumbled on the barrier of methodology. For example, the introduction of the concept of a cognitive map necessitated considerable research about methodologies for recovering the content of such maps before the concept could become part and parcel of geographic thinking. The early attempts to use Lynch's (1960) methods of compiling sketch maps were criticized because they lacked the metric geometry needed to allow cartographic interpretation (i.e. they lacked scale and a frame of reference for checking direction and orientation, and were dependent on graphic skills). The 'mental maps' methodology used by Gould (1966) proved to be stated preference surfaces produced by trend surface analysis of communal rankings of places rather than externalizations of stored spatial memories. Demko and Briggs (1970) and Johnston (1972) showed the difference between these 'stated' (i.e. ideal or preferred) evaluations and 'revealed' (or actual) behaviors in the context of predicting migratory movements.

The need to find appropriate methods for externalizing information stored in the mind was investigated from a variety of viewpoints using statistics, diagrams, verbal descriptions, model building, and metric and non-metric scaling procedures (procedures which were summarized in Kitchin's (1996) CMap software). While sketching and describing were direct procedures for externalizing and representing spatial knowledge, other methods recovered 'latent' spatial structures via indirect methods of recovering these 'spatial products'. Experimentation with these methodologies dominated behavioral research in the 1970s and much of the 1980s. While some potential researchers were 'turned off' by this methodological emphasis, it was essential in that it enabled reliable and valid access to spatial information stored in long-term memory, and facilitated its representation and analysis in ways not previously possible. Dissolving the methodological barriers by constructing relevant experimental designs and procedures made possible the next critical steps in behavioral research.

Summary

Experiences which pass sensory thresholds and of which we consequently become aware are encoded, stored, and used in mental manipulations to assist in the ongoing process of interacting with and living in an everyday environment. Thus, researchers have pursued answers to questions such as: what relationships exist between objective reality and the world constructed inside our heads? How can we determine the nature of the relationship between a human in the world and the world in a human? What influences whether a message emanating from a complex external world is accepted as one worth storing in long-term memory? How can we determine what people know about the world they live in? How can we represent, analyze, and interpret the world existing in our minds? Do we construct experiments that are designed (like those in psychology) to find out how we think and what we think about? Or do we follow geographic tradition by observing revealed actions, activities, and behaviors, suggesting reasons for why they took place? How best can we represent information obtained from individuals in forms so that others can compare their own constructions with them? And what contribution in performing these tasks is made to the accumulation of geographic knowledge?

Many of these questions have surfaced in the writings of behavioral geographers at various times over the last 40 years. The fact that they have surfaced time and again shows some constancy in the philosophy of learning and investigation and points to the seriousness of the problems behind these musings.

The philosophies that at times have underlain behavioral research and thinking in geography have included (and still include): the classic empiricism and scientific methods of positivism; the philosophies of mind of various cognitive thinkers; the transactionalists and the interactional constructivists; and the realists, the rationalists, and the naturalists. Each has contributed to defining research questions, research methodologies, and research interpretations. Each has provided a base for discussing the nature of reality, the confound between subjective and objective realities, the mind–body problem, and modes of interpretation of findings. We have seen how rationalism and empiricism helped develop the hazards/risks stream, how realism and naturism helped the landscape aesthetics stream, and how transactionalism, interactional constructivism and structuralism helped define the cognitive behavioral stream. In particular, the latter focused attention on cognitive behavioral processes, encouraged the search for primary data, and encouraged the use of a wide variety of quantitative and qualitative methodologies, experimental designs, and analytical and inferential procedures.

But the essence of trialing these various philosophies boils down to how one interprets the concept of reality. The epistemologies that, to me, have given the greatest insights into developing knowledge about human–environment relations have been the interactional-constructivist and transactional approaches. These emphasize the importance of mind in making sense of objective reality, and also allow treatments ranging from controlled laboratory experimentation with people undertaking stringently controlled tasks, to the qualitative interpretation of subjectively acquired information about actions, beliefs, and emotions relating to specific or general environmental settings.

To label this wide variety of ways of thinking and reasoning with a simple 'ism' (such as 'behavioralism') does not do justice to the richness and complexity of the behavioral approach. In fact, there is not one 'behavioral approach', but many. This diversity makes behavioral research in geography appealing and potentially productive. With the 'which ism' categorization question out of the way, this stream of geospatial thinking and reasoning could be a major focus in the future search for geographic knowledge.

References

Couclelis, H. and Golledge, R.G. (1983) 'Analytic research, positivism, and behavioral geography', *Annals of the Association of American Geographers*, 73: 331–9.

Cox, K.R. and Golledge, R.G. (1969) *Behavioral Problems in Geography: A Symposium*. Evanston, IL: Northwestern University Press.

Demko, D. and Briggs, R. (1970) 'An initial conceptualization and operationalization of spatial choice behavior: a migration example using multidimensional unfolding', *Proceedings, Canadian Association of Geographers*, 1: 79–86.

Downs, R.M. (1970) 'The cognitive structure of an urban shopping center', *Environment and Behavior*, 2: 13–39.
Downs, R. and Stea, D. (1973) *Image and Environment: Cognitive Mapping and Spatial Behaviour*. Chicago, ILL: Aldine.
Gale, S. and Olsson, G. (1979) *Philosophy in Geography*. Boston, MA: Reidel.
Gleeson, B.J. (1996) 'A geography for disabled people?', *Transactions of the Institute of British Geographers*, 21: 387–96.
Golledge, R.G. (1979) 'Reality, process, and the dialectical relation between man and environment', in S. Gale and G. Olsson (eds), *Philosophy in Geography*. Dordrecht: Reidel, pp. 109–20.
Golledge, R.G. (1981) 'Misconceptions, misinterpretations, and misrepresentations of behavioral approaches in human geography', *Environment and Planning A*, 13: 1325–44.
Gould, P. (1966) 'On mental maps'. Paper presented at the Community of Mathematical Geographers, Michigan University, Ann Arbor, MI.
Hanson, S. and Hanson, P. (1993) 'The geography of everyday life', in T. Gärling and R.G. Golledge (eds), *Behavior and Environment*. Amsterdam: Elsevier, pp. 249–69.
Imrie, R. (1996) 'Ablist geographies, disablist spaces: towards a reconstruction of Golledge's "Geography and the Disabled"', *Transactions of the Institute of British Geographers*, 21: 397–403.
Johnston, R.J. (1972) 'Activity spaces and residential preferences: some tests of the hypothesis of sectoral mental maps', *Economic Geography*, 48: 199–211.
Kitchin, R.M. (1996) 'Exploring approaches to computer cartography and spatial analysis in cognitive mapping research: CMAP and MiniGASP prototype packages', *Cartographic Journal*, 33: 51–5.
Lee, H.N. (1973) *Percepts, Concepts, and Theoretic Knowledge*. Memphis, TN: Memphis State University Press.
Lewin, K. (1951) *Field Theory in Social Science*. New York: Harper.
Lowenthal, D. (ed.) (1967) *Environmental Perception and Behavior*. Chicago, IL: Department of Geography, University of Chicago.
Lynch, K. (1960) *The Image of the City*. Cambridge, MA: MIT Press.
Moore, G.T. and Golledge, R.G. (eds) (1976) *Environmental Knowing: Theories, Research and Methods*. Stroudsburg, PA: Dowden, Hutchinson and Ross.
Piaget, J. (1950) *The Psychology of Intelligence*, trans. M. Piercy and D. Berlyne. London: Routledge and Kegan Paul.
Piaget, J. and Inhelder, B. (1967) *The Child's Conception of Space*. New York: Norton.
Sholl, M.J. (1987) 'Cognitive maps as orienting schemata', *Journal of Experimental Psychology: Learning, Memory, and Cognition*, 13: 615–28.
Simon, H.A. (1957) *Models of Man*. New York: Wiley.
Tolman, E.C. (1948) 'Cognitive maps in rats and men', *Psychological Review*, 55: 189–209.
Wolpert, J. (1964) 'The decision process in a spatial context', *Annals of the Association of American Geographers*, 54: 537–58.

7 STRUCTURATION THEORY: AGENCY, STRUCTURE AND EVERYDAY LIFE

Isabel Dyck and Robin A. Kearns

Introduction

In this chapter we make a case for structuration theory in geographical analysis. Rather than a theory to be applied, sociologist Anthony Giddens (1984) sees it as useful in providing 'sensitizing' concepts in analysis of the constitution of the individual and society. Geographers' engagement with structuration theory, initially in the 1980s, signalled the intense exploration of ways that social theory could inform understandings of the 'sociospatial dialectic' (cf. Soja, 1980). As historical materialism vied with humanism in theorizing the recursiveness of social relations and spatial structures (see for example Gregory and Urry, 1985), concepts of structuration provided an attractive entry point for geographers entering the agency–structure debate prominent at the time. Giddens' particular formulation of structuration theory problematized human agency and dismantled the notion of a 'macro'/ 'micro' dichotomy in the process. Through this theorizing, he urged a conceptualization of the *contextuality* of social life that sat well with contemporary geographical concerns and was taken up notably by Gregory (1981; 1982; 1989), Pred (1984) and Thrift (1985). This enduring problematic has been taken up in different guises through the social sciences; here we make the case for the ongoing relevance of structuration theory in geographical work.

Attention to Giddens' way of integrating agency and structure in 'bridge building' between humanism (see Chapter 3) and Marxism (see Chapter 5) initially remained primarily within debates in cultural and social geography, but notions of structuration now inform, explicitly or implicitly, work in various subfields of geography. Interest in the complex relationship between human agency and the constraints of structure brings common ground to the domain of human geography inquiry; while this problematic is taken up variously through different theoretical perspectives, Giddens' highly focused explication provides a strong foundational statement from which to examine processes of enablement and constraint. Geographers working with notions of structuration have emphasized the *spatiality* of such processes. In this chapter we use examples from our work to show the value of structuration theory in pursuing the ongoing interest in the mutual influences of society and space in forging everyday geographies. The chapter proceeds as follows: we first set out central concepts of Giddens' structuration theory taken up in geographical inquiry; we then briefly signal other work by geographers using concepts of structuration, before going on to describe in some detail how our own work has been informed by structuration theory; lastly, we comment on structuration theory's continuing utility and its reworking in the context of other contemporary social theory that informs the agenda of human geographical inquiry. In the course of the chapter we provide research examples to 'bring to life' some of the theoretical concepts that are presented.

Giddens' Structuration Theory

Structuration theory, as developed by Giddens (1976; 1979; 1981; 1984), essentially views society as neither existing independently of human activity nor being a product of it. Rather, this theory suggests an inherent spatiality to social life. For Giddens the problem of order, central to the sociological project of the time, was not one of discovering an underlying pattern of social life but rather a concern for how social systems are bound together in time and space. This concern was in an attempt to provide a non-functionalist theory of society in response to an orthodoxy that was unable to integrate face-to-face interaction with institutional analysis. He intended to sensitize social analysis through emphasizing the knowledgeability of the individual agent in the reproduction of social practice, the time–space contextuality of social life and the hermeneutic or interpretive nature of analysis.

The notion of the duality of structure is central to Giddens' structuration theory, in which neither the human agent nor society is regarded as having primacy. This duality is a recursive process in which 'structure is both medium and outcome of the reproduction of practices' (1981: 5) that are themselves fuelled by both the intended and the unintended consequences of human conduct. The concepts of social system and social structure are integral to the notion of the duality of structure. Social systems are, essentially, regularized relations between individuals and groups, comprising routinely reproduced social practices situated in time and space. These social systems are grounded in the knowledgeability of actors and contain structured properties as understood through the concept of structure.

Structure is regarded as 'rules and resources', which only exist temporally when 'presenced' by actors; that is, when drawn upon as stocks of knowledge in day-to-day activity. Structure, then, exists only through the concrete practices of human agents, recognized as competent and knowledgeable, who reproduce social life through their routinized day-to-day encounters. Institutions, from this position, are viewed as chronically reproduced rules and resources. Resources include both physical environments and social relations within such environments. Furthermore, rules are not static, but may be amended due to the negotiable quality of meanings, evaluations and even power. An important element of the notion of structuration is the major role played by the unintended outcomes of human activity, as well as those that are intended. Together these feed back into 'structure' and further influence day-to-day activities as 'the unacknowledged conditions of further acts' (1984: 8). 'Constraints', therefore, are not externally imposed on the flow of action; instead, the structural components of society that are 'embedded in an enduring way in institutions' are both enabling and constraining (1983: 78). The discovery of structure exposes both constraint and empowerment.

A fundamental aspect of the duality of structure is Giddens' understanding of human agency and the contextuality of social life. Agency is based on the idea that the individual is a perpetrator of events and he or she could have acted differently. The issue is not that 'agency' is a given quality, but *how* it is possible for human beings to act as agents. Here, Giddens distinguishes between discursive consciousness – what people can put into words about their actions – and practical consciousness – that is, what actors know about how to do things in a variety of contexts of social life, but may not be able to put into words. He sees the routine and reflexive application of practical consciousness, by knowledgeable practitioners, in the chronic constitution and reconstitution of social life. Actors may not know the meanings of rules but can use them skilfully in interaction with the possibility of transformation, for rules and resources are not static but the media of production and reproduction of practices. This transformative capacity is of enormous importance in understanding the notion of social power in relation to human agency, for social change can be generated through social practice as it grows out of the everyday activities of individuals. The strategic conduct of human agents, however, does not take place under conditions of their own choosing. Unintended consequences become conditions bounding further action, and the binding of social systems through their extension over time and space suggests that the structured properties of social systems may be beyond the control of individual actors. This extension, system integration, is achieved through technological means such as letters, print, the telephone and, more recently, the explosion of electronic communication.

The contextualization of human interaction in time and space is central to Giddens' thinking, as suggested in this statement:

> **the settings and circumstances within which action occurs do not come out of thin air; they themselves have to be explained within the very same logical framework as that in which whatever action described and 'understood' has also to be explained. It is exactly this phenomenon with which I take structuration to be concerned. (1984: 343)**

As Thrift (1985) points out, the recognition of the contextuality of action in time and space is not to argue for localism, but rather to advocate a concern with how social systems change over time and space. He interprets Giddens' view as one that comprehends the spatial aspects of social experience by understanding the intermingling of 'presence' and 'absence' in everyday life, or in other words the continual interplay of agency and structure over time and space. Here the notion of locale is significant. Locales occur at all physical scales, from a room in a house to the territories demarcated by nation-states, and are not just points in space, but rather have features that are used 'in a routine manner to constitute the meaningful content of interaction' (Giddens, 1985: 272). Further, such locales are 'regionalized', meaning that regularized social practices in a given location may be 'zoned' through legislation or informally shared understandings, in time and space. For example, the separation of home from workplace is a form of regionalization, just as the internal divisions of halls, rooms and floors of a home are zoned according to their use temporally and by type of activity. What becomes important is the implication of regionalization for power relations, as certain social practices may be more or less visible; examples include the sequestration, through spatial separation, of insanity and crime (see Gleeson, 1999). Following from this, the environment can logically be seen as a matrix of locales, or settings for encounters, which contain particular combinations of resources that may be drawn upon in action. Resources include physical attributes and people in a locale, but also refer to stocks of knowledge. It is important to note within this conceptualization that locales are not 'givens' but are created – for human agency designates human beings as makers of their milieu, albeit within unequal power relations. What Giddens' work emphasizes is that there is a transformative capacity to all human action. Power is generated through the expansion of social systems and structure over time and space that are simultaneously experienced and drawn upon as the 'rules and resources' of particular locales.

Giddens' Structuration Theory and Empirical Work

It is important to note that Giddens views his explication of structuration as providing 'sensitizing concepts' for informing research, rather than a set of concepts to be applied; as

Fincher (1987) suggests, individual concepts cannot readily be extricated from Giddens' extensive treatment and total approach of structuration theory. Indeed, geographers investigating a variety of topics have employed different emphases in working with notions of structuration. Gregory (1982) and Pred (1984), for example, worked with Giddens' critical engagement with time geography in tracing how the material contingencies of everyday life are fundamental to understanding place as unfolding process. An associated attempt to achieve a firmer binding of structure and agency occurred through the work of Michael Dear and Adam Moos (Moos and Dear, 1986; Dear and Moos, 1986) who were among the first geographers to both unravel the language of structuration theory *and* attempt to apply it in an empirical study (in their case, of mental health care and the so-called 'boarding house ghetto'). Giddens' ideas later emerged as a catalyst in a heated arena of debate that reformulated medical geography into 'health geography' through embracing a sociocultural framework (Dorn and Laws, 1994; Kearns, 1993). The value of embracing the tension between structure and agency, whether invoked through the language and concepts of structuration theory, or through a more generalized recognition of the complex links between the individual and society that operate at different, layered scales, has continued to inform recent work on culture and health (Gesler and Kearns, 2002). Other contemporary usage occurs in migration research where the concept of 'structurated patriarchy', drawing directly on Giddens' work, is used in analyses of the gendering of migration processes and outcomes (Halfacree, 1995). A specific focus on rules and resources, within the context of the core structuration concept of the 'duality of structure', is used in theorizing community activism in opposition to school closures in Ontario, Canada (Phipps, 2000). Phipps developed a typology of rules and resources that could be useful for communities to use in activism, and found structuration theory vital to a reinterpretation of educational and facility-closure literatures.

In our own work we have also been influenced by Giddens' conceptualization of the integration of agency and structure in one framework. Specific concepts from Giddens' structuration theory provided an analytical lens for the first author's work with women with young children, and the emphasis on the recursiveness of agency and structure continues to inform her later work with immigrant women. The second author's recent work in health geography also carries the legacy of notions of structuration in analysing relationships between health, place and health care where the ongoing importance of taking account of both structure and agency is recognized.

Structuration theory, mothering work, and creating 'safe space' for children

In the early 1980s women with young children were entering the paid workforce in unprecedented large numbers. In my research I was interested in the suburb as a domestic workplace, and in how women worked practically and 'morally' through the conflicts arising from participating in paid employment while continuing to be the prime carers of children. In the language of structuration, I was interested in where and how knowledge, in the form of information, was shared and circulated; that is, where and how understandings, rules and stocks of knowledge were both modified and reaffirmed in this context of social change. The locales of everyday mothering work could be expected to be laden with culturally and gender-specific meanings, to be 'presenced' through social interaction. The sensitizing concepts of structuration theory helped me explicate not just how 'context' was a backdrop against which the lives of the women in the study took place, but how it was actively implicated in the ways they practised motherhood in a particular locality.

In-depth interviews and diaries kept by study participants were used to explore various aspects of the women's everyday routines – their actions, concerns and interpretations of their mothering work as well as the locales in which they took place. I then sketched these data utilizing Giddens' time–space maps, which showed where women went, when, with whom and why (see 'Anna's day', Box 7.1). This mapping revealed the opportunities and constraints of the everyday locales of mothering work – homes, schoolyards, streets, parks, transportation modes – that helped me examine the interplay of agency and structure, including the capacity of women to transform the meaning and practices of mothering. That is, in the language of structuration, I observed the regionalization of locales, the rules and resources that were drawn upon, and the unintended consequences of the women's actions for how they were able to combine mothering work and paid labour. Under a rubric of 'what's best for the children', practical examples of combining waged and domestic labour were shared by women and the meaning of a 'good' mother was renegotiated. The spaces of the street, parks, schoolyards – where women waited for their children to come out of school – and preschool parenting education sessions were all locales where such 'mothering talk' took place. Further, it was through such encounters in the shared spaces of mothering work that conditions allowing women to remain 'good mothers' while participating in waged labour were created. For instance, women created flexibility in their use of time and space through making arrangements for exchanging childcare and babysitting and constituting the street as a safe space for children through shared 'surveillance' of street play. In short, children could be left under the care and watchful eyes of other mothers while a woman was away from home (Dyck, 1989; 1990; 1996).

This work on suburban women, with its fine-grained focus on the locales of routinized quotidian life, and informed by sensitizing concepts of structuration, permitted an exploration of the circulation and negotiation of meanings and knowledge. The study brought insight into how an array of ideas entered stocks of knowledge and became part of the complex recursive constitution of spaces, identities, and transformation – as well as reproduction – of cultural norms. Agency and structure were commingled; women's everyday locales, and the routines of which they are a part, operated as sites for both reproduction and change in mothering work and social identity.

Gender and migrant spaces

As in the study of suburban mothers, the first author's later work with immigrant women shows them to be skilled, knowledgeable agents negotiating cultural knowledge in various locales – the home, workplaces, immigrant education programmes, and neighbourhood spaces such as parks. A study concerning immigrant women's management of health and illness found that traditional healing and biomedical knowledges were negotiated and sometimes integrated in a way that suggested women were pragmatists in the way they used traditional medicine and folk remedies in dealing with illness, using these approaches instead of, or in addition to, biomedical strategies according to how they best fitted in with the circumstances of their everyday lives (Dyck, 1995). This work contests the commonly held notion that cultural beliefs act as barriers to some minority populations' use of western medicine. Another study is also showing how notions of femininity and motherhood are being reworked in specific, neighbourhood

BOX 7.1 ANNA'S DAY

Anna has two children, seven and four years old. While she had worked for short periods of time since having children, she had decided not to return to paid employment until both her children were attending school full-time. Her time–space map (Figure 7.1) illustrates her daily routine, indicating the places and sequencing of her activity on a day that she recorded and described in an interview. The home (represented by the middle column of boxes) is the base from which her trips start and to where she returns. The other columns of boxes indicate both everyday locales of activity, such as the school and preschool, and those that may be less regular but constitute the routinized activity of mothering work and home provisioning. The constant features of Anna's days during the week are the hours of her husband's employment and the children's school and preschool schedules. As can be seen from the time–space map, her more discretionary use of time and space is also centred around the children's activities, such as soccer and swimming lessons, or other domestic activity, such as shopping for everyday needs and taking the family pet to the veterinarian. The mapping indicates the busyness of her days, as does her comment, 'If it's not one thing it's another. I find it hard sometimes to keep track of what I need to do each day.' While her activity may seem mundane from the 'outside', the discussion of structuration theory you have read indicates how this everyday organization and use of time and space is the very stuff of cultural continuity *and* transformation!

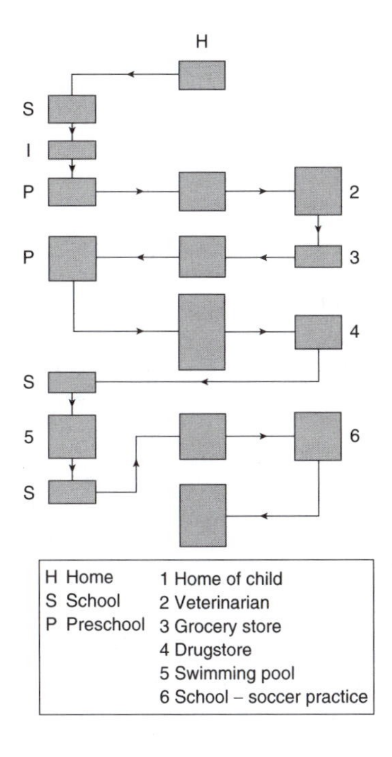

Figure 7.1

locales; here immigrant women negotiate not only the taken-for-granted knowledge they bring from their countries of origin, but also the unfamiliar 'ways of doing' things they encounter in Canada (Dyck and McLaren, 2002; McLaren and Dyck, 2002). As in the study of mothering work, these two studies demonstrate that women's everyday practices, and the knowledge that guides these, are constituted through interactions in the everyday locales making up quotidian life where agency and structure 'play out'.

In this recent work insights of Giddens' theory of structuration prevail, but through a different lens and a different set of sensitivities. For instance, a gap in Giddens' work was a marginalization of culture and gender. This is not to say, however, that Giddens' structuration theory has no currency for work concerning immigration and migration processes and outcomes. Halfacree (1995), for example, sees structuration theory useful in analysing the gendering of migration through the notion of 'structured patriarchy' together with a biographical approach (see also Halfacree and Boyle, 1999). Here Giddens' duality of structure is explicitly drawn on in demonstrating how differences in women and men migrants' labour market participation, framed by women's usual status as the secondary migrant, reconstitute structures of patriarchy which are 'both the medium and outcome of the gendering of "labour migration"' (Halfacree, 1995: 173).

Notions of structuration, health and place

The second author's work illustrates the ways in which structuration theory is useful in analysing various dimensions of the relationships among health and place. He traces an interest in structuration ideas to his master's thesis in which he grappled with the connections between individual behaviour, institutional influences and community politics in a rural New Zealand locality where there were proposals for a large-scale irrigation scheme (Kearns, 1982). These concerns with the links between geographic scales and domains of action travelled with him to Canada where he joined an emergent tradition of mental health care research at McMaster University. To date, this work had embraced two, largely separate, traditions of political-economic and behavioural work which loosely map onto the terms 'structure' and 'agency'. In other words, while some studies of mental health care systems and users operated at high levels of process and outcome, emphasizing power and class (structure), others dealt with the (largely statistical) detail of who was thinking or doing what, when (agency) (e.g. Dear, 1981; Dear and Taylor, 1982). Dear and Taylor's (1982) work attempted to bring the political-economic and behavioural perspectives together in research on why some communities reject the prospect of group homes for mentally ill people, and was followed by Dear's collaboration with Moos in thinking through the relationship between agency and structure. Working alongside people who were applying this theory inevitably impacted the second author's work. He was both fascinated with, and concerned for, the fragmented lives of people with long-term psychiatric illnesses whose world had become the so-called 'boarding house ghetto'. Disenchanted with both the generalization of large statistical samples and the depersonalization of political-economic explanations for stigmatized poverty, he sought direct experience of the world of the drop-in centre and boarding house as a way of informing the questions posed in, and the analysis of, a small-scale survey (Kearns, 1987). Ideas of sense of place were central to the inquiry, but this construct itself was reworked to include considerations of structure and agency. For too much humanistic thinking threatened to leave people's humanity unrealistically detached from societal and

BOX 7.2 DAVID'S HOUSING

David lives in a boarding house on a street only a few blocks away from the poorer end of Main Street in the downtown area. He was diagnosed with paranoid schizophrenia seven years ago in the small town he grew up in and was hospitalized in the regional city. When discharged, he stayed in town, as that is where the best community mental health care service is, and where other former patients he got to know during his numerous stays in hospital live. He also feels safer in the city, away from family friends and people who knew him as a teenager. His room is dark and musty, and barely large enough for a bed, a cupboard and a hotplate for cooking. It is one of 18 rooms in the large rambling dwelling that has known better days. David has to leave the boarding house each morning after breakfast and cannot return until the afternoon. It is just the rules of the house, he says. There are not many options by day. He walks a lot. He could spend time at the 'Care Centre', a 'drop-in' run by the local mental health association. But David chooses to spend more time at the local doughnut shop, where the owners turn a 'blind eye' to how long he takes to finish a coffee. His choices are limited, but he does have some agency: he opts to hang out in a less stigmatized place than among the other regulars. Yet in other ways his life is highly structured at a number of levels. First, at the household scale, rules cast him out on the street to find places of respite by day. Second, at the urban level, bylaws have meant that boarding houses such as his are clustered into a few city streets, enhancing his chances of networking among former patients but heightening their visibility and contributing to the stigma of mental illness. Third, and beyond the specifics of David's here and now, the 'place' of the mentally ill has been structured through generations of societal misunderstanding and marginalization in the health care system such that David's day is characterized by coping rather than creativity.

material influence, and, in effect, *displaced* from all that holds us *in place*.

In other health geography work the pioneering engagement with Giddens' ideas by Moos and Dear has been followed, for instance, by Baer et al. (2000) who use Gidden's notions of space and time constraints to trace the influences on physician recruitment and retention. Indeed as geographical work on health has embraced the blurred boundaries of social and cultural geography, ideas drawn from structuration theory retain their currency whether or not the specific language of Giddens' explication of structuration is used. For instance, in an effort to understand the influence of high housing costs on family wellbeing, the notion of 'discounting health' was developed to signal the way in which expenditure on not only health care but also health practices is constrained by the fixed costs of housing. Within a sample of racialized Pacific Island migrant families, in-depth interviews revealed that agency remained and decisions were constrained not only by structures of housing and welfare provision, but also by cultural proscription (e.g. church donations) (Cheer et al., 2002). Elsewhere, in the rural Hokianga district of New Zealand, resistance to health reforms and the potential loss of a locally owned health care system offered an opportunity to tease out the dynamics of structure and agency. In particular, the agency of 'social entrepreneurs' was found to be crucial in challenging and transforming institutional structure that was anathema to what Giddens might have labelled the long durée of community tradition and history (Kearns, 1998). Other research shows that just as health and place cannot be divorced in understanding health and illness experience, so too health care and its consumers – potential or actual – should not be separated analytically, for each constitutes and structures the other (Kearns and Barnett, 1997).

In sum, health geography has been more strongly affected by the general influence, than by the concepts or language of structuration. It has sensitized health geographers to structure–agency ideas such that it is now seldom that individual health experience or structural preconditions for health and health care will be researched or theorized without reference to each other. In our view an 'internalized' structuration perspective now prevails and this is evident in more recent health geographies in which structural constraints such as the 'contract culture' collide with individual aspirations towards community development (Kearns and Joseph, 2000).

Continuations and Conclusions

Giddens' work has been valuable to geographers in bringing together the social and the spatial through a conceptualization of the contextuality of social life that admits the human agent and structure in an integrated framework. Structuration concepts initially informed a moment in cultural geography when social theory was seriously engaged in the course of formulating the questions and theoretical and methodological approaches of human geography (Gregson, 1987). The notions of locale and regionalization were particularly important in addressing issues of scale and the concept of place. It is perhaps the link between Giddens' conceptualization of time–space and the desire to move beyond the underlying positivism of Hägerstrand's time geography that acted as a catalyst for the first consideration of Giddens' work in the 1980s. It is also noteworthy that personal connections can lead to transformations within professions. In this respect, the collegial connections between Derek Gregory and Anthony Giddens at Cambridge were arguably of great influence in facilitating the flow of ideas from sociology to geography. Certainly the pioneering work on the relationships among agency, structure and the recursive constitution of people and places has been fundamental to a progression of inquiry in human geography. However, the rapid adoption of postmodernist ideas within human geography's cultural turn has diverted a concentrated focus on notions of structuration. 'Third wave' concerns with difference, identity, and uncertainty infuse work at the critical edge of the discipline.

Our current work is located in this climate of inquiry. So where do notions of structuration now fit? Poststructural concerns with difference and identity have tended to deflect attention away from material conditions and the 'rules and resources' through which structure is realized, but the tricky problematic of agency–structure remains. We have noted earlier some recent work in geography that directly draws on concepts from Giddens' structuration theory. In our own work, notions of structuration provide a necessary yeast to the dough: while we do not explicitly draw on Giddens' concepts we recognize the ongoing theoretical importance of the tension between structure and agency in constructing explanations of everyday geographies. Nevertheless, the globalization of goods and services, transnational movements of people at unprecedented rates, enormous advances in communication technology, and economic restructuring have shifted, if not erased, the agency–structure problematic.

This has been reformulated and extended through an engagement with issues of power relations that stretch over time and space globally, and create the conditions of western industrial 'multicultural societies' that are now more prominent than in the early and mid 1980s. Since the 1980s the significance of gender and 'whiteness' as structured relations of power, and the instability of meanings, have become of central concern in understanding western postindustrial societies; these are dimensions of structuration that were not addressed in Giddens' original formulation of a non-functionalist theory of society. Yet the insights from Giddens' formulation of structuration theory remain invaluable in retaining the crucial importance of materiality and issues of scale in analyses. Notions of structuration also retain their currency in understanding how the spaces/locales of everyday life are both constitutive of and constituted by meanings that are reproduced or reworked in small-scale ways. Grappling with the structure–agency debate also endures in work concerned with the way that individual and collective action is invariably constrained, yet holds the capacity to be transformative at the level of community politics and the experience of place. Recent work drawing specifically on structuration theory suggests there is considerable opportunity to employ its constituent notions in new ways as a world predicated on the interlocking scales of global and local processes continues to unfold.

References

Baer, L.D., Gesler, W.M. and Konrad, T.R. (2000) 'The wineglass model: tracking the locational histories of health professionals', *Social Science and Medicine*, 50: 317–29.

Cheer, T., Kearns, R.A. and Murphy, L. (2002) 'Housing policy, poverty and culture: "discounting" decisions among Pacific peoples in Auckland, New Zealand', *Environment and Planning C: Government and Policy*, 20: 497–516.

Dear, M.J. (1981) 'Social and spatial reproduction of the mentally ill', in M.J. Dear and A.J. Scott (eds), *Urbanisation and Urban Planning in Capitalist Society*. London: Methuen, pp. 481–97.

Dear, M.J. and Moos, A.I. (1986) 'Structuration theory in urban analysis: 2. Empirical application', *Environment and Planning A*, 18: 351–73.

Dear, M.J. and Taylor, S.M. (1982) *Not On Our Street*. London: Pion.

Dorn, M. and Laws, G. (1994) 'Social theory, body politics and medical geography', *The Professional Geographer* 46: 106–10.

Dyck, I. (1989) 'Integrating home and wage workplace: women's daily lives in a Canadian suburb', *The Canadian Geographer*, 33: 329–41.

Dyck, I. (1990) 'Space, time and renegotiating motherhood: an exploration of the domestic workplace', *Environment and Planning D: Society and Space*, 8: 459–83.

Dyck, I. (1995) 'Putting chronic illness in place: women immigrants' accounts of their health care', *Geoforum*, 26: 247–60.

Dyck, I. (1996) 'Mother or worker? Women's support networks, local knowledge and informal childcare strategies', in K. England (ed.), *Who Will Mind the Baby? Geographical Perspectives on Child Care and Working Mothers*. London: Routledge, pp. 123–40.

Dyck, I. and McLaren, A.T. (2002) 'Becoming Canadian? Girls, home and school and renegotiating feminine identity'. RIIM Working Papers (02–12), Vancouver.

Fincher, R. (1987) 'Social theory and the future of urban geography', *The Professional Geographer*, 39: 9–12.

Gesler, W.M. and Kearns R.A. (2002) *Culture/Place/Health*. London: Routledge.

Giddens, A. (1976) *New Rules of Sociological Method*. London: Hutchinson.

Giddens, A. (1979) *Central Problems in Social Theory*. Berkeley, CA: University of California Press.

Giddens, A. (1981) *A Contemporary Critique of Historical Materialism, Vol 1. Power, Property and the State*. Berkeley, CA: University of California Press.

Giddens, A. (1983) 'Comments on the theory of structuration', *Journal for the Theory of Social Behavior*, 13: 75–80.

Giddens, A. (1984) *The Constitution of Society*. Cambridge: Polity.

Giddens, A. (1985) 'Time, space and regionalisation', in D. Gregory and J. Urry (eds), *Social Relations and Spatial Structures*. London: Macmillan, pp. 265–95.

Gleeson, B. (1999) *Geographies of Disability*. London: Routledge.

Gregory, D. (1981) 'Human agency and human geography', *Transactions of the Institute of British Geographers*, n.s. 7: 254–6.

Gregory, D. (1982) *Regional Transformation and Industrial Revolution*. London: Macmillan.

Gregory, D. (1989) 'Presences and absences: time–space relations and structuration theory', in D. Held and J.B. Thompson (eds), *Social Theory of Modern Societies: Anthony Giddens and His Critics*. Cambridge: Cambridge University Press, pp. 185–214.

Gregory, D. and Urry, J. (eds) (1985) *Social Relations and Spatial Structures*. London: Macmillan.

Gregson, N. (1987) 'Structuration theory: some thoughts on the possibility of empirical research', *Environment and Planning D: Society and Space*, 5: 73–91.

Halfacree, K. (1995) 'Household migration and the structuration of patriarchy: evidence from the U.S.A.', *Progress in Human Geography*, 19: 159–82.

Halfacree, K. and Boyle, P. (1999) 'Introduction: gender and migration in the developed world', in P. Boyle and K. Halfacree (eds), *Migration and Gender in the Developed World*. London: Routledge, pp. 1–29.

Kearns, R.A. (1982) 'Irrigation at Maungatapere: individuals, community and institutions'. MA thesis, Department of Geography, University of Auckland.

Kearns, R.A. (1987) 'In the shadow of illness: a social geography of the chronically mentally disabled in Hamilton, Ontario', PhD dissertation, Department of Geography, McMaster University.

Kearns, R.A. (1993) 'Place and health: towards a reformed medical geography', *The Professional Geographer*, 45: 139–47.

Kearns, R.A. (1998) '"Going it alone": community resistance to health reforms in Hokianga, New Zealand', in R.A. Kearns and W.M. Gesler (eds), *Putting Health into Place: Landscape, Identity and Wellbeing*. Syracuse, NY: Syracuse University Press, pp. 226–47.

Kearns, R.A. and Barnett, J.R. (1997) 'Consumerist ideology and the symbolic landscapes of private medicine', *Health and Place*, 3: 171–80.

Keaens, R.A. and Joseph, A.E. (2000) 'Contracting opportunities: interpreting the post-asylum geographies of Auckland, New Zealand', *Health and Place*, 6: 159–69.

McLaren, A.T. and Dyck, I. (2002) '"I don't feel quite competent here": immigrant mothers' involvement in schooling'. RIIM Working Papers (02–12), Vancouver.

Moos, A.I. and Dear, M.J. (1986) 'Structuration theory in urban analysis: 1. Theoretical exegesis', *Environment and Planning A*, 18: 231–52.

Phipps, A.G. (2000) 'A structuration interpretation of community activism during school closures', *Environment and Planning A*, 32: 1807–23.

Pred, A. (1984) 'Place as historically contingent process: structuration and the time-geography of becoming places', *Annals, Association of American Geographers*, 74: 279–97.

Soja, E.W. (1980) 'The socio-spatial dialectic', *Annals, Association of American Geographers*, 70: 207–25.

Thrift, N.J. (1985) 'Bear and mouse or bear and tree? Anthony Giddens's reconstitution of social theory', *Sociology*, 19: 609–23.

8 REALISM AS A BASIS FOR KNOWING THE WORLD

Andrew Sayer

The name 'realism' might suggest it's a philosophy that claims to have, unlike all the others, a 'realistic' view of the world, a privileged access to some absolute truth about that world. Well, sorry, but that's *not* what realist philosophy claims.[1] In fact its most fundamental doctrine should make us wary of such simplistic views of knowledge and truth. The most basic idea of realist philosophy is that the world is whatever it is largely independently of what particular observers think about it, and not simply a product of the human mind. People used to think the world was flat, but when they started thinking it was round, we don't imagine that the earth itself changed shape too. Its shape was unaffected by our ideas about it. At the same time, our ideas about the world are always constructed in terms of various ways of seeing – perceptual schemata, concepts and discourses. We cannot step outside these to see the world directly as it is, for we need these schemata etc. in order to see and think. So the world *exists* largely independently of our knowledge of it, but our *descriptions* of it do not, for they clearly depend on available knowledge. Our accounts of the world are always made in terms of available discourses and these may vary in their ability to make sense of it, as they did in the case of different discourses on the shape of the earth. There is an obvious caveat to make here, to do with the fact that knowledge is itself part of the social world, but I'll come to that in a moment.

Most people – including most researchers – are realists at least some of the time, though some researchers may be reluctant to acknowledge this and try to follow other approaches. Like all the 'isms' in this book, realism can be defined as much through what it opposes as what it asserts.

Non-realists fail to make a distinction between the world and our knowledge of it, and so end up imagining either that the discourses or types of knowledge are simple reflections of it (positivism), or that conversely the world is a product of our knowledge (idealism). Both these views make it difficult to see how knowledge can be fallible. For realists, the fallibility of knowledge suggests that the world is not just whatever we care to imagine. When we make mistakes, are surprised by events, or crash into things, we sense the 'otherness' of the world, its independence from our ideas about it. The implications of this are double-edged: on the one hand, this otherness or independence of the world implies that the task of developing ideas that can make sense of it is going to be inherently *difficult*; on the other hand, the very fact that we can often realize when we have got things wrong, through getting some negative feedback from the world, implies that distinguishing among the various properties of the world is not impossible. The very fact that we can successfully do so many things through our practical interventions in the world suggests that the knowledge informing those interventions has at least some 'practical adequacy'. The flat earth theory was quite practically adequate for many activities, even though we might want to say now that it was untrue. The round earth theory is more practically adequate than this, enabling us to do new things like putting satellites into orbit, but it's not perfect (the world seems not to be perfectly spherical).

Many readers may find this basic realist idea of the independence of the world from our thought about it too obvious to be worth labouring, and most advocates of other 'isms' will accept it when pressed, though they often forget it, or conflate the world with our conceptions of it. However, it's time to address that caveat. What about social phenomena? Aren't they 'socially constructed', and therefore *dependent* on our ways of thinking? Yes they are, but not in a way that contradicts our basic realist proposition. The UK Conservative Party is a social construction, but it is not *my* social construction; and though I may 'construe' – not construct – it in various ways, it would exist even if I didn't construe it at all. The discipline of geography is a social construction, a product of the interactions of many academics over long periods of time. Without that process of social construction it would not exist as an object. But this looks like a more difficult case: surely, the discipline is *our* construction, that is, the social construction of people like us. But here's the qualification to the basic realist point: I said the world can exist *largely* independently of any particular observers. Your doing a geography degree and my writing this piece makes only a little difference to the discipline of geography, for it is something that is already constructed, though it continues to be reproduced and transformed by many others.[2]

If we are not to reduce social construction to mere wishful thinking – the social world is whatever we care to suppose or construct it as – we have to bear in mind three points. First, we have to ask *who* is doing the constructing? In most cases it is not the geographer or social scientist, who is merely doing some 'construing' or observation and interpretation. Even if our research has some influence on the social phenomena under study – perhaps as the result of our interviews influencing our interviewees – the changes are at any time usually small, and they presuppose that there is something independent of the researcher that can be changed, thus confirming the basic realist point.

Second, we have to think about social construction as a process over time. Once things have been constructed, they gain a degree of independence from their constructors and from subsequent observers.

Third, we have to remember that attempts at construction always use materials – not only physical materials in this case, like concrete, but ideational materials like people's beliefs and habits – and the attempts succeed or fail according to how they make use of the particular properties of those materials. I may imagine myself to be a brilliant entrepreneur but fail to establish a successful firm because my attempts at constructing one failed to take into account the properties of the forms of social organization and activities I was trying to marshal. I may have underestimated the skills needed or be unable to pay workers enough to get them to work for me, or I may fail to stimulate any demand for the product. At any specific time these properties exist largely independently of how I care to think about them. Social constructions, like the knowledge that informs them, are *fallible*. Even when they succeed, they usually turn out to differ from what their authors or constructors had intended. This volume is a social construction but it may turn out differently from what the editors had wanted.

So yes, of *course* social phenomena are socially constructed, but if we are to avoid imagining that this amounts to some kind of automatically successful collective wishful thinking, and if we are to avoid confusing other people's constructions with our own construals of them, we need to remember the simple realist point about the independence of objects (including other people's ideas) from our thoughts about them.

This still allows us to acknowledge that ideas or discourses, ways of thinking, are extremely important and influential in shaping societies and their geography. Discourses of racial superiority, for example, had major effects in the shaping of many countries, legitimizing colonialists' seizure of indigenous land, and domination. Ideologies of gender restrict women's movement in public space. Recent shifts in political discourses from emphases on state support to individual entrepreneurship have influenced regional policy. All these discourses have had effects, but not only the effects they intended or comprehended: they also tend to have produced unintended consequences, including resistance.

This power of discourses is something that poststructuralists emphasize too, but they often lapse into a form of idealism in which it seems that anything can be constructed on the basis of any discourse, as if all discourses were infallible and hence all-powerful. People are indeed shaped by discourses, but as with any form of social construction, just how they are shaped depends on their properties – properties that at any particular time exist independently of or prior to the forces then impinging on them. If that sounds difficult, think of ordinary things that can be shaped. Wood is easier to shape in lasting ways than air because wood offers particular forms of resistance and has particular kinds of susceptibility. People are not infinitely plastic, and in so far as they can be shaped this presupposes that they have certain properties – certain powers and resistances. If we say simply that they are 'socially constructed', we are liable to misunderstand how the shaping of people and institutions at time *t* is constrained and enabled by their properties, which exist largely independently of the current shaping but were influenced by social forces at an earlier time *t*–1.

Like poststructuralists, we should be struck by the enormous cultural variety of human societies – something that ought to be particularly clear to geographers. But not just any object exhibits such variety; Russian sulphur dioxide is the same as British sulphur dioxide because sulphur dioxide lacks the properties that enable it to take on cultural forms that people (and perhaps some higher animals) possess. Culture doesn't produce variety on its own, but requires beings who are capable of being culturally influenced. Poststructuralists tend not to notice this but, under the influence of a kind of sociological or cultural imperialism, try to reduce everything to culture and discourse. Geography, as a good materialist subject, taking seriously people's physical as well as cultural properties and needs, ought to be able to see through and resist this imperialism. When someone says something is 'socially constructed', always ask 'by whom, and of what, and with what effects?'

These, then, are the most basic ideas of realism. Let's now move on to more specific ones that have implications for how we study geographical and other phenomena.

The first concerns causation. There was a long debate in geography, which ran for most of the twentieth century, between advocates of 'idiographic' and 'nomothetic' views of the subject. The former argued that subjects like geography and history could study only unique events and places and were therefore unlike science in being resistant to generalization. By contrast, the nomothetic school thought those disciplines could indeed find regularities and laws of behaviour and hence become sciences in their own right, if they only looked for them. The nomothetic geographers saw the idiographic approach as unscientific, lacking in theory or systematic method, and producing unique results lacking in wider application, so that each research monograph amounted to no more than another book on the shelf, in fact little more than mere opinion. A nomothetic approach would supposedly enable geographers gradually to accumulate knowledge of laws of behaviour, by systematically testing hypotheses about regularities. This, it was hoped, would enable them to generalize and build up a science of geography through which one could explain and predict by reference to theories. This was most clearly evident in the positivist approaches of the quantitative revolution.[3]

What has emerged in the last two or three decades in geography differs from *both* these models, and realism illuminates why this is the case. Realists argue for a different conception of science, one that doesn't rest upon the discovery of empirical regularities. Early attempts to find regularities have only thrown up approximate and temporary ones. The tendency of spatial interaction to decline with distance, for example, is only approximate and unstable. The regularity is itself spatially and temporally variable. More importantly, on its own, it is not very interesting. It becomes more interesting when one focuses on the qualitative differences between various kinds of interaction, their purposes and effects, and on what kinds of things produce those regularities.

There has definitely been a considerable increase in self-conscious theorizing about geographical phenomena, but most of it is about *how to conceptualize them* rather than merely for the purpose of generating hypotheses about regularities or patterns among events. There are things which geographers and others make generalizations about, but they expect them to vary somewhat over space and time. Thus, for example, in economic geography there has been a lot of interest in how to conceptualize globalization: researchers have evaluated different concepts of globalization and allied phenomena such as global media, cultural flows, and 'neoliberalism'. They know that while these phenomena are widespread, they are unlikely to be absolutely constant over time and space but will take various forms. To be sure there are some interesting patterns or regularities, of at least an approximate and temporary kind, but researchers have also necessarily had to take seriously certain unique events and objects, such as changes in world trade policies or changes of government. In either case – that is whether they are dealing with common patterns or unique events – they are more interested in the nature of the phenomena, the mechanisms producing them and their effects, than in whether they form regularities or not. Unique events, no less than common ones, are produced or caused by something and they have real effects. In either case we need to know how this happens, and finding regularities is not strictly necessary for this. In other words, the obsession with establishing such regularities – which is one of the characteristics of positivism – is not necessary for explaining things.

By 'cause' we mean simply that which produces (or perhaps blocks) change. A cause is not, as positivism assumes, a consistent regularity between one event and another, such that one is inclined to say, 'if A, then B'. A cause is a mechanism that produces change. All objects – including people, institutions and discourses – have particular 'causal powers', that is, things they are capable of doing, such as a person's power to breathe or speak. They also have particular 'causal susceptibilities' such as an individual's susceptibility to certain changes. Thus people are susceptible to cultural conditioning and peer group pressure, whereas lumps of rock are not. Many of these powers and susceptibilities are acquired through socialization, and while some of them need to be exercised or performed frequently to persist, some can persist unexercised for long periods of time, indeed some may never be exercised. Thus, fortunately, most of us rarely exercise or develop our powers to be violent.

Whether these powers or susceptibilities are activated depends on conditions whose presence is 'contingent', by which I mean not dependent but 'neither necessary nor impossible'. These conditions, which form the context of any given action, have their own causal powers and susceptibilities. Thus, for example, you could terminate your studies here and now but whether you do so depends on many circumstances, which may or may not be present. Furthermore, when a causal power or susceptibility is activated, the particular consequences depend on the context, that is on the presence and placement of the conditions in which it is activated. For example, an important element in the operation of housing markets is the formation of new households, as when children leave home or people begin cohabiting or get married, or separate or divorce. But their success in getting housing depends on the housing stock, on what's available and on the price. Thus, the consequences of actions are 'context-dependent', so that the same action may produce different results in different contexts. The implication of this is that we should not assume that the same cause will always produce the same effects. Quite often the effects will vary according to context, producing *ir*regularities. Stable, exact regularities occur only when processes and contexts are stable, and it is significant that in natural sciences, such regularities usually arise only under experimentally controlled conditions, which are unavailable in social science.

However, the absence of these regularities need not prevent us from explaining what is happening. What we need if we are to understand the operation of causal mechanisms is careful conceptualization of the objects concerned so as to identify their powers, and likewise of their contexts, together with a primarily qualitative analysis of how they operate, if and when they do. Quantitative analysis may also be useful, but the language of mathematics lacks any notion of causation, of forcing, producing, qualitatively changing or shaping. The important questions for getting at causation are ones which ask: what is it *about* this object (person/group/institution/structure/process/discourse) which enables it to do this thing? What is it *about* urbanization that produces new kinds of social relation and experience? What is it *about* multinational firms and their contexts which enables them to change their internal geographies?

There is something else that is crucially important about social phenomena in addition to how they are caused and what they are capable of causing. This something else concerns meaning. Here, the situation in human geography differs from that in physical geography. The latter, confronting objects like rocks, soils and slopes, has to develop a battery of concepts for distinguishing them – concepts like 'periglacial', 'moraine' and 'chernozem'. These concepts simultaneously help physical geographers to understand both the physical world and each other through a shared set of meanings. These meanings are *external* to their objects. The stuff we call 'moraine' does not have a discourse about itself; it does not have its own language and self-understanding. But human geographers are primarily interested in things that do have such understandings. Concepts and meanings in society are not merely externally descriptive of people and institutions but are 'constitutive' of what they are. The meaning of a geography degree or a seminar or marriage or gender is not simply external to its object, as the concept 'moraine' is to its object, but internal; what these things *are* depends on what they mean to the people involved in them, because they reproduce and transform them on the basis of their understanding of them. A mass of people walking down a street in the same direction may be a funeral procession, a political demonstration, a carnival, or football supporters on their way to a match. While there might be subtle differences among these in appearance and action, what marks the event as one thing rather than another is the actors' understanding of what they are doing, and what their intentions are. Unless we understand these things, we could observe the outward behaviour forever and still not know what was going on.

This is something that positivism does not recognize, even though positivists' own understanding of the social world and their ability to function within it unknowingly presupposes a knowledge of the constitutive meanings of the phenomena they observe and measure. Many of these meanings are so familiar we may not even notice them. Yet they are usually historically, geographically and culturally fairly specific, and they need analysing rather than taken for granted. This is something that humanistic geographers have emphasized, but I and other realists would argue all human geographers and social scientists need to do this; it is not something that need be taken seriously only by those interested in specifically 'cultural' phenomena like religions or lifestyles. Economic phenomena presuppose mutual understandings – e.g. common ideas about the function and value of money – about what it is to observe a contract, to be employed and so on, and these ideas are geographically and historically variable. Thus, the obligations of employers to employees vary between (and within) western and East Asian capitalism, for example.

So social phenomena are 'concept-dependent', and some, like political ideas, or the discipline of geography, are directly conceptual. Thus women's and men's life courses and patterns of mobility are strongly differentiated by behaviour based on gendered norms about what women and men ought and ought not to do. To understand how their use of private and public space has changed we need to study how the meanings of what it is to be a man or a woman have changed. Regardless of whether we find regularities or irregularities, the actions will depend at least in part on what they mean for actors.

To acknowledge that social phenomena are intrinsically meaningful or concept-dependent, however, is not to endorse the content of what people believe. People may act towards one another on the basis of misunderstandings as well as understandings (for example, imagining that gender differences in things like housework and paid work derive from our genes), though a misunderstanding is itself a kind of understanding in that it is meaningful. Hence to understand how men and women use space differently, we must certainly attend to what they think and assume about what men and women should 'naturally' do, and cite this in our explanation of their behaviour, but it does not mean we have to endorse these ideas. If we were to accept that these socially learned differences were innate, we would have misunderstood them.

It is for this reason that social science, including human geography, stands or needs to stand in a critical relation to its object of study, for it must be prepared not only to acknowledge the real effects of common-sense understandings held by people it studies, but to contradict them where they seem to be based on falsehoods.[4] It is also for this reason that I would advocate not merely a realist approach but a *critical* realist one. More generally, what would be the point of the study of society if it failed to evaluate everyday understandings, that is, if it failed to evaluate what people already knew just from living in society?

I said earlier that realists stress the importance of conceptualization in science – of worrying about how to conceptualize the objects we study instead of accepting everyday definitions and rushing on to gather data in the hope that we can get something meaningful by flushing the data through a statistical package so as to find some regularities, without bothering with what they mean or stand for. We can now see, I hope, that conceptualization is important not only for allowing the qualitative analysis that is necessary for identifying causal powers and susceptibilities and the manner of their exercise, but for getting at the meanings that actions and other social phenomena have for actors themselves – the meanings which make them those kinds of phenomena rather than others.

Remember that many social phenomena are ambiguous in terms of what they mean to actors. We therefore need to try to get at those ambiguities. Is this Remembrance Day march just about mourning the war dead, or also about celebrating the military? In what ways are people ambivalent about the places in which they live? Does migration indicate the freedom of the powerful or the desperation of the asylum seeker?

One more thing on causation and meaning in society. Although a lot of philosophers have assumed them to be radically different and opposed, realists do not. If a cause is simply that which produces change, than meanings can be causal too, for we usually communicate – share meaning – in order to produce some change. I am trying to reason with you so as to cause you to change your mind about approaches to geography. When someone asks me a question or challenges what I say, they produce (i.e. cause) a change in my behaviour by making me respond to them. When social scientists say discourses are 'performative', or produce effects, this is just another way of saying they are causal or 'causally efficacious'.[5]

I have already made a few comparisons between realism and positivism, humanism and poststructuralism. What about another influential development in social and geographical research – feminist approaches? In relation to feminism, there are points of both similarity and difference, and such is the diversity of feminist thought that there are some realist feminists.[6] First of all, realism is a philosophy of social science, not a social theory; roughly speaking it is about how to approach social science, rather than a particular theory about what society is, or a theory looking at society from the point of view of a particular social group, as feminism does. However both feminism and realism share a critical stance towards their objects of study; they agree that social science needs to be critical of its objects of study. Many feminists emphasize the way that social science reflects the social position of its authors or researchers, and hence has largely reflected a white, middle-class viewpoint. Realists would agree that this has happened, and that this has tended to restrict and distort social research. Some feminists have argued that, by contrast, members of oppressed groups have a privileged standpoint such that they are able to see things that others cannot. As a realist I would say that while this is possible, those in position of power are also able to see other things that the dominated cannot see. What is problematic in this situation is not only that the male view is selective, but that there are patriarchal relations of domination at all.

Some feminists, and indeed many non-feminists, argue that social scientific knowledge is not objective, as often assumed, but subjective, reflecting the views and values of the social researcher and his or her position in the social field. This is often a thoroughly confused position. First, all knowledge is indeed subjective in the sense that it requires subjects, beings who can think and make decisions. This is a precondition of, not an obstacle to, the development of objectivity in the sense of learning about the world. Second, two different senses of 'objective' are widely confused: objectivity in terms of value neutrality is quite a different thing from objectivity in terms of truth.[7] Although our values can sometimes make us see only what we want to see, they don't *necessarily* do this. Just because you feel that something is good or bad, it doesn't follow that you can have only a false (or practically inadequate) view of it. We can sometimes acknowledge unpalatable facts. Sometimes strong views about what is good or bad can enable you to see things that others who are less concerned fail to see; this is surely exemplified by feminism's exposure of gender itself. Realists would be closer to those feminists and others[8] who have argued for a position – sometimes termed 'strong objectivity' – that we are more likely to achieve objective or adequate knowledge of the world if we reflect carefully upon the difference that our particular subjective standpoint makes to the kind of knowledge we develop (Harding, 1991; Haraway, 1991). Are researchers projecting the special circumstances of their own situation onto those they study? How can they try to correct for this? A greater social diversity among researchers would help. So too would recognizing that social research involves a social relationship between researcher and researched.

The necessity of a subjective dimension to all knowledge doesn't mean that debates in social research are reducible to clashes of subjective views among which there can be no adjudication or resolution. It doesn't entail a relativist position in which truth is merely relative to your point of view. If they genuinely are debates rather than just different language games, they must be debates over something in common; there must be some thing over which they disagree, which, as realists insist, has at least some independence from those views. 'Malestream' views are wrong not simply because they are male, but in so far as they misrepresent the world, for example, by failing to note gender differences. A feminist view is unlikely to make that particular mistake, but by the same token, in as much as feminist views are right, it is not because those views are associated with women but because they adequately represent the world, for example, by explaining gender differences.

Note that I'm trying to steer a course between two hopeless extremes here: one extreme which imagines beliefs can be neatly sorted into those that are simply true in some absolute sense and those that are simply completely false; and another (called relativism, which is more literally 'hopeless') which imagines that we can never even sort out *some better* or more adequate beliefs from others. The otherness of the world from our knowledge of it makes it hard to see what it would mean to say that we knew the absolute truth about it, for there will always be alternative ways of describing it. But the obdurate nature of the world, its failure to do everything we imagine or want it to do, suggests we can at least sometimes sort out better or more adequate/true ideas about it from less adequate/true ones.

You might have hoped for something stronger from realism – perhaps a magic key that would enable you to distinguish what is 'realistic' from 'unrealistic' – but as I hope to have shown, things aren't that simple. But nor is the situation hopeless. We can still make progress. Many researchers are realists in practice, at least some of the time, even though they may not recognize themselves as such. When they worry about how to conceptualize the things they are studying, in order to decide what is and is not attributable to those things themselves; when they ask, 'What is it *about* this object (person, group, institution, practice, structure, etc.) that enables it to do that?'; when they study how mechanisms or processes of change work over time and space without expecting them necessarily to produce regularities; when they try to elicit people's understandings of their actions and situations, while retaining a critical distance towards those understandings as regards their adequacy; they are practising realists. You probably do some of these things sometimes. As a realist, I urge you to try to do them all the time.

NOTES

1 It *is*, however, what 'literary realism' claims, as in 'realist' novels or films, which (absurdly) purport to represent the world directly 'as it is', without any mediation of concepts, discourses or perceptual schemata.

2 Even in this case, mere observation on its own is unlikely to change anything. Change requires communication and practical interventions.

3 The quantitative revolution and spatial analysis were only partly and inconsistently positivist,

presumably because positivism is strictly impractical.

4 This critical stance is not something that humanistic geographers tend to be happy with.
5 These are not causes on the positivist cause–effect regularity model, but that model cannot distinguish mere regularities or coincidental correlations from changes which the cause event brought about. Note also that if I fail in persuading you, it doesn't mean I wasn't trying; activation of causal mechanisms (arguing, reasoning) doesn't guarantee regular effects.
6 For example, New, Davies.
7 A third meaning of objective is 'pertaining to objects'. This again is different from 'objective' as in 'true knowledge' or 'Value-free knowledge'.
8 Something like this was argued by the French sociologist Pierre Bourdieu (2000).

References

Bourdieu, P. (2000) *Pascalian Meditations.* Cambridge: Polity.

Davies, C.A. (1998) *Reflexive Ethnography.* London: Routledge.

Haraway, D. (1991) *Simians, Cyborgs and Women: The Reinvention of Nature.* London: Free Association.

Harding, S. (1991) *Whose Science, Whose Knowledge? Thinking from Women's Lives.* Oxford: Oxford University Press.

New, C. (1998) 'Realism, Deconstruction and the Feminist Standpoint', *Journal for the Theory of Social Behaviour,* 28(4): 349–72.

New, C. (2003) 'Feminism, Deconstruction and Difference', in J. Cruickshank (ed.), *Critical Realism: The Difference it Makes.* London: Routledge.

9 POSTMODERN GEOGRAPHIES AND THE RUINS OF MODERNITY

David B. Clarke

Introduction

The term *postmodern* burst unceremoniously onto the geographical scene in the mid to late 1980s (though the word was used much earlier in other contexts). Against the sober promises of realism to rethink science for human geography (Chapter 8), and the wise counsel structuration theory offered in attempting to resolve the long-standing family feud between structuralists and humanists (Chapter 7), the reckless, dizzying antics of postmodernists seemed to throw reason itself into doubt – and to care about the fact only as much as someone who has ceased to believe in God continues to worry about Satan. And they had a point, too: 'what proof is there that my proof is true?' asks Lyotard (1984: 24) in *The Postmodern Condition*. This disarmingly simple question throws scientific certainty into spiralling doubt, yet also makes you feel – if it's *that* easy to undermine – there's little point in worrying about it unduly anyway (Doel, 1993). Science is a matter of faith just as much as anything else. It has its believers, its unbelievers – and its fair share of guilt trips, too. Such is the non-committal credo of the postmodernist (though one would hardly guess it from the uptight sermons *The Dictionary of Humanist Geography* delivers to enlighten its Ley readers: Johnston et al., 2000: s.v. postmodernism, postmodernity).

The very first thing to say about the postmodern wor(l)d is that it's *inherently confusing*. (And postmodernists are wont to perform cheap tricks like that, so we now no longer know if we're talking about *words* or *worlds*, and are left feeling thoroughly disoriented regarding their relationship – which we once took for granted as a straightforward matter of *representation*.) If this prevents us from offering a clear-cut definition at the outset, the following eloquent montage of meanings should set us on track:

> **Postmodernity means many different things to many different people. It may mean a building which arrogantly flaunts the 'orders' prescribing what fits and what should be kept strictly out to preserve the functional logic of steel, glass and concrete. It means a work of imagination that defies the difference between painting and sculpture, styles and genres, gallery and street, art and everything else. It means a life that looks suspiciously like a TV serial, and a docudrama that ignores your worry about setting apart fantasy from what 'really happened'. It means licence to do whatever one may fancy and advice not to take anything you or others do too seriously. It means the speed with which things change and the pace with which moods succeed each other so that they have no time to ossify into things. It means attention drawn in all directions at once so that it cannot stop on anything for long and nothing gets a really close look. It means a shopping mall overflowing with goods whose major use is the joy of purchasing them. It means the exhilarating**

> freedom to pursue anything and the mind-boggling uncertainty as to what is worth pursuing and in the name of what one should pursue it. (Bauman, 1992: vii)

'Postmodernity is all these things and many others,' Bauman adds, 'But it is also – perhaps more than anything else – a *state of mind*.' Similarly, Eco suggests that 'postmodernism is not a trend to be chronologically defined, but, rather ... a *Kunstwollen*, a way of operating' (1985: 66). Let's simply log this thought for now.

A second thing to say about the postmodern wor(l)d is that people are always coming up with conceptual schemata that are supposed to make everything clearer – the overall effect of which is often to make things *even more confusing*. Ordering schemes are *always* confusing, as Foucault (1974: xv) famously made clear with reference to a passage in Borges on the classification of animals offered by 'a certain Chinese encyclopaedia' – to which Perec retorts with the following list, gleaned from genuine government documents:

> (a) animals on which bets are laid, (b) animals the hunting of which is banned between 1 April and 15 September, (c) stranded whales, (d) animals whose entry within the national frontiers is subject to quarantine, (e) animals held in joint ownership, (f) stuffed animals, (g) etcetera (this etc. is not at all surprising in itself; it is only where it comes in the list that makes it seem odd), (h) animals liable to transmit leprosy, (i) guide-dogs for the blind, (j) animals in receipt of significant legacies, (k) animals able to be transported in the cabin, (l) stray dogs without collars, (m) donkeys, (n) mares assumed to be with foal. (1999: 197)

This 'perfectly astonishing miscellaneity' (1999: 167) is just as evident in the customary distinctions made between (1) 'postmodern*ity*', a historical epoch or period, (2) the process of 'postmodern*ization*', (3) 'postmodern*ism*' as a cultural current or aesthetic style, (4) et cetera:

> Underwriting much of the literature ... has been a conceptualization of postmodern*ity* as an epoch progressively instituted in space–time in accordance with a logic of de-differentiation, which explodes the relative autonomy of the distinct spheres of economy, polity, civil society, and so on, through the negatory process of postmodern*ization*; and from which a correlative cultural logic or 'structure of feeling' (postmodern*ism*) may be directly discerned. (Doel and Clarke, 1997: 145)

Such schemata invariably break down. Take 'dedifferentiation', for instance. The basic idea is that *modernity* witnessed the *separation* of life into distinct spheres that had previously been inseparable (i.e. unified), allowing for their rational organization, monitoring, and surveillance. The way in which modern western society developed thus allowed the question of what makes 'economic sense', say, to be meaningfully disentangled from what is 'aesthetically pleasing' or 'morally right'. This had all kinds of (often contradictory) unintended consequences – but the modern process of differentiation has now, it seems, entered into a kind of *postmodern* 'reversal'. This is not a straightforward process of 'reintegration' and 'reparation', however, but rather a 'leaking' of different modern logics across into formerly distinct modern spheres: political concepts (e.g. 'rights') being applied to the market (e.g. 'consumer rights'); the aesthetic driving the political or the economic (think of media-driven electoral campaigns or the consumer society); and so on. With the short-circuiting of formerly distinct spheres, 'We witness ... the complete *decentring of society*. Ex-centric, dis-integrated, dis-located, dis-juncted, deconstructed, dismantled, disassociated, discontinuous, deregulated ... de-, dis-, ex-. These are the prefixes of today. Not post-, neo-, or pre-' (Tschumi: 1994: 225). So, if this is the case, how apt is the notion of a *separate* cultural logic, postmodern*ism*? Such contradictions are par for the course because postmodernity is contradictory at heart. It's no longer strictly 'modern' – but it's not something entirely new and different either, for modern categories can still be seen at work everywhere: 'the term "postmodern" implies contradiction of the modern without transcendence of it' (Kuspit, 1990: 60). Or perhaps it's 'the continuation of the Modern and its transcendence' (Jencks, 1986: 7). Such contradictions are, in fact, without possibility of resolution.

A third thing to say is that *geographers have added to the confusion more than most* (Dear and Flusty, 2001; Minca, 2002) – for which we may be justly proud! One of the most famous books on postmodernity was written by a geographer: Harvey's *The Condition of Postmodernity* (1989). But this, as we shall see, is actually a repudiation of the whole 'postmodern' affair from a Marxian perspective – more akin to the work of Habermas (1983), Jameson (1984; 1991) or, especially, Callinicos (1989) than to Lyotard (1984), whose title Harvey mirrors (or is that parodies?). Likewise, Soja's *Postmodern Geographies* (1989) makes the case that postmodernity involves the 'reassertion of space' as modernity's obsession with history begins to crumble under its own weight – reworking Jameson's (1984: 83) argument that postmodernity ruptures the 'capacities of the individual human body to locate itself ... and cognitively to map its position' in a fragmented world of mut(il)ated space. Whilst there is something to this, one senses that Soja's thesis is unlikely ever to be entirely convincing. Space and time are never completely separable. So it's far more likely that we'll be able to speak of 'modern space–time' and 'postmodern space–time' than to suggest some linear historical narrative whereby 'time and history' give way to 'space and geography' (Clarke, 2003). Finally, Dear's *The Postmodern Urban Condition* (2000) is a later, more sober addition to the line-up of book-length treatments of postmodern geography. As with Soja (1989; 1996), however, Los Angeles is once more taken to be the epicentre of the postmodern universe (cf. Davis, 1985; Gordon and Richardson, 1999). When Soja (2000: xvii), anticipating the criticisms of those who are sick and tired of hearing the LA story (Elden, 1997), says that 'what's been happening in Los Angeles can also be seen taking place in Peoria, Scunthorpe, Belo Horizonte, and Kaohsiung, with varying intensities … and *never* in exactly the same way', I begin to suspect that he may not have been to Scunthorpe and wonder why he denegates his claim, stressing a variety of intensities and highlighting '*never*'.

So, '*never*' trust anyone's arguments about postmodernity. And be especially cautious when it comes to geographers. Humanists, Marxists, and many others have used and abused the term willy-nilly (and who can blame them? – it's in the spirit of the word, after all!). The purpose of *this* chapter, though, is to try to stamp some sense onto the notion, and to see how useful it might be – now that modish overuse of the word has finally faded away and we can at last get down to business.

Irresistible Force Meets Immovable Object

To make headway here, we need to get to grips with what a better understanding of the postmodern might entail. I've just noted that the line one gets from Marxists, humanists, and the like is typically unreliable. They all have their axes to grind and there are many for whom 'pomo' remains a four-letter word. So, let's go straight to the horse's mouth and hear Lyotard's (1984) argument.

The most frequently quoted line of *The Postmodern Condition* appears in the Introduction, where Lyotard states, 'I define *postmodern* as incredulity towards metanarratives'

(1984, xxiv). By *metanarrative* (literally, 'big story') Lyotard means the kind of 'overarching principles' that *legitimated* a certain kind of *modern* discourse – a discourse which claimed to be capable of disclosing the *truth*, hence guaranteeing the value and utility of *knowledge*-based action to society at large. And let us note that scientists, radicals, politicians, moralists and countless others have consistently claimed the right to act on the basis of the truth. Lyotard intends to argue that this kind of situation no longer holds good; that we now find ourselves facing 'the obsolescence of the metanarrative apparatus of legitimation' (1984: xxiv). Nike logic: *Just do it!* This, in a nutshell, *is* the postmodern condition: we've lost faith in the 'grand narratives' of modernity, just as we once lost faith in God. (At this point someone will ask who 'we' are supposed to be, since most of us continue to act *as if* we believe. Good question – but for now, simply suspend your disbelief: you can see what Lyotard's getting at. See also Lyotard, 1992.)

Lyotard indicates that he's using the term '*modern* to designate any science that legitimates itself with reference to ... some grand narrative, such as the dialectics of Spirit, the hermeneutics of meaning, the emancipation of the rational or working subject, or the creation of wealth' (1984: xxiii). What he's suggesting is that *before* modernity, stories were simply stories – fables or myths – which allowed us to get along in the world. Modernity, however, purported to introduce a new state of affairs. It proclaimed its superiority by declaring that *its* stories were *more* than stories: they could guarantee the truth and produce the reality effects that would demonstrate as much (modern medicine might provide a specific example). They would ultimately lead humanity on the road to emancipation (with knowledge serving as a means to an end) and to a full and final understanding of reality (knowledge, here, serving as its own, 'speculative' end). Now, for Lyotard, there's no question that knowledge could ever *really* provide access to the real and emancipate humanity. Such metanarratives were simply invoked as a way of science legitimating itself – of precluding other, potentially competing 'language games' from measuring up against its declarative statements of authenticity. In fact, we only ever have access to reality through language; through particular language games. We can never dive down beneath language and see directly how well it maps on to reality. But modernity got us all thinking that we could. Indeed, modernity might well be defined in terms of an overriding belief in the power of knowledge to grant privileged access to the truth. And on this score, perhaps, modernity *still* sounds all well and good.

Modernity's dream of universal knowledge might sound all well and good – until one recognizes some uncomfortable home truths. The first is simply that the qualities of science in the information age bear little resemblance to the original blueprint, and continue to accelerate away from the initial modern vision at an alarming rate: 'postmodern science is discontinuous with the science that preceded it' (Lechte, 1995: 99). The proliferation of knowledge production has led to an ever-more complex set of incommensurate and incomparable knowledges, to the extent that no one can any longer regard this as a temporary state of affairs, *en route* to some future final grand synthesis. The second thing to appreciate is that modernity's dream of universal knowledge could very easily turn into a totalitarian nightmare – and, in fact, did. It is not only in science fiction novels that modernist visions take on a sinister pall. After Auschwitz, modernity as a whole is flung into crisis – for the Holocaust was not some kind of reversion to premodern barbarity, but a systematic deployment and implementation of *modern* rationality. It even *exemplified* modernity (Bauman, 1989; Lyotard, 1990; Clarke et al., 1996). For Adorno (2003), Auschwitz induced a *crisis of representation* within modernity: 'It is impossible to speak of the Holocaust and impossible to keep silent' (Easthope, 2002: 112). So, both the *inaccuracy of the description* and the *evidence of its perversion* have served to undermine those self-same 'grand narratives' that once held out the hope of a happy future in utopia. The upshot is the *delegitimation of universal knowledge.* Competing language games are all we now have. But there is not necessarily anything to lament in this state of affairs. As Lapsley and Westlake joyously conclude, 'the postmodern condition can embrace inventive pluralism and proliferate resistance to existing forms of oppression' (1988: 208). Only those caught up in the unhealthy grip of modernist nostalgia will see the absence of metanarratives as one big *problem* rather than an overwhelming *opportunity* – which, like Nietzsche's death of God, opens up 'Our new infinite' (1974: § 374).

Or will they? Lyotard's principal claim for postmodernity is 'incredulity towards metanarratives'. One does not have to delve too far within the covers of *The Postmodern Condition* to realize that the venerable tradition of *historical materialism* looks suspiciously like a prize-winning specimen of a now outmoded modern metanarrative. For Harvey (1989), this is all a bit rich – to say the least. As a card-carrying Marxist, he can see perfectly well where the condition of postmodernity has sprung from. His unfailing Marxian method offers a clear vantage point for spying out its origins. And no one is going to get away with declaring his vantage point null and void, not least when it offers a perfect view of why they should want to do so in the first place! Here, then, is where the irresistible force meets the immovable object. Lyotard (1984) claims that modernist metanarratives have had their day: no one bothers to believe in such nonsense any more. Harvey (1989) demonstrates that his own favoured (Marxian) 'metanarrative' still works perfectly well, thank you very much – and is quite capable of exposing the sort of vacuous discourse that would wish to undermine its credibility (namely postmodernism). Harvey's response to Lyotard – like that of Habermas (1987), who indicates just how much of modernity's positive potential remains unfulfilled – is underpinned by the serious concern that, in practice, postmodernism 'delegitimates' social critique (and progressive politics) rather more than it 'delegitimates' venture capitalism (and neoconservatism). Postmodernity is, perhaps, 'nothing more than the cultural clothing of late capitalism' (Harvey, 1987: 279).

The polemical tone Harvey strikes in *The Condition of Postmodernity* is highly effective in portraying a whole panoply of French philosophers as a jilted generation of disaffected erstwhile radicals. For example, it makes one question the identity of the collective 'we' Lyotard implicitly invokes in *The Postmodern Condition*: jet-setting, globe-trotting intellectuals fit the bill rather more than the workers of the world, it seems. It also sounds a resounding rallying cry for continued mobilization around the Marxian metanarrative that Lyotard has supposedly ruled out of court – though Lyotard (1988: 92) himself has stressed that his 'strong drift away' from Marxism stemmed from the suspicion that 'by keeping to the question of realism, of true or false knowledge' one occludes the possibility of sensing all those other injustices, which are not permitted to appear within a given frame of reference. It is perhaps unsurprising, therefore, that Harvey earned more than a few bruises from people he might have expected to recruit as allies; he received a decidedly frosty reception for characterizing feminism as a 'local' struggle (Deutsche, 1991; Massey, 1991; Morris, 1992; Harvey, 1992). Harvey's call to 'keep the faith' provoked such a backlash by castigating postmodernism as inherently reactionary, for failing to bear witness to a 'postmodernism of resistance' that might open up a space for *other* voices, apart from class alone (Bondi, 1990; Bondi and Domosh, 1992; Soja and Hooper, 1993). Yet despite all the hubbub, *The Condition of Postmodernity* was also widely fêted, not least for its magisterial overview of the changed experience of space and time accompanying postmodernity. Harvey deftly develops Jameson's (1984) thesis that postmodernism has had fundamentally *disorienting* effects, ranging over everything from the incredible shrinking world wrought by 'time–space compression' (Kirsch, 1995) to changes in the built environment that have become the talk of the postmodern town (Ellin, 1996). All these hallmarks of postmodern culture, Harvey maintains, can – and should – be traced back to the logic of capitalism.

Despite some slight differences between Harvey (1989) and Jameson (1984), the two are united by a common method. For Jameson (1984), postmodernism is 'the cultural logic of late capitalism'. Harvey's (1987) 'cultural clothing' idea clearly follows suit – but whereas Jameson draws on Mandel (1975) to explain how we've become cast adrift in 'hyperspace', Harvey (1989) sees postmodern culture as a reflection of a new regime of 'flexible accumulation' and its attendant 'mode of regulation' (Aglietta, 1979). These differing inflections aside, both authors offer restatements of the classic Marxian 'base–superstructure' metaphor. Marx reasoned that 'The mode of production of material life conditions the social, political and intellectual life process in general'; the 'economic structure of society' is 'the real foundation, on which arises a legal and political superstructure and to which correspond definite forms of social consciousness' (1971: 20). This metaphor is justly notorious for the range of differing interpretations that have been put on it – such as Althusser's infamous remarks about economic determination in the last instance, where 'the lonely hour of the last instance never comes' (1969: 111). Harvey, like Jameson, clearly regards the base–superstructure model as foundational for understanding the way in which capitalism has spawned postmodern culture: 'The odd thing about postmodern cultural production is how much sheer profit-seeking is determinant in the first instance' (1989: 336). Whilst Harvey might have a point – Bauman (1993a) raises much the same issue in *Postmodern Ethics* – I'm afraid I think there is something seriously wrong with Harvey's dogged insistence that the conjugation of 'economy' and 'culture' necessarily accords to a particular logic (Amin and Thrift, 2004).

No matter how far one might sympathize politically with Harvey, there are grave problems with maintaining a staunchly modernist stance under postmodern conditions, where once-separate spheres have done away with themselves as specific determinations. Even Jameson (1981) adopts a more reflexive conception of the Marxist (meta)narrative than Harvey is willing to admit (though as Easthope (1999: 146) wryly observes, 'what is left of Marxism if it becomes merely the greatest story ever told' is difficult to fathom). Likewise, however much one might worry that Lyotard's arguments delegitimate social critique, Lyotard can hardly be held personally to blame. As Huyssen puts it, 'No matter how troubling it may be, the landscape of the postmodern surrounds us. It simultaneously delimits and opens our horizons. It's our problem and our hope' (1984: 52). Lyotard is an astute observer of modernity's meltdown, and if he occasionally fiddles whilst Rome burns, he nonetheless puts his finger on something vitally important – to which Harvey's response seems, sadly, little more than wishful thinking. My diagnosis may well be wrong – but if Harvey is in denial, wishful thinking is *really* not going to help. Failing to recognize something for what it is can be comforting: it absolves one from the task of straining to detect the unfamiliar amongst the outlines of the familiar. But equally, ignorance is not necessarily bliss. Not recognizing something for what it is can be extremely hazardous. It grants one a decidedly false sense of security. This, I would contend, is the supreme danger of Harvey's take on postmodernity.

Let us close this section by pressing this point a little further, since it will ultimately allow us to formulate a better understanding of the nature of postmodern geographies. Here, I take the liberty of quoting myself at length:

> [For Harvey,] the postmodern needs only to be brought to account before the established precedent of historical-materialist analysis to be revealed for what it is. All fashionable pronouncements of 'incredulity towards metanarratives' (Lyotard, 1984, xxiv) dissolve into the comfortingly familiar form of old-fashioned ideological rhetoric once the Marxian metanarrative is revealed as still having teeth. The legitimacy of this argument lies in the cogency of its demonstrable truth, and Harvey's (1989) is indeed the most cogent demonstration one could hope for. The problem, however, is that the 'incredulity towards metanarratives' proceeds from the opposite direction: from the impossibility of finding anything solid enough into which to bite. The question of legitimacy is, moreover, simply an *unnecessary* defence in the face of incredulity – incredulity being what it is. The only real response to such postmodern objections is to reject them out of hand, which Harvey (1989) manages with considerable aplomb. The danger, however, is that this can easily amount to the theoretical equivalent of whistling through the graveyard ... Harvey's repeated declaration of 'business as usual' ... amounts to a blunt refusal to accept the novelty of the postmodern condition. Yet its novelty cannot be so easily refuted. (Clarke, 2003: 177)

To change the metaphor only slightly, the postmodern world is not solid enough for anyone to find the kind of foothold that Harvey wishes to find – in order to reach the luminous summits and enjoy unrestricted panoramic views (Doel, 2005). Here we must take our leave from Harvey, therefore, and consider what a range of other authors have said.

Modern Times, Postmodern Geographies?

What would a truly valuable conception of postmodern geography look like? It's difficult to say, because despite the wealth of books that promise to map out the field, they almost invariably adhere to the modernist principles we have already had cause to dismiss. One arguably gains a better sense of the postmodern world from so-called fictive literature, like Auster's *New York Trilogy* (1987), than from the majority of texts with 'postmodern' in the title (see Jarvis, 1998). Soja's (1989; 1996; 2000) LA trilogy is a case in point. It spins ever onward and outward from its audacious initial thesis – developed from the perennially unlikely coupling of an anti-structuralist Marxist (Lefebvre, 1991) and an anti-humanist post-Marxist (Foucault, 1986) – that modern times (the cumulative progression of history) have given way to postmodern spaces (the syncretic imbroglio of geography). However entertaining all this may be, it exemplifies the most persistent error committed by self-professed postmodern geographers, namely the theorization of 'modern' and 'postmodern' as epochal concepts, which take their place in a sequential narrative: we were once modern; we are now postmodern; how things have changed. But the future ain't what it used to be. As Lyotard duly warns us, 'The idea of a linear chronology is itself perfectly "modern"' (1992: 90). Instead, the '*Post modern* would have to be understood according to the paradox of the future (*post*) anterior (*modo*)' (Lyotard, 1984: 81). Just as we have never been modern (Latour, 1993), therefore, we will never have been postmodern.

One key text to provide us with a sophisticated sense of postmodern geography is Doel's *Poststructuralist Geographies* (1999).

Citing Lyotard and Thébaud (1985: 16) to the effect that the 'Postmodern is not to be taken in a periodizing sense,' Doel notes that most accounts of postmodern geography openly flaunt their misapprehension by constructing listy, 'two-column, contrapuntal differentiations of the modern and the postmodern, in which there is a shift of dominance from, say, "reason, unity and order" to "madness, fragmentation and disorder"' (1999: 68). In the face of such widespread misunderstanding, it is important to recognize that 'Postmodernism ... is not modernism at its end but in the nascent state, and this state is constant' (Lyotard, 1984: 79). Here, we should revive the thought we logged earlier: that postmodernism is, if anything, a 'way of operating' (Eco) or a 'state of mind' (Bauman) – in other words, a way of going on.

To get ourselves in this particular frame of mind, let us begin by clarifying that postmodernity is not in *opposition* to modernity. Not only is it *not a distinct epoch* coming 'after modernity' – which would imply *absolute* historical discontinuity (a conception in evidence wherever hyphenation finds favour: 'post-modern') – it is *not even a shift in dominance* (a *relative* discontinuity, which is what all those tedious two-column lists imply). This latter case, as Doel maintains,

> **is easily unhinged through a dialectical image of thought that emphasizes how the two apparently autonomous and self-sufficient sides of the diachronic break are in actuality two mutually constitutive aspects of a single synchronic structure. And since the negation of modernity always already belongs to modernity, the post-modern negation is an impostor bereft of either proper form or specific content. (1999: 68)**

Unless we are careful, therefore, 'postmodern' can all too easily prove to be a contradiction in terms; a meaningless buzzword. So what *does* postmodernity mean? As Doel insists, 'Postmodernity is not an epoch, but the ceaseless refusal, from within modernity, to silence and forget what cannot be represented and remembered within modernity' (1999: 69). This is a significant formulation, so let's hear it once again, this time from Lyotard: 'The postmodern would be that which, in the modern, puts forward the unrepresentable within representation itself' (1984: 81). The postmodern is, on this understanding, a kind of *sensitivity to the unrepresentable* (Farinelli et al., 1994; Olsson, 1991). It is a recognition that modernity, with its penchant for representation – for constructing the big picture, so that 'we' can see that there's a place for everything and that everything is in its place – necessarily involved the *denial* or *effacement* of whatever didn't fit. The modern was hooked on the Truth, and the 'Truth – insofar as it exists at all – is first and foremost pictured' (Hebdidge, 1988: 209; cf. Heidegger, 1977). The postmodern, in contrast, is about remaining open to the other, not closing off possibilities, not airbrushing the unrepresentable out of the picture or off the map. And to this extent it is inherently geographical – more sensitive to difference and differentiation, to (s)pace and (s)pacing, than geography ever was (Doel, 1999; Soja and Hooper, 1993).

On Doel's poststructuralist account, therefore, postmodern geography 'deconstructs the orderly lineaments of Euclidean, non-Euclidean, and *n*-dimensional spaces', so that rather than the unrepresentable other remaining trapped within a restrictive scheme of representation or field of vision, 'space resists unification and totalization and becomes dissimilatory – a conduit for difference, otherness, and heterogeneity' (1999: 70–1). Like Dadaism, postmodern space 'marks a disjunction, effecting a possible place of difference and alterity' (Easthope, 2002: 4). It aims to open up 'a gap in signification' – to resist the tendency to try 'to recuperate it into some form of coherent meaning' (2002: 4). It is clear, from this kind of portrayal, why Harvey's (1989) work reads as a arch-modernist attempt to conserve a *particular* point of view under the guise of a *universal* struggle. So, if postmodernism casts Marxism's modernism in a different light, this is not necessarily something to get uptight about. The postmodern is not at all inherently conservative: far from it, in fact.

The previously mentioned geographical expressions of discontent that greeted Harvey's (1989) portrayal of feminism as a strictly 'local' struggle, in contrast to the 'universal' history of class iniquity, reflected a range of positive engagements between postmodernism, feminism, postcolonialism, and an array of other movements relating to a variety of marginal*ized* voices (Benhabib, 1991; Bhabha, 1994; hooks, 1990; Hutcheon, 1988; Young, 1990). Indeed, emphasizing the cultural politics of difference roundly rejects the idea of privileging class. As Bhabha puts it, deliberately twisting Habermas's (1987: 348) words, differences 'no longer "cluster around class antagonism, [but] break up into widely scattered historical contingencies"' (1994: 171). Harvey's (1993) considered response to such points arguably scores a tactical rather than a strategic goal, highlighting consonant rather than dissonant political interests – but also points up once more the difficulties of maintaining his default position. The Marxian metanarrative may be a cracking good story, but maintaining its truth-value so doggedly serves merely to highlight its modernist streak. Yet perhaps there is also something too easy about academics simply privileging difference instead (Strohmayer and Hannah, 1992). This is the trenchant critique made by Easthope (2002) – not at all from a reactionary position but as a gritty theoretical intervention designed to point up that Marxism doesn't hold a monopoly on utopian wishful thinking.

Amidst all these debates, perhaps one of the most remarkable aspects of the postmodern debate has been its perspicacious capacity to reconsider the modern – and to recognize what modernity was actually about all along. A few pertinent remarks on modernity should reveal the Möbius-like topology of the (post)modern. From its entry into the centre of western public discourse in the seventeenth century, the term 'modern' took on a significance far greater than its ostensible meaning alone allows (Bauman, 1993b). It signalled something vital about a society that was affording the 'current' or 'new' an elevated significance – for the first time emphasizing the *possibilities of the future* over the previously esteemed *authority of the past.* A modern, future-oriented society, however, necessarily casts the present as inadequate, as in need of perpetual improvement: hence the fundamental dynamism of modernity as a social formation. Yet 'perpetual improvement' covers a multitude of sins. In consistently attempting to achieve *what might yet be achieved*, modernity waged a constant war against ambivalence. This war of attrition relied upon a powerful ordering zeal, which could not but define as 'irrational' whatever stood in the way of its projected future accomplishments (Bauman, 1991). With the uncanny knack of viewing nature as untidy possessed only by expert horticulturalists, modern powers sought to prune and trim society into shape, turning wild cultures into gardening cultures (Bauman, 1987). Indeed, as the Haussmanization of Paris (Harvey, 2003) or Robert Moses' impact on New York (Berman, 1982) impeccably attest, modernity was always a process of creative *destruction* – 'the immense process of the destruction of appearances ... in the service of meaning ... the disenchantment of the world and its abandonment to the violence of interpretation and history' (Baudrillard, 1994: 160).

To all intents and purposes, says Bauman, such characteristics continue to mark all present-day western and westernized societies – with one significant difference: 'if throughout the modern era the "messiness", ambivalence, and uncertainty inherent in social and individual life were seen as temporary irritants, to be eventually overcome by the rationalizing tendency, they are now seen as unavoidable and ineradicable – and not necessarily irritants' (1993b: 596). This, for Bauman, is what postmodernity finally means. It is, accordingly, a fully *modern* development. Postmodernity is modernity having irreversibly passed a critical threshold, where the impossibility of its goals has become all too clearly recognizable, but beyond which its momentum is unstoppable. If modernity is a juggernaut out of control (Giddens, 1990), therefore, postmodernity is 'modernity minus its illusions' (Bauman and Tester, 2001: 75) – 'the immense process of the destruction of meaning, equal to the earlier destruction of appearances' (Baudrillard, 1994: 161).

This is why it is necessary to see the postmodern as, first and foremost, a way of operating or state of mind; a way of carrying on. For whilst it is far from being detached from evidence-based claims about the state of the world, it is primarily an interpretation of the status of that world. And this interpretation is certainly a long way from any literal reading of the term as 'after modernity'. As Bauman reflects:

> The 'after' bit which the concept of 'postmodernity' entailed seemed to me suspicious from the start ... It seemed to me that the 'postmodern perspective' which allowed the scrutiny of modernity's failures and the debunking of many of its undertakings as blind alleys, far from being in opposition to modernity or growing on its grave, had from the start been its indispensable *alter ego*: that restless, perpetually dissentful voice that enabled modernity to succeed in its critical engagement with found reality and the many realities sedimented by that engagement. I liked Lyotard's quip: one cannot be truly modern without first being postmodern.
>
> The 'time of postmodernity' is to me the time in which the modern stance has come to know itself, and 'knowing itself' means the realization that the critical job has no limits and could never reach its terminal point; that, in other words, the 'project of modernity' is not just 'unfinished', but *unfinishable*, and that this 'unfinishability' is the essence of the modern era. (in Bauman and Tester, 2001: 74–5)

So much for the 'time of postmodernity', then – but what, finally, can we say about the 'space of postmodernity'? The paradoxical answer can only be that postmodern spaces and places as such are, in a sense, not going to look very different from modern spaces and places. Yet in another sense, to paraphrase Proust, *we have been given new eyes to see spaces and places in other ways*. This, I believe, is the lasting legacy of the postmodern debate in geography.

To demonstrate as much, I will offer just one, brief example: Flusty's virtuosic consideration of various forms of 'interdictory space' in – yes, you've guessed it! – Los Angeles:

> His [Flusty's, 1994] taxonomy of interdictory spaces identifies how spaces are designed to exclude by a combination of their functional and cognitive sensibilities. Some spaces are passively aggressive: space concealed by intervening objects or grade changes is 'stealthy', and spaces that may be reached only by means of interrupted or obfuscated approaches [are] 'slippery'. Other spatial configurations are more assertively confrontational: deliberately obstructed 'crusty' spaces surrounded by walls and checkpoints; inhospitable 'prickly' spaces featuring unsittable benches in areas devoid of shade; or 'jittery' space ostentatiously saturated with surveillance devices. (Dear, 2000: 146–7)

This is a wonderful example of the postmodern capacity to see with new eyes. It is certainly possible to offer more thoroughgoing theorizations of postmodern space–time – in terms of its *fluidity* (Bauman, 2000) for instance; or the *obscene proximity* of pornogeography, 'the total promiscuity of things' (Baudrillard, 1993: 60). But this more modest example admirably serves our purpose. It gives a perfect flavour of the *possibilities* of postmodern geography.

The End

So, we must conclude. 'Must we conclude? It is by no means certain that any definite conclusion can be reached. Indeed, some are emphatic that it cannot' (Benko, 1997: 27). Well, perhaps not, then – except to say that this high-speed summary of the development, significance, and legacy of geography's postmodern turn has hinged on the attempt to cut through the confusion the term 'postmodern' initially delivered, especially regarding its 'ironic' stance (Rorty, 1989). Let us therefore close by noting the deadly seriousness of this irony, particularly in relation to the end of modernity (Vattimo, 1988). Baudrillard cites as an epigram an idea from Canetti's (1986) novel, *The Human Province*:

> **A painful thought: that beyond a certain precise moment in time, history is no longer *real*. Without realizing it, the whole human race suddenly left reality behind. Nothing that has happened since then has been true, but we are unable to realize it. Our task and our duty now is to discover this point or, so long as we fail to grasp it, we are condemned to continue on our present destructive course. (1987: 35)**

Now, Baudrillard takes the idea that we have dropped out of history entirely seriously. But he insists that Canetti is mistaken in thinking that we could ever go back and discover the point at which it happened, let alone put things back on track. Passing beyond the end, we realize that the end itself no longer means anything, and never meant anything in the first place. The end was always an illusion. It was one of modernity's illusions. The means modernity developed in pursuit of its illusory ends were often painful, but were somehow judged to be worth it (Freud, 1955). The ends seemingly justified the means. Today, however, those means persist entirely in the absence of the ends (Agamben, 2000).

References

Adorno, T.W. (2003) *Can One Live after Auschwitz? A Philosophical Reader*, ed. R. Tiedemann. Palo Alto, CA: Stanford University Press.

Agamben, G. (2000) *Means without Ends: Notes on Politics*. Minneapolis: University of Minnesota Press.

Aglietta, M. (1979) *A Theory of Capitalist Regulation: The US Experience*. London: Verso.

Althusser, L. (1969) *For Marx*. Harmondsworth: Penguin.

Amin, A. and Thrift, N. (eds) (2004) *The Cultural Economy Reader*. Oxford: Blackwell.

Auster, P. (1987) *The New York Trilogy*. London: Faber.

Baudrillard, J. (1987) 'The year 2000 has already happened', in A. Kroker and M. Kroker (eds), *Body Invaders: Panic Sex in America*. New York: New World, pp. 35–44.

Baudrillard, J. (1993) *Baudrillard Live: Selected Interviews*, ed. M. Gane. London: Routledge.

Baudrillard, J. (1994) *Simulacra and Simulation*. Ann Arbor, MI: University of Michigan Press.

Bauman, Z. (1987) *Legislators and Interpreters: On Modernity, Post-Modernity and Intellectuals*. Cambridge: Polity.

Bauman, Z. (1989) *Modernity and the Holocaust*. Cambridge: Polity.

Bauman, Z. (1991) *Modernity and Ambivalence*. Cambridge: Polity.

Bauman, Z. (1992) *Intimations of Postmodernity*. London: Routledge.

Bauman, Z. (1993a) *Postmodern Ethics*. London: Routledge.

Bauman, Z. (1993b) 'Modernity', in J. Krieger, M. Kahler, G. Nzongolap-Ntalaja, B.B. Stallins and M. Weir (eds), *The Oxford Companion to the Politics of the World*. Oxford: Oxford University Press, pp. 592–6.

Bauman, Z. (2000) *Liquid Modernity*. Cambridge: Polity.

Bauman, Z. and Tester, K. (2001) *Conversations with Zygmunt Bauman*. Cambridge: Polity.

Benhabib, S. (1991) *Situating the Self: Gender, Community and Postmodernism in Contemporary Ethics*. Cambridge: Polity.

Benko, G. (1997) 'Introduction: modernity, postmodernity and the social sciences', in G. Benko, and U. Strohmayer (eds), *Space and Social Theory: Interpreting Modernity and Postmodernity*. Oxford: Blackwell, pp. 1–44.

Berman, M. (1982) *All That Is Solid Melts into Air: The Experience of Modernity*. London: Verso.

Bhabha, H.K. (1994) *The Location of Culture*. London: Routledge.

Bondi, L. (1990) 'Feminism, postmodernism and geography: a space for women?', *Antipode*, 22: 156–67.

Bondi, L. and Domosh, M. (1992) 'Other figures in other places: on feminism, postmodernism, and geography', *Environment and Planning D: Society and Space*, 10: 199–213.

Callinicos, A. (1989) *Against Postmodernism: A Marxist Critique*. Cambridge: Polity.

Canetti, E. (1986) *The Human Province*. London: Picador.

Clarke, D.B. (2003) *The Consumer Society and the Postmodern City*. London: Routledge.

Clarke, D.B., Doel, M.A. and McDonough, F.X. (1996) 'Holocaust topologies: singularity, politics, space', *Political Geography*, 15: 457–89.

Davis, M. (1985) 'Urban renaissance and the spirit of postmodernism', *New Left Review*, 151: 106–13.

Dear, M.J. (2000) *The Postmodern Urban Condition*. Oxford: Blackwell.

Dear, M.J. and Flusty, S. (eds) (2001) *The Spaces of Postmodernity: Readings in Human Geography*. Oxford: Blackwell.

Deutsche, R. (1991) 'Boys town', *Environment and Planning D: Society and Space*, 9: 5–30.

Doel, M.A. (1993) 'Proverbs for paranoids: writing geography on hollowed ground', *Transactions of the Institute of British Geographers*, 18: 377–94.

Doel, M.A. (1999) *Poststructuralist Geographies: The Diabolical Art of Spatial Science*. Edinburgh: Edinburgh University Press.

Doel, M.A. (2005) 'Dialectical materialism: stranger than friction', in N. Castree and D. Gregory (eds), *David Harvey: A Critical Reader*. Oxford: Blackwell, pp. 55–79.

Doel, M.A. and Clarke, D.B. (1997) 'From Ramble City to the Screening of the Eye: *Blade Runner*, death and symbolic exchange', in D.B. Clarke (ed.), *The Cinematic City*. London: Routledge, pp. 140–67.

Easthope, A. (1999) *The Unconscious.* London: Routledge.
Easthope, A. (2002) *Privileging Difference.* Basingstoke: Palgrave.
Eco, U. (1985) *Reflections on* The Name of the Rose. London: Secker and Warburg.
Elden, S. (1997) 'What about Huddersfield? Edward W. Soja, *Thirdspace: Journeys to Los Angeles and Other Real-and-Imagined Places*', *Radical Philosophy*, 84: 47–8.
Ellin, N. (1996) *Postmodern Urbanism.* Oxford: Blackwell.
Farinelli, F., Olsson, G. and Reichert, D. (eds) (1994) *Limits of Representation.* Munich: Accedo.
Flusty, S. (1994) *Building Paranoia: The Proliferation of Interdictory Space and the Erosion of Spatial Justice.* West Hollywood, CA: Los Angeles Forum for Architecture and Urban Design.
Foucault, M. (1974) *The Order of Things: An Archaeology of the Human Sciences.* London: Routledge.
Foucault, M. (1986) 'Of other spaces', *Diacritics*, 16: 22–7.
Freud, S. (1955) *Civilization and its Discontents.* London: Hogarth.
Giddens, A. (1990) *The Consequences of Modernity.* Cambridge: Polity.
Gordon, P. and Richardson, H.W. (1999) 'Los Angeles, City of Angels? No, City of Angles', *Urban Studies*, 36: 575–91.
Habermas, J. (1983) 'Modernity: an incomplete project', in H. Foster (ed.), *The Anti-Aesthetic: Essays on Postmodern Culture.* Seattle, WA: Bay, pp. 3–15.
Habermas, J. (1987) *The Philosophical Discourse of Modernity: Twelve Lectures.* Cambridge: Polity.
Harvey, D. (1987) 'Flexible accumulation through urbanization: reflections on postmodernism in the American city', *Antipode*, 19: 260–86.
Harvey, D. (1989) *The Condition of Postmodernity: An Enquiry into the Origins of Cultural Change.* Oxford: Blackwell.
Harvey, D. (1992) 'Postmodern morality plays', *Antipode*, 24: 300–26.
Harvey, D. (1993) 'Class relations, social justice and the politics of difference', in M. Keith and S. Pile (eds), *Place and the Politics of Identity.* London: Routledge, pp. 41–66.
Harvey, D. (2003) *Paris, Capital of Modernity.* London: Routledge.
Hebdidge, D. (1988) *Hiding in the Light: On Images and Things.* London: Routledge.
Heidegger, M. (1977) 'The age of the world picture', in *The Question Concerning Technology and Other Essays.* London: Harper and Row, pp. 115–54.
hooks, b. (1990) 'Postmodern blackness', *Postmodern Culture*, 1. http://jefferson.village.virginia.edu/pmc/text-only/issue.990/hooks.990, accessed 9 May 2005.
Hutcheon, L. (1988) *A Poetics of Postmodernism: History, Theory, Fiction.* London: Routledge.
Huyssen, A. (1984) 'Mapping the postmodern', *New German Critique*, 33: 5–52.
Jameson, F. (1981) *The Political Unconscious: Narrative as Socially Symbolic Act.* London: Methuen.
Jameson, F. (1984) 'Postmodernism, or the cultural logic of late capitalism', *New Left Review*, 146: 53–92.
Jameson, F. (1991) *Postmodernism, or the Cultural Logic of Late Capitalism.* Durham, NC: Duke University Press.
Jarvis, B. (1998) *Postmodern Cartographies: The Geographical Imagination in Contemporary American Culture.* London: Pluto.

Jencks, C. (1986) *What is Postmodernism?* London: Academy.
Johnston, R.J., Gregory, D., Pratt, G. and Watts, M. (2000) *The Dictionary of Human Geography*, 4th edn. Oxford: Blackwell.
Kirsch, S. (1995) 'The incredible shrinking world? Technology and the production of space', *Environment and Planning D: Society and Space*, 13: 529–55.
Kuspit, D. (1990) 'The contradictory character of postmodernism', in H.J. Silverman (ed.), *Postmodernism: Philosophy and the Arts*. London: Routledge, pp. 53–68.
Lapsley, R. and Westlake, M. (1988) *Film Theory: An Introduction*. Manchester: Manchester University Press.
Latour, B. (1993) *We Have Never Been Modern*. Cambridge, MA: Harvard University Press.
Lechte, J. (1995) '(Not) belonging in postmodern space', in S. Watson and K. Gibson (eds), *Postmodern Cities and Spaces*. Oxford: Blackwell, pp. 99–111.
Lefebvre, H. (1991) *The Production of Space*. Oxford: Blackwell.
Lyotard, J.-F. (1984) *The Postmodern Condition: A Report on Knowledge*. Manchester: Manchester University Press.
Lyotard, J.-F. (1988) *Peregrinations: Law, Form, Event*. New York: Columbia University Press.
Lyotard, J.-F. (1990) *Heidegger and 'the Jews'*. Minneapolis: University of Minnesota Press.
Lyotard, J.-F. (1992) *The Postmodern Explained to Children: Correspondence 1982–1985*. London: Turnaround.
Lyotard, J.-F. and Thébaud, J.-L. (1985) *Just Gaming*. Minneapolis: University of Minnesota Press.
Mandel, E. (1975) *Late Capitalism*. London: Verso.
Marx, K. (1971) *A Contribution to the Critique of Political Economy*. London: Lawrence & Wishart.
Massey, D. (1991) 'Flexible sexism', *Environment and Planning D: Society and Space*, 9: 32–58.
Minca, C. (ed.) (2002) *Postmodern Geography: Theory and Praxis*. Oxford: Blackwell.
Morris, M. (1992) 'The man in the mirror: David Harvey's *Condition of Postmodernity*', *Theory, Culture and Society*, 9: 253–79.
Nietzsche, F. (1974) *The Gay Science*. New York: Random House.
Olsson, G. (1991) *Lines of Power/Limits of Language*. Minneapolis: University of Minnesota Press.
Perec, G. (1999) *Species of Spaces and Other Pieces*. Harmondsworth: Penguin.
Rorty, R. (1989) *Contingency, Irony, and Solidarity*. Cambridge: Cambridge University Press.
Soja, E.W. (1989) *Postmodern Geographies: The Reassertion of Space in Critical Social Theory*. London: Verso.
Soja, E.W. (1996) *Thirdspace: Journeys to Los Angeles and Other Real-and-Imagined Places*. Oxford: Blackwell.
Soja, E.W. (2000) *Postmetropolis: Critical Studies of Cities and Regions*. Oxford: Blackwell.
Soja, E.W. and Hooper, B. (1993) 'The spaces that difference makes: some notes on the geographical margins of the new cultural politics', in M. Keith and S. Pile (eds), *Place and the Politics of Identity*. London: Routledge, pp. 183–205.

Strohmayer, U. and Hannah, M. (1992) 'Domesticating postmodernism', *Antipode*, 24: 29–55.
Tschumi, B. (1994) *Architecture and Disjunction*. London: MIT Press.
Vattimo, G. (1988) *The End of Modernity: Nihilism and Hermeneutics in Post-modern Culture*. Cambridge: Polity.
Young, I.M. (1990) 'The ideal of community and the politics of difference', in L.J. Nicholson (ed.), *Feminism/Postmodernism*. London: Routledge, pp. 300–23.

10 POSTSTRUCTURALIST THEORIES

Paul Harrison

'All truth is simple' – Is that not a compound lie?
(Friedrich Nietzsche, 1990[1888]: 33)

Like all 'isms', 'poststructuralism' is an awkward term and one which continues to generate more confusion, frustration, argument and outright anger than most. One of the main reasons for this is that it is often very unclear quite to whom or what the term refers. Unlike the positivists of the Vienna Circle for example (see Chapter 2) or indeed the vast majority of Marxists (Chapter 5), feminists (Chapter 4) and realists (Chapter 8), the figures, works and views gathered under the title 'poststructuralism' are for the most part *not* self selecting; they did not and have not signed up to a manifesto and do not share a credo. Indeed – and as we shall see below – considered at its broadest the term 'poststructuralism' simply describes the state of contemporary continental philosophy, a lineage of philosophical figures and texts stretching back approximately 220 years. In this context the term 'poststructuralism' refers to a more or less loosely grouped collection of texts and philosophers which came to prominence in France during the 1960s.

Clearly 50 years of vibrant and challenging theory and thought cannot be summed up in a chapter such as this, nor should one rush to attribute one view to what is a diverse and still developing body of thought. Having said this, at the outset of this chapter I want to suggest three things which give poststructuralism its unique place within the continental tradition. First, there has been a revival of ontological questioning; poststructuralism marks a return to and a revitalization of 'first philosophy' – of basic foundational questions of being – though more often than not by treating such foundational questions as contingent historical issues. Second, and following on, poststructuralism is radically anti-essentialist; for poststructuralism, meaning and identity are effects rather than causes. Third, and following on again, it has become increasingly clear over the past two decades that poststructuralism has a major ethical aspect, particularly in its concern for radical otherness and difference. If these points sound somewhat obscure, hopefully this chapter will go some way to clarifying them; however I cannot recommend strongly enough that anybody who is interested should engage with the primary literature itself.

The chapter is divided into four main sections. The first of these gives a brief introduction to poststructuralism's anti-essentialism via reflecting on the method of genealogy. The second section moves on to give a somewhat schematic history of poststructuralism, situating it in the wider continental tradition. The third section looks briefly to the future and to the political and ethical demands on poststructuralism. The final section considers a number of further readings, and reviews two key poststructural geography works. Running throughout the chapter is a rather blunt critique of positivist approaches and of the traditional philosophical and theoretical attempt to escape from history and situation – from the 'original difficulty of life' as the American philosopher John Caputo (2000) puts it – the main purpose of which is to simply raise questions concerning the nature of certain assumptions about verification, representation and truth.

The Cold Truth of Genealogy

Talking about the giving of definitions the German philosopher Friedrich Nietzsche (1844–1900) claimed that only that which has never had a history can be defined with any certainty. What does this mean? Nietzsche is suggesting that if someone, a philosopher perhaps, thinks that all we need for understanding are clear definitions – that one can get to the essence, identity or meaning of something by summing it up in abstract or logical terms – they are badly mistaken:

> There is their lack of historical sense, their hatred even of the idea of becoming, their Egyptianism. They think they are doing a thing an *honour* when they dehistoricize it, *sub specie aeterni* [from the viewpoint of eternity] – when they make a mummy of it. All that philosophers have handled for millennia has been conceptual mummies; nothing actual has escaped from their hands alive. They kill, they stuff, they worship, these conceptual idolaters – they become a mortal danger to everything when they worship. (1990: 45)

According to Nietzsche, when we think we are talking about essences, facts or things-in-themselves – anything that seems or should be simple, obvious, unproblematic or clear – we forget that everything has a history. Nietzsche is suggesting that those who believe Truth is something that, like money or salvation, can be gained or possessed, those who believe 'even to the point of despair, in that which is' (1990: 45), fail to grasp the shifting, *becoming*, nature of existence. While a concept seems to identify something certain and immutable, something common, like 'good' and 'evil' perhaps, or 'human' and 'animal', or 'truth' and 'error', or simply 'positivism' and 'poststructuralism', it is rather the sedimentation of a history of mutations and conflicts over definition, the strata of which outline attempts to wrestle control of the term's meaning. A concept has all the unity of a mole's burrow and, like the needle vibrating on a speedometer, a word or a sign marks an ongoing relationship between forces. Nietzsche is warning us to be particularly wary of *transcendental* claims, be they religious or philosophical-scientific: of claims which whisper the reassuring salvation of 'now and forever' – *sub specie aeterni* – under their breath; of claims which would direct you to Truth with a capital T, Reality with a capital R, the Good with a capital G. When you hear claims like this Nietzsche advises to start sniffing around, as for all this sweetness something is rotten somewhere; a moral lesson is being instructed as identity is confirmed and contingency disavowed: 'This workshop where *ideals are manufactured* seems to me to stink of so many lies' (1998: 47). As Caputo writes:

> Whatever is called 'Truth' and adorned with capital letters masks its own contingency and untruth, even as it masks the capacity for being-otherwise. For our being human spins off into an indefinite future about which we know little or nothing, which fills us with little hope and not a little anxiety, a future to come for which there is no program, no preparation, no prognostication. (2000: 36)

This is the 'cold truth' of genealogy and of poststructuralism: its truth without Truth: its secret which is not a secret; its foundation which is an abyss.

The French historian of ideas Michel Foucault (1926–1984), perhaps the most well known of poststructuralists, was deeply influenced by Nietzsche's work and by the insights that could be gained by genealogy in particular. Foucault practised genealogy as a historical-philosophical method, producing studies of prisons and punishment (1977a) and of human sexuality and subjectivity (1978; 1988; 1990). These studies were detailed genealogies of the various ways in which bodies and minds were and are historically constituted. In his 1971 essay 'Nietzsche, genealogy, history' Foucault writes that:

> Genealogy is grey, meticulous, and patiently documentary. It operates on a field of entangled and confused parchments, on documents that have been scratched over and recopied many times. (1977b: 139)

We tend to think that the purpose of historical investigation is to trace the development of a phenomenon, be it morality, sexuality or punishment, by working back towards its hidden source or origin in order to discern the underlying principle or cause – be it climate change, human nature, the civilizing urge of a people, a crisis in the mode of production, the will to power of an individual or their troubled relationship with their mother. And indeed this is how much historical research has been and is conducted. However, such research both assumes and sets out to discover some extrahistorical or transcendental structure or mechanism standing behind, guiding and shaping phenomena, some mechanism which allows words to keep their meaning, desires to always point in one direction and ideas to retain their logic. Genealogy does not seek to discover such a source or secret; rather 'the genealogist needs history to dispel the chimeras of the origin' (1977b: 144). The aim is not recovery or restitution but dispersion. The aim of genealogy is not to institute a despotic aspatial and ahistorical *sub specie aeterni* in our thought and methods but to catch a glimpse of life as it takes flight and to be obligated by this movement; to understand history as a productive, differential field. Thus for Foucault:

> it is no longer an *identity* that we need to recover ... but a *difference*. It is no longer a positive ideal that needs to be restored but simply a certain capacity to resist identities that are imposed upon us just to set free our capacity to invent such new identities for ourselves as circumstances allow. (Caputo, 2000: 34)

If Foucault refuses to posit transcendental or metaphysical causes to events, to give a simple narrative to history and an explanation to the present, how does he construct his accounts? Again Foucault follows Nietzsche, stepping back from reason and towards sensation, from theory to practice, from brain to nose. For Foucault, all we have to hold onto is the body in all its unforeseeable mutability:

> The body is the inscribed surface of events (traced by language and dissolved by ideas), the locus of a dissociated Self (adopting the illusion of a substantial unity), and a volume in perpetual disintegration. Genealogy ... is thus situated within the articulation of the body and history. Its task is to expose a body totally imprinted by history and the process of history's destruction of the body. (1977b: 148)

Thus in his studies of disciplinary techniques and confessional practices Foucault traces the linking of networks of *biopower* across the west: the articulation of technoscientific discourses and practices which call forth, shape, identify, classify, regulate and judge bodies. In particular he focuses upon the creation of docile and productive bodies which underlies the rise of capitalism: bodies trained and sorted through the emergent network of schools, workshops, prisons, barracks and hospitals in order to fit into the new machinery of production.

Yet if Foucault argues that nothing is fundamental, if his argument is radically anti-essentialist and if knowledge and truth are themselves inseparable from power, how can he make the historical claims he does and how are we to judge his work? Surely there is a performative contradiction here or, as the German philosopher Jürgen Habermas (1929–) (1991) puts it, a 'crypto-normativism': in offering a philosophical-historical critique surely Foucault must be appealing to some external critical standards as well as to some standard of truth and reason, if only implicitly. For many Foucault's writing demonstrates a double gesture common to most if not all poststructuralist thought: the denial of any external standard of reason and truth on the one hand while attempting to critique and convince on the other. And hence the accusations of irrationalism and nihilism which so often accompany poststructuralism. Foucault's work is certainly problematic in a number of respects; however, on this key issue of the status of critique – and looking ahead to our third section – it is worth focusing briefly on his essay 'What is Enlightenment?' in which he writes of the situation of his own work and of wider tasks of modern philosophy and theory.

Foucault opens by commenting that: 'Modern philosophy is the philosophy that is attempting to answer the question raised so imprudently two centuries ago: *Was is Aufklärung?* [*What is Enlightenment?*]' (1984: 32). It was the German philosopher Immanuel Kant (1724–1804) who first asked this question in an essay written in 1784. Foucault notes what could be called the standard interpretation of Kant's essay: that Kant understood the Enlightenment as an exit from our self-imposed immaturity, from our willing acceptance of 'someone else's authority to lead us in areas where the use of reason is called for' (1984: 34), examples being submission to military discipline, political power or religious authority. In place of such orders to 'obey without thinking' Kant suggests the alternative 'Obey, and you will be able to reason as much as you like', and as an example he suggests paying one's taxes while being able to argue as much as one likes about the system of taxation (1984: 36). Foucault notes that Kant is actually proposing some sort of contract, that this is an issue of politics as much as of science: 'what might be called the contract of rational despotism with free reason' (1984: 37) wherein there is the free use of reason but only within certain prescribed limits. Foucault breaks off from his reading of Kant's essay at this point and begins to focus less on what it *says* and more on what it *shows*. For Foucault the crucial and radical point in the essay is how Kant takes the *present moment* as the object of his critical reflections. Against the contract or settlement of reason within the limits of an arbitrary reasonableness, in this instant the critical questioning is incessant: 'What difference does today introduce in respect to yesterday?' (1984: 34). Here the Enlightenment is understood not as a threshold over which we pass once moving from subservience to freedom within certain limits, as one would graduate from school to work, but rather as a process of continual questioning of such thresholds. In particular for Foucault it is a questioning of the geohistorical constitution of ideas, concepts and values which underpin the most apparently unquestionable and normal of attitudes and assumptions; 'maturity' is the unending process of the *production* of autonomy and freedom, not another settlement into another despotism. Thus Foucault's is a 'practical critique ... a *critical ontology of ourselves*, which opens the possibility of being otherwise by calling into question through reflecting on how we have become what we are' (Owen, 1999: 602). In this sense poststructuralism produces *immanent* critiques, critiques in the absence of an overarching explanatory schema. For Foucault the Enlightenment is 'a set of political, economic, social, institutional and cultural events' that linked the 'progress of truth and the history of liberty in a bond of direct relation' and formulated 'a philosophical question that remains for us to consider' (1984: 43): *what is Enlightenment? How have we become what we are? What are the uses of reason today?* The problem and question of reason is to be treated historically and not metaphysically. In this way Foucault rejects the '"blackmail" of the Enlightenment' (1984: 43) found in critiques of his work such as those given by Habermas; one does not have to be 'for' or 'against' the Enlightenment as if some clash of cultures or civilizations were in the offing. Foucault's analysis suggests 'that one has to refuse everything that might present itself in the form of such a simplistic and authoritarian alternative' (1984: 43).

What is Poststructuralism?

What of poststructuralism then? As noted above, the term 'poststructuralism' names one of the most recent phases of continental philosophy. Following Foucault's lead we can say that the history of continental philosophy extends for roughly 220 years, beginning with the publication of Kant's critical philosophy in the 1780s. The British philosopher Simon Critchley provides a useful provisional rundown of the phases and figures since that time, as shown in Table 10.1.

Like Foucault, Critchley argues that it is the reaction to Kant's philosophy which continues to inform current debate over (and misunderstandings of) poststructuralism. As we saw in the previous section, Kant's philosophy crystallized the Enlightenment by claiming the sovereignty of reason. One route from this point to the present is via what has become known as the 'analytic tradition' in philosophy – which gave rise to logical positivism and its offspring. On the reading of analytic philosophy Kant's contribution means that the focus of thinking and theory should be first and foremost on *epistemological* questions, i.e. questions concerning validity, verification, and evidence, at the service of reason. As the Viennese positivist Otto Neurath put it:

> **The representatives of the scientific world-conception stand on the ground of simple human experience. They confidently approach the task of removing the metaphysical and theological debris. (quoted in Critchley, 2001: 96–7)**

However this is not the only route from the Enlightenment to the present. As Critchley describes, for many the entire project of the German Enlightenment suffered an 'internal collapse' soon after it was proposed:

> **The problem can be simply described: the sovereignty of reason consists in the claim that reason can criticize all our beliefs ... But if this is true – if reason can criticize all things – then surely it must also criticize itself. Therefore there has to be a *meta-critique* if the critique is to be effective. (2001: 19–20)**

And yet Kant's philosophy was unable to provide such a metacritique; in particular it could not link up theory and practice, reason and experience, understanding and sensibility, nature and freedom, the pure and the practical. Taking this last dualism – the pure and the practical – we can return to Nietzsche's comments near the start of this chapter. As we have seen, Nietzsche takes the reified air of purity to task for its disavowal of the contingent, the sensible, the mutable and the becoming, its disavowal of the 'original difficulty of life' with which we started. Rather than taking the high road to pure forms and essences, continental thought took the low path becoming, *pace* Martin Heidegger (1889–1976), a radical *onto-hermeneutics*. Cutting lower perhaps than Neurath thought possible, to the layer where most spades are turned, this tradition asked about

Table 10.1 Phases and figures in continental philosophy

1 German idealism and romanticism and its aftermath (Fichte, Schelling, Hegel, Schlegel, Novalis, Schleiermacher, Schopenhauer)
2 The critique of metaphysics and the 'masters of suspicion' (Feuerbach, Marx, Nietzsche, Freud, Bergson)
3 Germanophone phenomenology and existential philosophy (Husserl, Max Scheler, Karl Jaspers, Heidegger)
4 French phenomenology, Hegelianism and anti-Hegelianism (Kojève, Sartre, Merleau-Ponty, Levinas, Bataille, de Beauvoir)
5 Hermeneutics (Dilthy, Gadamer, Ricoeur)
6 Western Marxism and the Frankfurt School (Lukács, Benjamin, Horkheimer, Adorno, Marcuse, Habermas)
7 French structuralism (Lévi-Strauss, Lacan, Althusser), poststructuralism (Foucault, Derrida, Deleuze), postmodernism (Lyotard, Baudrillard) and feminism (Irigaray, Kristeva)

Source: Critchley, 2001: 13

> 'the *a priori* conditions not only for the possibility of sciences which examine entities as entities of such and such a type ... but also for the possibility of those ontologies themselves which are prior to the ontical sciences and which provide their foundations. (Heidegger, 1962: 31)

Of course for many – in particular many within the analytic tradition – such questioning is pointless, either having been answered or being unanswerable. As Ludwig Wittgenstein (1889–1951) commented in the *Tractatus Logico-Philosophicus* – a work which was and often still is taken as a programme for logical positivism – 'The world is all that is the case' (1961: #1), meaning that we can answer all our speculative questions and propositions by verifying them against the world, all other propositions being either analytic, in that they are self-referential and logical, or nonsense. Hence the final line of Wittgenstein's great work, perhaps one of the most famous in recent philosophy: 'What we cannot speak about we must pass over in silence' (1961: 7). Any proposition or claim that is not analytic or cannot be verified is not strictly speaking a proposition or a claim at all but an opinion and, as such, something which you should perhaps keep to yourself. And yet it is precisely the danger contained in this view which motivates so much continental philosophy: the danger of the reduction and limitation of truth to questions of representation, calculation, measurement and correspondence, outside which everything else is condemned as mere opinion. However, doesn't this mean that continental philosophy and poststructuralism in particular are simply bad psychology dressed up as philosophy – that *bête noire* of the US culture wars, pseudoscience?

Not all of these worries are misplaced; certainly there are works labelled poststructuralist which are awry in their arguments and convictions. Yet just because poststructuralism engages explicitly with the contexts, conditions of possibility, *a prioris* and *aporias* of phenomena such as truth and error, presence and absence, subjectivity and objectivity, testimony and fiction, representation and the sublime, correspondence and communication, that does not make it pseudoscience. The founder of phenomenology (see Table 10.1), Edmund Husserl (1859–1938), saw philosophy's role as the investigation into the nature and being of the lifeworld (*Umwelt*) from which theoretical and scientific thought emerges and within which it finds its significance and meaning. The phenomena, being or givenness of the lifeworld is the condition of possibility for subjectivity and objectivity alike, being presupposed in any definition of either; quite simply there is nothing outside (this) context. As Adriaan Peperzak writes:

> reason cannot prove its own beginnings. At least some beliefs, perceptions, feelings must be accepted before we can begin arguing. In order to avoid all arbitrariness, we must find out which basics, instead of being 'subjective' in the subjectivist sense of the word, are so fundamental that they deserve our respect and even trust. (2003: 3)

While for many poststructuralism is the epitome of contemporary nihilism, such a view is a serious failure to engage with its context and history for, with Nietzsche, poststructuralism is dedicated to *resisting* nihilism. Nothing evacuates potential meaning faster and in a more underhand manner than a simple presentation of the facts as if this were all that could or needs to be said. Nihilism is not due to being irrational or somehow 'against' the Enlightenment; rather it stems from the valuation of a purely calculative understanding of truth irreversibly cut off from context – ideals separated from their workshop. This is to say not that truth is *purely* contextual or that it is 'situated', 'local' or 'relative', but simply that reason finds its rationale only in the 'original difficulty of life'. If we forget to critically question our onto-hermeneutic situation we cannot begin to answer meaningfully the questions '"*why* in this way and not otherwise?", "*why* this and not that?", "*why* something rather than nothing?"' (Heidegger, 1998: 134): *sub specie aeterni* is the herald of an unresponsive and irresponsible despotism.

The Promise of Poststructuralism

Poststructuralism is often characterized as being an overly *negative* critical stance. In its radical anti-essentialism and anti-foundationalism, poststructuralism seems unable to offer any recognizably progressive programme, be it in terms of knowledge or politics. It is for this reason that many accuse poststructuralists of both epistemological and political relativism. To counter such views, in this section I want to bring out the *affirmative* nature of poststructuralism and to do so by turning briefly to the writing of French-Algerian philosopher Jacques Derrida (1930–) and to the mode of thought with which his name has become synonymous: deconstruction.

We saw above how through his use of genealogy Foucault attempted to bring to the fore the extreme historical and geographic contingency of apparently transcendental forces, entities and concepts. Derrida employs a similar technique though his investigations tend to focus upon key concepts within the philosophical, religious and political traditions of the west. For example his book *Politics of Friendship* (1997) considers various articulations of the concept of politics and the political from Plato (427–347 B.C.) through to Emmanuel Levinas (1906–1995). Derrida shows how throughout this history the concept of politics and of democracy in particular

> rarely announces itself without some sort of adherence of the state to the family, without what we could call some sort of *schematic* of filiation: stock, genus or species, sex ... blood, birth, nature, nation. (1997: x)

Democracy's great force and claim are that it treats all individuals singularly and without prejudice; however Derrida's analysis claims that in being always defined in terms of masculine friendship, democratic thought and practice consistently fall short of and indeed systematically withdraw from such obligations. Derrida demonstrates how the political imagination of the west has great difficulty in imagining ways of being together and the social relationship *per se* otherwise than as a reciprocal relationship between similar men. On the last page of the book Derrida asks the question which has motivated the study:

> **is it possible to think and to implement democracy, that which would keep the old name 'democracy', while uprooting from it all those figures from friendship (philosophical and religious) which prescribe fraternity: the family and the androcentric ethic group? Is it possible, in assuming a certain faithful memory of democratic reason and reason *tout court* ... not to found, where it is no longer a matter of *founding*, but to open to a future, or rather to the 'to come', of a certain democracy? (1997: 306)**

From this quote it should be clear that Derrida's conceptual investigation is not a simple nihilistic assault on the concept of democracy; rather, and like all deconstructive readings, it is an attempt to open the concept up to the possibility of being thought otherwise. Derrida is suggesting that we do not need a new foundation of the political – a new programme or blueprint – as such foundations unavoidably put to work transcendental or metaphysical presuppositions. Rather we need an opening of the concept of the political beyond its current imagination and conceptualization:

> **The idea of such [deconstructive] analysis is not to level democratic institutions to the ground but to open them to a democracy to come, to turn them around from what they are at present, which is the prevention of the other ... Preparing for the in-coming of the other, which is what constitutes radical democracy – that is what deconstruction *is*. (Caputo, 1997: 44)**

Since the early 1990s the deeply ethical nature of poststructuralist theory and of deconstruction in particular has come to the fore. In Derrida's writing this ethical impulse often revolves around the idea of the 'to come' (*l'à-venir*). For Derrida and for poststructuralism *per se* there 'can be no future as such without radical otherness, and respect for this radical otherness' (Derrida in Derrida and Ferraris, 2001: 21). Importantly 'radical otherness' is not otherness considered in terms of an identity but rather that which 'defies anticipation, reappropriation, calculation – any form of pre-determination' (2001: 21). While we cannot but anticipate the future, prepare, make plans, analyse, reckon and strategize, and we would be highly irresponsible not to do so, Derrida suggests that the rationale for such rationalization can only lie in our 'relationship' to the incalculable 'to come' of the future. We calculate because of the incalculable; a future is possible for us because of our relation to a future always to come but never present. In this sense our plans, calculations and rationalizations have the form first and foremost not of grounded propositions, of proofs or certitudes, but of pledges and promises. As Derrida comments: 'From the moment I open my mouth I promise' (1987a: 14) – even if like Wittgenstein in the *Tractatus* it is to say that there is nothing else to say – for all thought and thinking 'requires a *yes* more "ancient" than the question "what is?" since this question presupposes it, a *yes* more ancient than knowledge' (1992a: 296). Without a 'relationship' to the unknown in the form of an affirmation and a promise, knowledge is not possible. Hence, and to return to the example of *Politics of Friendship*, Derrida comments that

> **democracy remains to come; this is its essence in so far as it remains: not only will it remain indefinitely perfectible, hence always insufficient and future, but, belonging to the time of the promise, it will always remain in each of its future times, to come: even when there is democracy, it never exists, it is never present. (1997: 306)**

Hence deconstruction is already pledged, already engaged, already obligated and, at the same time, always to come, always promised, always possible.

Poststructuralism and Geography

Throughout this chapter I have been keen to stress that the term 'poststructuralism' covers a complex and diverse set of writings and ideas and as such it has had many routes into human geography. In the companion chapter to this one, John Wylie (Chapter 27) provides a concise and well-informed overview of poststructuralism's influence in geography; rather than replicate this account I want to focus on just a few selected and, I believe, indicative texts which the reader interested in poststructuralism and geography may want to consider.

To begin with it is worth noting a number of texts by (mainly) non-geographers which either come from or can serve to inform a poststructuralist understanding of and approach to space. Edward S. Casey's *The Fate of Place: A Philosophical History* (1997) is particularly useful for its long view of thinking on place and space. In a similar vain Jeff Malpas' *Place and Experience: A Philosophical Topology* (1999) is an interesting attempt by a contemporary philosopher to engage in the issues around space. While both Casey and Malpas have a distinctive phenomenological tenor to their discussions, David Farrell Krell's *Architecture: Ecstasies of Space, Time and the Human Body* (1997) is more influenced by Derrida's post-phenomenological deconstructive approach, though it is perhaps less successful than his other works. Covering similar topics, though at a less frenetic pace, are Robert Mugerauer's *Interpreting Environments: Tradition, Deconstruction and Hermeneutics* (1995), Karsten Harris' *The Ethical Function of Architecture* (1998) and John Rajchman's *Constructions* (1998). In many ways Derrida's key contribution for geographers is the idea of *spacing*, and as such, Derrida's own writing on the topics of space and place is minimal and somewhat oblique to say the least. However his substantive writing on the concepts of cosmopolitanism and hospitality are significant (1992b; 2000; 2001; 2002), as are his reflections on language, translation, exile and distance (1987b; 1998). An excellent example of deconstruction in a context relevant to geography is David Campbell's *National Deconstruction: Violence, Identity and Justice in Bosnia* (1997). As for Foucault, a great deal of his work can be understood at least in part as an endeavour of 'spatial history'; along with those primary texts noted previously, see in particular Stuart Elden's *Mapping the Present: Heidegger, Foucault and the Project of Spatial History* (2001), the geographer Chris Philo's (1992) important paper, and the essays collected in Joanne P. Sharpe et al.'s *Entanglements of Power: Geographies of Domination/Resistance* (2000). Some of the best writing in poststructuralism and space has come from a feminist perspective; of particular note in this context is the work of Elizabeth Grosz, including her books *Space, Time and Perversion: Essays on the Politics of Body* (1995), *Architecture from the Outside: Essays on Virtual and Real Space* (2001) and her influential *Volatile Bodies: Towards a Corporal Feminism* (1994). Of import for geographers is Donna Haraway's work on rethinking human–nature relations; see in particular her landmark collection of essays *Simians, Cyborgs and Women: The Reinvention of Nature* (1991). Finally, while not explicitly on the topics of space and place, of interest to many geographers in recent years have been Brian Massumi's *Parables for the Virtual: Movement, Affect, Sensation* (2002), William E. Connolly's *Neuropolitics: Thinking, Culture, Speed* (2002) and Michael Hardt and Antonio Negri's radical poststructuralist manifesto for the twenty-first century *Empire* (2001) and its sequel *Multitude* (2004). While all the texts listed here will help any would-be poststructuralist in gaining a feel for the broad contours of discussion and debate, it should come as no surprise that these texts can be quite detached from and seemingly lacking a context in current geographical thought. Again I would refer the interested reader to Wylie's overview for a more extensive context. Here I want to concentrate on two contributions by geographers which attempt to bridge this gap: Marcus Doel's *Poststructuralist Geographies: The Diabolical Art of Spatial Science* (1999) and Sarah Whatmore's *Hybrid Geographies: Natures Cultures Spaces* (2002).

While certainly not the first to enquire into poststructuralism and geography or to be avowedly poststructuralist in outlook, Doel's book *Poststructuralist Geographies* nonetheless in many ways represents the culmination of a 'first wave' of explicitly post-structural theorization from *within* human geography. Doel's central claim is that poststructuralism is already a thought about the nature of space, in particular about the nature of space as an *event*. According to Doel geography has suffered from 'pointillism', an overly sure (and metaphysical) belief in the common-sense substantiality of 'what is' – of the kind critiqued by Nietzsche above – and this has led to a failure to think about the nature and work of difference and differentiation or *becoming*. For Doel the realization that everything must take place, must – as Doel puts it after Henri Lefebvre – 'undergo trial by space', demonstrates that space is not so much a stage or thing but a happening, an event. Things do not simply sit in space but rather are spaced out in various ways and constitutively so; however, at the same time this 'spaceing' or '(s)playing' disturbs and disrupts any claim or semblance of solidity, permanence, identity or transcendence. These insights, developed and explored through readings of Baudrillard, Deleuze, Derrida, Irigaray and Lyotard, lead to a critique of both humanist and Marxist approaches within the discipline, both of which are taken to task for their overly 'sedentary' modes of thought, their unfounded belief in the foundations of place and thereby their failing to 'let space take place'. Played out across a wide range of references and through a proliferation of quotes, jokes and neologisms, what emerges from Doel's account is an ambitious and provocative attempt to rethink and rephrase our basic apprehension of space and so of geography itself. *Poststructuralist Geographies* has been critiqued on a number of fronts, both by those broadly sympathetic to its arguments and by those opposed. There can be little doubt that Doel's book is a challenging read; it is densely written and presupposes some prior knowledge of the writers of which he makes use. For some this is symptomatic of elitist and exclusionary tendencies within poststructuralism; however, perhaps a more pointed criticism here is that Doel's book is indicative of 'the central failing of post-structuralism ... that the only thing it is capable of saying anything about is itself' (Bancroft, 2000: 122). Less rhetorically, the geographer Jeffery E. Popke makes a similar criticism of Doel's brand of poststructuralism in his article 'Poststructuralist ethics: subjectivity, responsibility and the space of community' (2003). While sympathetic to and welcoming of Doel's rethinking of space in terms of difference, Popke argues that Doel 'fails to offer us any means for thinking this opening in relation to responsibility and justice' (2003: 309).

Rather than being a poststructural meditation on the nature of space *per se* – or at least not overtly so – Whatmore's *Hybrid Geographies* takes as its central task a rethinking and exploration of the distinction and relationships between nature and culture. Through a series of investigations Whatmore explores the vitalistic 'topologies of wildlife' which emerge as this binary is deconstructed. Like Doel, Whatmore wishes to demonstrate the heterogeneous, processual and emergent complexity of the world which underlies and is hidden in our inherited representations and assumptions. Hence the word 'hybrid' in the title: for Whatmore it is relations which are important and not essences, again a shift from being to becoming. This approach leads to a 'mapping' of various decentred or networked phenomena, exploring how various 'things' – from elephants to genetically modified crops – are constituted through diverse encounters, processes and performances and as such are always partial, provisional and incomplete. As much as it embodies the affirmative anti-essentialism and anti-foundationalism of post-structural thought, Whatmore's writing also draws on distinctive feminist and environmentalist traditions; and it is perhaps the combination of these three strands which leads to a more explicit thinking of the ethical than that found in Doel. For Whatmore the exposure of the relational and open nature of existence demonstrates the need for a 'relational ethics' concerned with ways of living, folding and becoming with 'more-than-human' others. In many ways – and not unlike Doel – Whatmore's text performs this commitment, attempting to embody the vital and polyvalent ethos of which she speaks. A consequence of this joining of saying and doing is that *Hybrid Geographies* is another challenging read which, like Doel's work, can be disorienting. However this is a productive disorientation, aimed at dispelling arbitrary or divisive conceptual formulations and naive methodological and empirical strategies. Arguably Whatmore's relational ethics – while more explicit than Doel's – remains somewhat ambiguous, perhaps indicative less of a lack of thinking about otherness than of the conflict between an ontological monism and a passionate commitment to difference. Despite these comments *Hybrid Geographies* is at the cutting edge of geographical thought, combining the theoretical and the empirical seamlessly in an incisive and affirmative immanent critique.

Summary

The aim of this chapter has not been so much to give an overview of poststructuralism as to give a brief indication of some of its directions and concerns. In so doing many bodies of works have been passed over, not least those of Giorgio Agamben (1942–), Judith Butler (1956–), Maurice Blanchot (1907–2003), Jean Baudrillard (1929–), Hélène Cixous (1937–), Gilles Deleuze (1925–1995), Félix Guattari (1930–1992), Luce Irigaray (1932–), Julia Kristeva (1941–), Emmanuel Levinas, Jean-François Lyotard (1925–1998) and Jean-Luc Nancy (1940–), to name only the most prominent – all of whom could be classified as post-structuralist, at least at some stage in their work. The point here is not to try to overwhelm but rather to highlight again that 'post-structuralism' is an awkward term and often one of limited value. Still, within this diversity of works I wish to reaffirm and add to the three points raised in the introduction. In reviving onto-hermeneutic questioning (if only to reject it), poststructuralism critically affirms our uncanny lack of foundations and essence. In this radical anti-essentialism, post-structuralism denies any short cuts to simple truths and the construction of accounts which would seek to reduce the phenomena under investigation to either ahistorical or aspatial causes *or* to simply the effect of context. In so doing, poststructuralism presents a relational and open movement of thought, one which is permanently under revision, undergoing 'trial by space'.

The chapter has sought to engage a number of the main criticisms of poststructural thought. Most generally, I have tried to show how the apparent obscurity of what is named by the term 'poststructuralism' stems from the fact that poststructuralism is the tip of the iceberg of continental philosophy. Once this is understood, it should be clear that poststructuralism is not simply the 'latest fashion' or 'elitist jargon'; indeed, trying to engage the term without respect for or in ignorance – wilful or otherwise – of this context is responsible for the vast majority of such misunderstandings and misrepresentations. A more telling criticism of poststructuralist thought is that it can become overburdened by its historical self-consciousness such that all that is produced in its name are commentaries on its own canonical and marginal texts. More specifically the chapter has engaged with claims that poststructuralism is nihilistic, anti-Enlightenment, irrational, pseudoscience nonsense, and epistemologically and politically relativistic. Taken together these criticisms claim that due to its anti-essentialism and anti-foundationalism poststructuralist work is incapable of producing insightful analysis or socially progressive and morally informed critique. Indeed, many take post-structuralism as an extreme and frivolous form of linguistic scepticism, preoccupied with splitting etymological hairs and bewitched by undecidability and obscurantism. In this understanding, poststructuralism's apparent obsession with concepts and language betrays its apathy in the face of real-world problems. And yet this is to ignore the incessant *urgency* which pulls poststructuralism along, its unfounded commitment to difference; to the singular, the marginal, the exceptional, the 'to come'. While thinking under the influence of poststructuralism does not *guarantee* insightful analysis or progressive critique – how could any philosophy, ontology or epistemology? – neither does it rule them out in advance. Indeed if any philosophy, ontology or epistemology claimed such a *guarantee* it would be the *least* insightful and progressive thought, being itself nothing more than an efficiency, a programmability, a calculability, or an automation of thinking.

In the end it is perhaps poststructuralism's general refusal to provide 'simple truths' and options which condemns it in the eyes of many; and yet, as Derrida writes, 'There is no moral or political responsibility without this trial and this passage by way of the undecidable' (1988: 116). Only a thought and a thinking which seeks to prepare and maintain a relationship with the unknown can be called thinking. As Nietzsche wrote:

> **I mistrust all systematizers and avoid them. The will to system is a lack of integrity. (1990: 35)**

References

Bancroft, A. (2000) 'Review of Marcus Doel's *Poststructuralist Geographies*', *Social and Cultural Geography*, 1: 120–2.

Campbell, D. (1997) *National Deconstruction: Violence, Identity and Justice in Bosnia.* London: University of Minnesota Press.

Caputo, J.D. (1997) *Deconstruction in a Nutshell: A Conversation with Jacques* Derrida. New York: Fordham University Press.

Caputo, J.D. (2000) *More Radical Hermeneutics: On Not Knowing Who We Are.* Bloomington, IN: Indiana University Press,

Casey, E.S. (1997) *The Fate of Place: A Philosophical History.* London: University of California Press.

Connolly, W. (2002) *Neuropolitics: Thinking, Culture, Speed.* London: University of Minnesota Press.

Critchley, S. (2001) *Continental Philosophy: A Very Short Introduction.* Oxford: Oxford University Press.

Derrida, J. (1978) *Of Grammatology.* London: Johns Hopkins University Press.

Derrida, J. (1987a) 'How to avoid speaking: denials', in S. Budick and W. Iser (eds), *Languages of the Unsayable: The Play of Negativity in Literature and Literary* Theory. Stanford, CA: Stanford University Press, pp. 3–70.

Derrida, J. (1987b) *The Post Card: From Socrates to Freud and Beyond.* London: University of Chicago Press.

Derrida, J. (1988) *Limited Inc.* Evanston, IL: Northwestern University Press.

Derrida, J. (1992a) 'Ulysses gramophone: hear say yes in Joyce', in D. Attridge (ed.), *Acts of Literature.* London: Routledge, pp. 253–309.

Derrida, J. (1992b) *The Other Heading: Reflections of Today's Europe.* Bloomington, IN: Indiana University Press.

Derrida, J. (1997) *Politics of Friendship.* London: Verso.

Derrida, J. (1998) *Monolingualism of the Other: Or, The Prosthesis of Origin.* Stanford, CA: Stanford University Press.

Derrida, J. (2000) *On Hospitality.* Stanford, CA: Stanford University Press.

Derrida, J. (2001) *On Cosmopolitanism and Forgiveness.* London: Routledge.

Derrida, J. (2002) 'Hospitality', in *Acts of Religion.* London: Routledge, pp. 358–420.

Derrida, J. and Ferraris, M. (2001) *A Taste for the Secret.* Cambridge: Polity.

Doel, M. (1999) *Poststructuralist Geographies: The Diabolical Art of Spatial Science.* Edinburgh: Edinburgh University Press.

Elden, S. (2001) *Mapping the Present: Heidegger, Foucault and the Project of Spatial History.* London: Continuum.

Foucault, M. (1977a) *Discipline and Punish.* London: Penguin.

Foucault, M. (1977b) 'Nietzsche, genealogy, history', in D.F. Bouchard (ed.), *Language, Counter-Memory, Practice: Selected Essays and Interviews.* Ithaca, NY: Cornell University Press, pp. 139–64.

Foucault, M. (1978) *History of Sexuality: An Introduction.* London: Penguin.

Foucault, M. (1984) 'What is Enlightenment?', in *The Foucault Reader.* London: Penguin, pp. 32–50.

Foucault, M. (1988) *History of Sexuality, Volume 2: The Use of Pleasure.* London: Penguin.

Foucault, M. (1990) *History of Sexuality, Volume 3: The Care of the Self.* London: Penguin.

Grosz, E. (1994) *Volatile Bodies: Towards a Corporal Feminism.* Bloomington, IN: Indiana University Press.

Grosz, E. (1995) *Space, Time and Perversion: Essays on the Politics of Body.* London: Routledge.

Grosz, E. (2001) *Architecture from the Outside: Essays on Virtual and Real Space.* Cambridge, MA: MIT University Press.

Habermas, J. (1991) 'Taking aim at the heart of the present', in D. Hoy (ed.), *Foucault: A Critical Reader.* Oxford: Blackwell.

Haraway, D. (1991) *Simians, Cyborgs and Women: The Reinvention of Nature.* New York: Free Association.

Hardt, M. and Negri, A. (2001) *Empire.* London: Harvard University Press.

Hardt, M. and Negri, A. (2004) *Multitude: War and Democracy in the Age of Empire.* London: Penguin.

Harris, K. (1998) *The Ethical Function of Architecture.* Cambridge, MA: MIT Press.

Heidegger, M. (1962) *Being and Time.* Oxford: Blackwell.

Heidegger, M. (1998) *Pathmarks.* Oxford: Oxford University Press.

Krell, D.F. (1997) *Architecture: Ecstasies of Space, Time and the Human Body*. Albany, NY: SUNY.
Malpas, J. (1999) *Place and Experience: A Philosophical Topology*. Cambridge: Cambridge University Press.
Massumi, B. (2002) *Parables for the Virtual: Movement, Affect, Sensation*. London: Duke University Press.
Mugerauer, R. (1995) *Interpreting Environments: Tradition, Deconstruction and Hermeneutics*. Austin, TX: University of Texas Press.
Nietzsche, F. (1990) *Twilight of the Idols/The Anti-Christ*. London: Penguin.
Nietzsche, F. (1998) *On the Genealogy of Morals*. Oxford: Oxford Paperbacks.
Owen, D. (1999) 'Power, knowledge and ethics: Foucault', in S. Glendenning (ed.), *The Edinburgh Encyclopaedia of Continental Philosophy*. Edinburgh: Edinburgh University Press, pp. 593–604.
Peperzak, A. (2003) *The Quest for Meaning: Friends of Wisdom from Plato to Levinas*. New York: Fordham University Press.
Philo, C. (1992) 'Foucault's geography', *Environment and Planning D: Society and Space*, 10: 137–62.
Popke, J.E. (2003) 'Poststructuralist ethics: subjectivity, responsibility and the space of community', *Progress in Human Geography*, 27: 298–316.
Rajchman, J. (1998) *Constructions*. Cambridge, MA: MIT Press.
Sharpe, J.P., Routledge, P., Philo, C. and Paddison, R. (2000) *Entanglements of Power: Geographies of Domination/Resistance*. London: Routledge.
Whatmore, S. (2002) *Hybrid Geographies: Natures Cultures Spaces*. London: Sage.
Wittgenstein, L. (1961) *Tractatus Logico-Philosophicus*. London: Routledge.

11 ACTOR-NETWORK THEORY, NETWORKS, AND RELATIONAL APPROACHES IN HUMAN GEOGRAPHY

Fernando J. Bosco

What is Actor-Network Theory?

There are many ways to describe actor-network theory (ANT) and much has been written about the perspective in recent years. ANT is an approach to sociology that has its origins in studies of the sociology of science and technology, mostly associated with the work of Bruno Latour, John Law, and Michel Callon, among many others, beginning in 1980s. ANT is about uncovering and tracing the many connections and relations among a variety of actors (human, non-human, material, discursive) that allow particular actors, events and processes to become what they are. The original concern of ANT was with understanding the construction of scientific knowledge as a product of all kinds of things (e.g. academic journals, a test tube in a lab, academic presentations, skills embodied in a scientist) coming together in different ways through a process of 'heterogeneous engineering' (Law, 1992). Since its original focus on the construction of scientific knowledge, ANT has gone much further and transcended disciplinary fields (Law, 1991; Law and Hassard, 1999). Today, scholars, including geographers of course, follow a similar route – that of tracing heterogeneous associations among things – to understand the construction of the social in general.

Any discussion of ANT requires some basic understanding of poststructural thinking, but ANT does not equal poststructuralism. Rather, ANT is one of several perspectives that share some of the anti-essentialist sentiments of poststructural thinking. In effect, ANT can be positioned together with other theoretical and methodological efforts in the social sciences that are attempting to better capture the complexity of our world today. As Law and Urry (2002) have recently argued, current social science methods are not well equipped to understand the realities of the twenty-first century. According to them, social science to date has dealt 'poorly with the fleeting – that which is here today and gone tomorrow' and it has also dealt 'poorly with the distributed – that is to be found here and there but not in between – that which slips and slides between one place and the other' (2002: 9). ANT is thus an attempt to deal with the complex and the elusive.

How can ANT tackle the analysis of many disparate research concerns? How can ANT claim to uncover the relations that give rise to such different arrangements of things as yuppie restaurants in New Zealand (Latham, 2002) and global terrorist networks? ANT (Etringer and Bosco, 2004) is a framework that suggests that knowledge, agents, institutions, organizations, and society as a whole, are *effects*, and that such effects are the result of relations enacted through heterogeneous networks of humans and non-humans. ANT tells us that such effects are just as much a part of the network as the actors that we are interested in studying. As Law (1999: 4) explains, actors are *network effects* that take the attributes of the entities which they include; entities and things are produced or 'performed' by and through relations. Thus, when one reads about 'performativity' in ANT, what scholars mean is that things are produced by relational effects, facilitated and enacted through networks.

The actor-network approach tells us that such effects can be analyzed and understood by tracing the networks they form and the changing relations that emerge and develop once heterogeneous things (both human and non-human) become related to each other. One of the most controversial points advanced by ANT is that the 'actors' in such heterogeneous networks of relations (again, both humans and non-humans) have the capacity to act. This requires some clarification, for if one attempts to interpret this statement in the language of more traditional social theory, one could be led to believe that everything in the world (a government official, an empty building, a law, a tree) had 'agency'. In other words, one could be led to assume that everything in the world was equivalent to an *intentional* individual actor. From the perspective of social theory, this would not make much sense.

However, ANT asks us to think about actors and agency differently. Specifically, and according to Latour, from the perspective of ANT an actor is something better described as an *actant*, 'something that acts or to which activity is granted by others ... [An actanct] implies *no* special motivation of *human individual* actors, or of humans in general. An actant can literally be anything provided it is granted to be the source of action' (1996: 373, original emphasis). It is not as if ANT does not see a difference between humans and non-humans. Rather, as Adam explains, when we work with ANT 'humans and non-humans are treated symmetrically in our descriptions of the world, especially with regard to the agency of non-humans, an agency which may or may not have been designed or inscribed by human actors' (2002: 332).

Latour's discussion and definition of actants take us back to the discussion of network effects, a key notion in ANT. From the perspective of ANT, agency is a distributed effect and the result of relations enacted through networks of heterogeneous things and materials. Agency is, in fact, *decentered* – i.e. it is not specifically centered or located in humans or in anything else (see Whatmore, 1999). For example, from the perspective of ANT, I would no longer be a geographer with the ability to write papers and produce knowledge if my computer, my colleagues, my books, my job, my professional network, and everything else in my life that allows me to act as what I am were taken away from me. ANT tells me that if this were to happen to me, I would become something different than what I am right now. Law (1992) makes this same case (about his identity as a sociologist being dependent on a particular actor network) and summarizes such a decentered notion of agency by defining *agency as network*. Thus, from such a point of view, things such as scientific knowledge, the government of a nation, and even what counts as a person are no other than network effects. According to Law, this implies that ultimately one is an agent because one 'inhabits a set of elements (including, of course, a body) that stretches out into the network of materials, somatic and otherwise, that surrounds each body' (1992: 3). Ultimately, once we conceptualize agency as a network, uncovering the heterogeneous 'actor networks' of associations allows us to explain the mechanics of power and organization in society, and to understand how different things (from knowledge to institutions to material artifacts and technologies) come to be, how they endure over time, or how they fail and exit from our lives and our world (Law, 1992).

Like other relational approaches in social theory (such as social-network analysis, discussed in relation to ANT in the next section), ANT is an attempt to overcome some of the most enduring dualisms in contemporary social theory, such as the structure–agency debate or the distinctions typically

BOX 11.1 ACTOR NETWORKS: TRACING 'TOPOLOGIES OF WILDLIFE'

What happens if we begin to analyze 'nature' and what we often consider to be 'wildlife' from the perspective of ANT? According to geographer Sarah Whatmore, the answer is that instead of seeing nature in relation to pristine, bucolic, and utopian images of animals and plants in the wild, one begins to see wilderness and wildlife as 'a relational achievement spun between people and animals, plants and soils, documents and devices in heterogeneous social networks which are performed in and through multiple places and fluid ecologies' (2002: 14). Whatmore constructs such a relational view of nature by tracing two different 'topologies of wildlife': the networks of ancient Roman games (involving animals chosen to compete in bloody public games against other animals and humans) and the contemporary species inventories built around the science of biodiversity (where, motivated by conservation efforts, certain animals are protected and their reproduction and use by humans are scientifically managed). Each of these two wildlife networks occupies its own time and place(s), but the building process behind each of them – the assembling of diverse elements through networks – is similar. For example, a leopard that used to take center stage in a Roman arena during gladiatorial games was merely one piece in an elaborate set of procurement networks involving military supply lines and political patronage that connected Rome with China, India and Africa. As Whatmore explains, the leopard that spectators saw in a Roman arena was the incarnation of a leopard somewhere in Africa. Leopards that made it to gladiatorial and wild animal combats were often starved, abused and/or diseased. Their characteristics had changed through their circulation in networks that involved their hunting and capture, transportation in wooden crates through land and over water (in carts and Roman merchant vessels), training, and storage in underground dens where they awaited their final performance day. By circulating through elaborate networks of heterogeneous elements, a leopard would become what the Romans called *leopardus* – a performance of wildlife. Similarly, wild animals today are part of complex and heterogeneous networks, not simply as 'unitary biological essences, but [as] a confluence of libidinal and contextual forces' (2002: 14). Currently, many wild animals are scientifically classified according to biodiversity principles in a system of taxonomies born out of, and very similar to, the colonial drive to subject the world to systematic scientific account. For example, through local, national and global institutions that regulate international wildlife trade, such as the Convention on the International Trade in Endangered Species, a particular kind of South American broadnose crocodile (known to locals as the *yacare*) has become classified as *Caiman latirostris* and as an endangered species. The naming and classification of the South American crocodile has effectively enrolled the animal in scientific networks of biodiversity that promote managed conservation approaches. Such approaches include the 'regulated sale of crocodile body parts [to] provide incentives to local people to ensure the species' survival' (2002: 27). This 'sustainable' use of the crocodile positions the animal in a complex network that includes the collection of eggs, their breeding in special stations where animals are marked and where hatching takes place, the return

BOX 11.1 (continues)

of some juvenile crocodiles to their habitat, and the breeding and fattening of others for commercial purposes. The process of becoming *Caiman latirostris* (a wild animal? an exotic leather?) continues as the network expands to include slaughtering and tanning. Finally, the skins are certified and enter the international reptile leather market. Through its enrolling in scientific networks of biodiversity, conservation and wildlife management, the broadnose crocodile paradoxically becomes the endangered *Caiman latirostris* and, at the same time, one of the most commonly traded crocodilian skins worldwide. Together, the examples of the leopard and the crocodile indicate how playing the part of and becoming a wild animal are the result of the enrollment of a multitude of human and non-human actors in actor networks that span the local to the global.

made between macro-level and micro-level analysis of social phenomena. But the main proponents of ANT ask that those working with the approach do not attempt to frame ANT under the typical categories of social theory. For example, reflecting on (and maybe lamenting) the way in which some scholars have been mobilizing the ideas of ANT in their research, Latour (1999) explained that the 'actor' (in *actor* network) is not supposed to play the role of 'agency', and that the 'network' (in *actor* network) is not supposed to play the role of structure or society. Instead, according to Latour, 'the social is a certain type of circulation that can travel endlessly without ever encountering either the micro-level or the macro-level' (1999: 19). ANT asks us that we do not think about hierarchies or categories, but rather think about constant circulations and flows.

In sum, ANT attempts to explain what happens in the world by exploring the myriad connections of actants in networks of associations. It does not make privileged distinctions between humans and non-humans. It asks us to consider what we see, the phenomena out there in the world that interest us and that we want to study, as performances, as the effects of relations that come about through connections and through networks. It tells us that to understand things, we must unearth the actor networks.

Networks in Geography and the Place of ANT

ANT is not exclusively a geographic theory, philosophy or perspective, since its origins lie outside the boundaries of contemporary human geography. Some could also say that there is nothing necessarily geographic about ANT – unless of course one is a geographer and therefore thinks geographically about the world in general. Thus, in thinking about ANT, geographers would also add that how the actor networks that give rise to knowledge, institutions and organizations come to be (and the reason for their success or their failure) is also related to a *spatiality* that is embedded in the actor networks. For example, in the topologies of wildlife traced by Whatmore (2002), the actor networks depended on (and were built upon) the specific geographic settings where wildlife could be performed, such as the Roman arena and the nature reserve. In other words, the spatiality of the actor networks is part of the explanation for their presence and it cannot be separated from or conceptualized outside the actor networks (Murdoch, 1997).

In recent years, many scholars have been thinking geographically about actor networks. In fact, ANT has informed the research projects of many human geographers and the perspective has quickly established its presence among other theoretical frameworks in the discipline, even displacing some older established ones. And the work of geographers has also been noticed by some of ANT's main proponents, who credit geographers with contributing to challenge conceptions of the world that rely too much on Euclideanism (Law, 1999). Evidence of the rising importance of ANT includes the increasing number of research articles (e.g. Murdoch and Marsden, 1995; Murdoch, 1997; 1998; Woods, 1998; Olds and Yeung, 1999; Dicken et al., 2001), special issues in well-established journals (e.g. Hetherington and Law, 2000 and their edited special issue in *Environment and Planning D: Society and Space*, 2000, entitled 'After Networks'), and books and book chapters in edited collections (e.g. Thrift, 1996; Whatmore, 1999; Latham, 2002; Whatmore, 2002), to name but a few. Similarly, the entry of ANT into the theoretical landscape of human geography is evident from the number of presentations in geography academic conferences in Europe and North America that invoke or acknowledge ANT as part of the framework that guides different research projects.[1] This is evidence of the increasing importance of ANT as an accepted and valuable approach in human geography. The main point is that, as exemplified by the number of books, articles and conference presentations, the conceptualization of networks provided by ANT has surpassed other more traditional ideas and theories about networks that have been around in geography for more than four decades, particularly in human geography.

Throughout the twentieth century, geographers have been interested in networks and their geographies. Networks permeate geographic thinking, even though geographers have different empirical concerns, think differently about networks because of different and many times conflicting theoretical perspectives, and pursue different analytical approaches to analyze the networks they are interested in. Networks as empirical phenomena and networks as a trope are both powerful notions that have entertained and informed geographers' research agendas across geographic subdisciplines. For example, transportation geographers and spatial analysts think about railroad and highway networks, about hub and spoke networks in airline traffic, and about Internet connectivity and flows, and attempt to model them mathematically. Economic geographers work on networks of production through studies of agglomeration of firms and other network-inspired concepts such as learning regions, untraded interdependencies and global commodity chains. Social, political, and cultural geographers are interested in global flows and connections among distant humans and approach such issues through studies of networks of ethnicity, immigrant flows, the creation of transnational identities, and the emergence of global social movements and resistance networks. A flurry of recent research on the World Wide Web, and the myriad connections and flows and cyber- and other alternative spaces that are being produced by the intersections of technology and society, has also been inspired by ideas about networks (perhaps most notorious here is Castells' 1996 notion of the 'network society' and 'space of flows', which has inspired much debate).

In some way, we should see the incorporation of ANT in geography as another in a series of approaches to analyzing and studying networks by geographers. ANT is perhaps the first approach since spatial network analysis in geography from the late 1960s and 1970s that makes networks the explicit focus of attention for the study of *all* phenomena. Let me clarify this. The spatial analysis of networks in geography developed in tandem with the quantitative revolution and attempted to analyze and model networks mathematically through methods such as graph theory. In the quantitative and spatial analysis view of networks, geographers were attempting to overcome the dichotomy between human and physical geography and thus argued that, for example, streams of migrants and streams of water within a drainage basin could be interpreted as networks and analyzed using common mathematical models (Hagget and Chorley, 1969).

In this sense, the network project of the quantitative revolution shared something with ANT. There was an attempt to analyze all geographic phenomena in terms of networks and an effort not to make a distinction between networks in human and physical geography. But the similarities ended there, really. ANT is very different from the spatial analysis of networks that was common in the 1960s, and is much more interested in power and on how things come to be – rather than on establishing precise measures of connectivity and the like. Additionally, ANT goes much further than traditional spatial analysis of networks and other network approaches in geography in calling for the theorization of society, nature, space and everything else from a relational perspective.

ANT's entry into geography also came at a time when social-network analysis, another influential network approach in geography, was beginning to be scrutinized more closely as a result of the cultural turn in geography and the social sciences. Many network concepts from the social-network perspective were (and still are) very popular among geographers. A good example of this is the concept of embeddedness, which in geography has always been thought about in terms of integration or participation in local, territorial networks.[2] Like ANT, social-network analysis and interest in processes such as embeddedness have not been exclusively geographic approaches. Rather, the literature on social networks comes from a federation of perspectives concerned with understanding the social in terms of a system of changing patterns of interactions occurring among actors who often participate in one or more social networks (Nohria and Eccles, 1992; Bosco, 2001). At this level, social-network analysis shares with ANT similar concerns with networks and relations and with overcoming the divide between micro- and macro-level analysis. But, as was the case with the first round of network analysis in geography, social-network analysis does not go as far as ANT because ANT is a more comprehensive approach.

Specifically, as the name of the approach implies, in social-network analysis the focus is on *social* networks. This concentration runs contrary to one of the most important lessons from ANT, that is, that what we typically see as 'the social' is constituted by relations among humans and non-humans, and that the divide between the natural and the social sciences should be broken down. Second, because of the focus on the social, agency in social-network analysis continues to be centered on the person and the human body, contrary to the account of agency offered by ANT discussed in the previous section and exemplified by Whatmore's (2002) accounts of how a South American *Caiman latirostris* or an African leopard are both subjects and objects of the networks they help fashion. Third, most studies in the social-network perspective are limited to analyses of structural forms, i.e. the forms and shapes of networks based on the number and types of connections which give a network different degrees of connectivity, centrality and so on. This is symptomatic of what Emirbayer and Goodwin (1994) have called the 'structural determinism' of social-network analysis as well as of what Law and Urry (2002) have called the Euclideanism of traditional social science, which is characterized by the presence of hierarchical orderings, divisions into levels of analysis and the use of metaphors of size, order, and proximity. This, again, runs contrary to ANT because ANT is not concerned with actual network structure as given by different types of connections. ANT is more concerned with accounting for how varieties of different actants (human and non-human) emerge out of different relations and give rise to constantly changing actor networks and different relations of power. Therefore ANT is more capable of accounting for fluidity and movement between the micro- and the macro-levels of analysis (Law and Urry, 2002). As a non-linear and non-representational approach (Thrift, 2000) ANT is superior to social-network analysis and other current relational approaches in being able to uncover the complex relations that shape our world.

In sum, ANT shares only a few similarities with the traditional spatial analysis of networks in geography and with social-network analysis. Overall, ANT offers a much more comprehensive view of networks and relations than either of those two approaches. ANT provides a framework that takes into account a multiplicity of actors and that permits accounting for the fluidity of multiple types of relations. ANT also provides a very different view of the spatiality of network relations, one that is not limited by a Euclidean view of space. However, some today would argue that ANT is analytically less precise in establishing why different kinds of network relations in a network might matter and that, as a result, ANT leaves us with a thin notion of spatiality. This is one often-cited limitation of ANT as an approach, and the reason why a few scholars argue that ANT might also be seen as an apolitical perspective. It is to a discussion of these issues then that this chapter now turns.

ANT, Spatiality, and Difference: Limitations or Advantages?

In recent years, several geographers have begun inviting us to see the spatiality of social relations not as fixed, essential, or reducible to considerations of distance or hierarchies of scales. Instead, we have been encouraged to see place and the connections between society and space relationally, as constantly produced and situated (Massey et al., 1999). As Amin (2002: 389) has recently put it, spatiality from a relational perspective is constituted through the folds and overlaps of different practices. The notion of spatiality that is emerging in human geography is very much inspired by network thinking, and the project of ANT has much to do with it. It is a notion of spatiality that is embedded in relational ideas about space and place as the result of interrelated processes rather than the product of Cartesian geometries and homogeneous considerations of society.

The fact that current thinking in human geography has begun to displace fixed, essentialized, and hierarchical notions of place and space is consistent with the project of ANT. The view of place as open and porous and as the product of the spatiality of networked relations can perhaps first be traced back to the notion of 'power geometry' introduced by Doreen Massey (1991; 1993; 1999a; 1999b). In recent years, relational ideas of spatiality have been further developed by other scholars, giving rise to spatial imaginings that highlight, as Whatmore put it, 'the simultaneity of multiple and partial space–time configurations of social life and the situatedness of social institutions, processes and knowledges' (1999: 31). This now includes geographic interpretations of ANT that direct our attention to the effects produced and generated by the fluidity or mutability of different spatial assemblages of actors (Murdoch, 1997; 1998; Sharp et al., 2000).

All relational approaches to spatiality also have implications for accounts of power relations. Power is enmeshed in networks of relations and has different expressions – dominating power, resisting power (Sharp et al., 2000) – and it is also related to the positions that individuals occupy in such networks (Sheppard, 2002). This indicates that relational understanding of power also cannot be separated from relational understandings of, for example, place. If places are constructed by the workings of intricate networks of social relations, then places should also be seen as solidifying intricate entanglements of power (Sharp et al., 2000). As a relational approach, ANT is very helpful here as well. From the perspective of ANT, thinking about power, networks, and place leads us to consider the ways in which different actors, who are placed and positioned differently in networks, are able to construct and enact different relations and forms of power, and to ponder how such relations and forms of power solidify both materially and discursively in places and flows through space–time.

ANT is an ideal framework to deal with the spatialities of power because ANT clearly directs our attention to the effects produced by the fluidity of spatial configurations of a variety of actors. Yet, some would argue that ANT as an analytical perspective can only be stretched so far in the treatment of the geographies of power and difference. ANT is an excellent framework to describe the complex and mutable composition of networks of heterogeneous actors. But what if our research interests also include the desire to recognize and evaluate *how* a range of types of relations in an actor network – connecting *different* actors/actants – are related to *different* outcomes and events? Some have argued that when it comes to recognizing different people and different voices and the politics associated with them (i.e. 'the other' or 'otherness'), ANT can almost be seen as a colonial framework (Lee and Brown, 1994). If everything is included, connected and ordered through actor networks, then ANT might leave 'no space outside the network … no room for alterity … nothing outside the relation it orders' (Hetherington and Law, 2000: 2). In other words, ANT might be valuable as a general framework that allows us both to think about configurations of power and to overcome the dualism between nature and society by following different configurations of relations and network effects. However, analytically ANT could also obscure *difference* (i.e. the different types of relations within and among different actors and processes) and thus could be considered an apolitical perspective. Such a limitation, however, is not unique to ANT and it is certainly dependent on how one approaches the perspective through research. For example, the task of tracing topologies of wildlife, as Whatmore has done with currently endangered animal species such as the *Caiman latirostris*, has important political and ethical implications because it demonstrates the ways in which animals become objectified under the auspices of wildlife management. Such a research project asks that we rethink the politics of current biodiversity and conservation efforts.

A critical view could be taken also in relation to ANT's conceptualization of spatiality. Some scholars have argued that ANT does not seem to recognize *explicitly* that different relations may have different spatial expressions. As a result, in some ANT-inspired research in geography the lines between networks and spatiality have been blurred extensively. For example, some geographers have argued that 'any assessment of spatial qualities is simultaneously an assessment of network relations' (Murdoch, 1997: 332) and offered views of space and scale that seem to dissolve into networks. Such moves can lead to vague specifications of spatiality and this has been one of the main criticisms directed towards ANT. As Sheppard (2002: 317) has explained, the concern with representing networks as non-hierarchical spaces where all the actors involved have significant power has resulted in a lack of attention to their internal differentiation. Relations in a network are not all the same, and their differences will inevitably produce different spatialities (Hinchliffe, 2000).

Geographers are certainly right in pointing out this tendency of ANT to erase difference within networks. The point should be taken, but it should also not be overemphasized and we should not dismiss ANT too quickly.[3] On the contrary, valuable insights into the constitution of different geographic processes and practices can be gained from the perspective; Lawrence Knopp's contribution in this volume (Chapter 20), in which he uses ANT to think about *placelessness*, is an excellent example. We should therefore approach ANT from a critical perspective and utilize its valuable insights to specify geographies more richly – without forgetting that ANT is inclined to utilize a vocabulary that evades recognitions of difference.

Summary and Conclusion

Geographers have brought ANT into their research agendas and made ANT one of the most sophisticated approaches to relational thinking available today. ANT has emerged as a perspective that allows the framing of research concerns about networks, relations among humans and non-humans, and spatiality through the lens of a flexible and malleable theory. Many see ANT as a poststructural approach to power relations that can be adapted, transformed, and modified to fit their research concerns and at the same time as an approach that can be 'grounded' and made politically relevant. It is an approach that encourages thinking about the fluidity of the social, the natural, and very importantly, of space, place and scale.

The flexibility and malleability of ANT at a perspective have created some confusion. At this point in time ANT means different things to different people and, despite the popularity of the approach, many remain very unclear as to what ANT is, what ANT can do and what it should not be called upon to do. Many think that ANT is about networks. While this is true, this chapter has shown that it is too simplistic and reductionistic to talk about ANT only in terms of networks without qualifying what is meant by 'networks' in more specific terms. There is even an ongoing debate among some of ANT's main proponents regarding the status of the approach. Bruno Latour (1999) has argued that ANT is not a finished theory; he has even claimed that the use of the words 'actor', 'network', and 'theory' to name the approach in English has contributed to the confusion about what the project of ANT is really about. Many other scholars are vocal about the limitations of ANT relative to other more traditional approaches in social theory, and criticize ANT for being apolitical and for offering a view of the world that ignores the real material consequences of actions of agents and institutions that are claimed to be actually operating 'out there' in the world. Yet, the project of ANT can have politically progressive implications, as indicated above in the discussion of Whatmore's (2002) topologies of wildlife.

In the end, many of those who are vocal about some of the problems with ANT cannot but acknowledge the utility of the perspective and be seduced by some of its main propositions. Thus many geographers continue to rework and reinterpret ANT by applying it to new contexts and new situations in different places – which after all is a very geographic thing to do. Overall, ANT is positioned to become an even more valuable approach in human geography as more scholars begin to adopt and refine the perspective through different research projects. ANT is a sophisticated way to deal with networks and relations among actors, objects, and processes that geographers typically care about in their research, from economic geography to urban, political and cultural geography. There is much room to further theorize and synthesize by using ANT as a point of departure, and therein lies the power and seduction of ANT for geographers today. We must engage ANT through research and let the challenge of tracing the actor networks and of unearthing complex relations begin.

NOTES

1 For example, the 2003 annual conference of the Association of American Geographers in New Orleans included 53 papers where 'network' was one of the keywords listed in the abstract. Of those papers, more than half dealt directly with or were inspired by actor-network theory (the other 20 or so papers dealt with networks in other ways and included both human and physical geography). A similar number of papers dealing with ANT were evident in the annual meeting of the same professional association in Los Angeles a year earlier.

2 Examples here include the work of economic geographers such as Amin and Thrift (1994) and Storper (1997), who focus on the territorial embeddedness of firms and markets in social and cultural networks of relations by placing emphasis on localities or regions.

3 Even Bruno Latour has recognized that ANT 'is a bad tool for differentiating associations [because] it gives a black and white picture, not a colored and contrasted one … [thus] it is necessary, after having traced the actor networks, to specify the types of trajectories that are obtained by highly different mediations' (1996: 380).

References

Adam, A. (2002) 'Cyborgs in the Chinese Room: boundaries transgressed and boundaries blurred', in J. Preston and M. Bishop (eds), *Views into the Chinese Room: New Essays on Searle and Artificial Intelligence*. New York: Oxford.

Amin, A. (2002) 'Spatialities of globalization', *Environment and Planning A*, 34: 385–99.

Amin, A. and Thrift, N. (1994) 'Living in the global', in A. Amin and N. Thrift (eds), *Globalization, Institutions, and Regional Development in Europe*. Oxford: Oxford University Press.

Bosco, F. (2001) 'Place, space, networks and the sustainability of collective action: the *Madres de Plaza de Mayo*', *Global Networks*, 1: 307–30.

Castells, M. (1996) *The Rise of the Network Society*. Oxford: Blackwell.

Dicken, P., Kelly, P., Olds, K. and Yeung, H. W.-C. (2001) 'Chains and networks, territories and scales: towards a relational framework for analyzing the global economy', *Global Networks*, 1: 89–112.

Emirbayer, M. and Goodwin, J. (1994) 'Network analysis, culture and the problem of agency', *American Journal of Sociology*, 99: 1411–15.

Ettlinger, N. and Bosco, F. (2004) 'Thinking through networks and their spatiality: a critique of the US (public) war on terrorism', *Antipode*, 36: 249–71.

Hagget, P. and Chorley, R. (1969) *Network Analysis in Geography*. New York: St Martin's.

Hetherington, K. and Law, J. (2000) 'After networks', *Environment and Planning D: Society and Space*, 18: 1–6.

Hinchliffe, S. (2000) 'Entangled humans: specifying powers and their spatialities', in J. Sharp, P. Routledge, C. Philo and R. Padisson (eds), *Entanglements of Power: Geographies of Domination/Resistance*. New York: Routledge.

Latham, A. (2002) 'Retheorizing the scale of globalization: topologies, actor networks and cosmopolitanism', in A. Herod and M. Wright (eds), *Geographies of Power: Placing Scale*. Malden, MA: Blackwell.

Latour, B. (1996) 'On actor network theory: a few clarifications', *Soziale Welt*, 47: 360–81.

Latour, B. (1999) 'On recalling ANT', in J. Law and J. Hassard (eds), *Actor Network Theory and After*. Oxford: Blackwell.

Law, J. (ed.) (1991) *A Sociology of Monsters: Essays on Power, Technology and Domination*. London: Routledge.

Law, J. (1992) 'Notes on the theory of the actor network: ordering, strategy and heterogeneity'. http://tina.lancs.ac.uk/sociology/soc054jl.html, accessed 14 August 2003.

Law, J. (1999) 'After ANT: complexity, naming, and topology', in J. Law and J. Hassard (eds), *Actor Network Theory and After*. Oxford: Blackwell.

Law, J. and Hassard, J. (eds) (1999) *Actor Network Theory and After*. Oxford: Blackwell.

Law, J. and Urry, J. (2002) 'Enacting the social'. Centre for Social Science Studies and Sociology Department, Lancaster University. http://www.comp.lancs.ac.uk/sociology/soc099jlju.html, accessed 19 August 2003.

Lee, N. and Brown, S. (1994) 'Otherness and the actor network: the undiscovered continent', *The American Behavioral Scientist*, 37: 772–90.

Massey, D. (1991) 'A global sense of place', *Marxism Today*, June: 24–9.

Massey, D. (1993) 'Power geometry and a progressive sense of place', in J. Bird, B. Curtis, T. Putnam, G. Robertson and L. Ticker (eds), *Mapping the Futures: Local Cultures, Global Change*. New York: Routledge.

Massey, D. (1999a) 'Philosophy and politics of spatiality: some considerations', *Geographische Zeitschift*, 87: 1–12.

Massey, D. (1999b) 'Imagining globalization: power-geometries of time–space', in A. Brah, M. Hickman and M. Mac An Ghaill (eds), *Global Futures: Migration, Environment and Globalization*. New York: St. Martin's.

Massey, D., Allen, J. and Sarre, P. (eds) (1999) *Human Geography Today*. Cambridge: Polity.

Murdoch, J. (1997) 'Towards a geography of heterogeneous associations', *Progress in Human Geography,* 21: 321–37.

Murdoch, J. (1998) 'The spaces of actor network theory', *Geoforum,* 29: 368–80.

Murdoch, J. and Marsden, T. (1995) 'The spatialization of politics: local and national actor-spaces in environmental conflict', *Transactions of the Institute of British Geographers*, 20: 368–80.

Nohria, N. and Eccles, R. (eds) (1992) *Networks and Organizations: Structure, Form, and Action*. Boston, MA: Harvard Business School Press.

Olds, K. and Yeung, H. (1999) '(Re)shaping "Chinese" business networks in a globalizing era', *Environment and Planning D: Society and Space*, 17: 535–55.

Sharp, J., Routledge, P., Philo, C. and Padisson, R. (eds) (2000) *Entanglements of Power: Geographies of Domination/Resistance*. New York: Routledge.

Sheppard, E. (2002) 'The spaces and times of globalization: place, scale, networks, and positionality', *Economic Geography*, 78: 307–30.

Storper, M. (1997) *The Regional World: Territorial Development in a Global Economy*. New York: Guilford.

Thrift, N. (1996) *Spatial Formations*. London: Sage.

Thrift, N. (2000) 'Afterwords', *Environment and Planning D: Society and Space*, 18: 213–55.

Whatmore, S. (1999) 'Hybrid geographies: rethinking the human in human geography', in D. Massey, J. Allen and P. Sarre (eds), *Human Geography Today*. Cambridge: Polity.

Whatmore, S. (2002) *Hybrid Geographies: Natures Cultures Spaces*. London: Sage.

Woods, M. (1998) 'Researching rural conflicts: hunting, local politics, and actor networks', *Journal of Rural Studies*, 14: 321–40.

12 POSTCOLONIALISM: SPACE, TEXTUALITY, AND POWER

Clive Barnett

Postcolonialism and the Critique of Historicism

As a field of academic inquiry, postcolonialism has its intellectual origins in the writings of a number of intellectuals who came to prominence in the middle part of the twentieth century, the period of intense anti-colonial struggles against formal European territorial control, especially in Africa and Asia (see Young, 2001). These include writers such as C.L.R. James, who recovered the forgotten history of Haitian rebellion in the French Revolution; Amilcar Cabral, the leader of the movement against Portuguese colonialism in Guinea and Cape Verde; and Aimé Césaire, a poet from French Martinique who became an important theorist of the Negritude movement, which asserted the value of previously denigrated African cultures. Each of these writers shared two common concerns. First, each emphasized that colonialism consisted of more than economic exploitation and political subordination; colonialism also involved the exercise of cultural power over subordinated populations. Culture is understood to have been wielded by colonialist powers to denigrate the traditions of non-western cultures, and to celebrate the superiority of particular versions of western culture.

If these writers understand culture to be an instrument of domination, then regaining control over the means of collective self-definition is regarded as an important strategy in the political struggle for emancipation. One good example of the analysis of this relationship between culture, domination, and resistance, is James' account of the history of cricket in the Caribbean. In *Beyond a Boundary* (James, 1963), the cricket field is refigured as an arena in which relations of racial superiority are asserted and subverted during colonialism, as well as one in which the continuing tensions between newly independent states and the former colonial power are played out after the end of formal colonialism. This leads us on to the second emphasis that this generation of anti-colonial writers share, which is a premonition that in so far as relations of colonial subordination are embedded in cultural systems of identity and representation, then the formal end of European colonialism would not necessarily mean the end of colonial forms of power. The clearest link between a generation of anti-colonial writers and the emergence of postcolonialism in the late 1970s and 1980s is, then, this shared concern with the conditions for the 'decolonization of the mind'. This process of decolonizing the mind is concerned with working through the embedded modes of reasoning, thinking, and evaluating that secrete assumptions about privilege, normality, and superiority (Sidaway, 2000).

The emphasis upon the destruction of non-western cultural traditions during colonialism might appear to imply that the work of decolonizing the mind requires the recovery and revaluation of these traditions. But this understanding of the cultural politics of postcolonialism can easily reinscribe a binary opposition between modernity and tradition that is itself a key ideological device used in the denigration of non-western societies. The invocation of 'authentic' traditions has, in fact, been one of the most problematic ways in which postcolonial elites have continued to wield political power over their citizens. A more complex way of understanding the relationship between the modern and the traditional is illustrated by the career of Ngûgî wa Thiong'o. His early novels were published in English under the name James Ngûgî, but in the 1970s he became involved in the production of popular theatre expressed in the most widely used indigenous language in Kenya, Gikiyu. Ngûgî was imprisoned because of this involvement, and out of this commitment emerged his decision to write original works in this language, rather than in English. In principle, this is an attempt to make his work available to local audiences, in a much broader way than is possible through the use of English (see Ngûgî, 1986). At the same time, however, Ngûgî's strategy is not straightforwardly aimed at recovering a lost tradition of indigenous, authentic narrative. It is, rather, more an act of postcolonial invention, fusing together genres and forms from different narrative traditions, both western and non-western. It is, then, a distinctive effort to inscribe an alternative modernity into global networks of cultural representation.

The most significant intellectual influence connecting anti-colonial writing to postcolonial theory is Franz Fanon. Fanon was born in French Martinique, and educated and trained in Paris. He spent much of his life working in Algeria at the height of the anti-colonial war between French and Algerian nationalists (the FLN) in the 1950s and early 1960s. Fanon came to identify strongly with the FLN struggle, and this infused his analysis of the psychological dimensions of colonialism. This is laid out in his two classic works. *Black Skins, White Masks* (Fanon, 1991) is an analysis of the impact of racism on the subjective identities of both dominant and subordinate groups. *The Wretched of the Earth* (Fanon, 1967) is one of the classics of modern political thought, a manifesto for the liberation of oppressed peoples around the world. One reason why this book is important is because of its prescient critique of the ideology of anti-colonial nationalism. Fanon suggested that nationalist ideologies were an essential element of anti-colonial struggle, but foresaw that once formal, political independence was won, this same ideology risked becoming a new mechanism for elites to exercise power over dissenters or marginalized populations. This critique of ideologies of nationalism is one crucial link between Fanon's work and that of various writers central to the emergence of postcolonial theory since the 1980s. Another link is a more directly theoretical one. Fanon was not just a practising psychiatrist, an experience that infused his analysis of the personal and group psychologies of both colonizers and the colonized. His writing was also informed by the main lines of modern continental philosophy, including Hegel's account of the master–slave dialectic, Marxian analysis of political struggle, and psychoanalytic theories of subjectivity. It is this last dimension in particular that makes Fanon such an important reference point for postcolonial theory: this line of work is concerned with rethinking the cultural legacies of colonialism and imperialism through a psychoanalytical vocabulary of subject formation.

One of Fanon's strongest assertions was that the so-called 'developed' or 'First World' was, in fact, the product of the 'Third World'. By this, he meant that it was through the exploitation of non-Europeans that the wealth, culture, and civilization of the West were built. This was more than an empirical observation, however. It was meant as a challenge to a whole way of understanding the dynamics of historical development. One way in which European colonial and imperial expansion was legitimized was through a claim that European culture was the prime mover of historical progress itself. Non-European cultures were denigrated as being either historically backward or, worse, wholly outside history. This same pattern of thought persists in central categories of twentieth-century social science, including ideas of *modernization*, of *development*, and of *developed* and *less developed*. All of these ideas presume one particular set of cultural values and practices as the benchmark against which to judge all others. In so far as they presume an idealized model of European history as the single model for other societies to emulate, these notions are often described as Eurocentric. Eurocentrism combines a strong sense of the particularity of European culture with a strong claim to the universality of these values. The seeming contradiction within the claims for the superiority of particular cultural values which are nonetheless held to be valued precisely because of their supposed universalizability is finessed by the projection of a linear model of historical progress onto the spaces of different societies. On this understanding, the assumption is that Europe is the core region of world history, out of which spreads all important innovations – science, capitalism, literature, and so on (see Blaut, 1993). This combination of cultural particularism and universalization therefore works through the spatialization of time: different parts of the world were ranked as being at different stages of a process of historical progress that assumed a single path of development, or modernization. This pattern of thought is known as historicism.

The biggest challenge of postcolonialism, as a tradition of critical thought, lies in questioning the legacies of this historicist way of thought (see Young, 1990). It is this critique of historicism that Fanon presaged in his work, by arguing that the history of the West was a not a hermetically sealed story of secularization, modernization, and accumulation. Rather than thinking of colonialism and imperialism as marginal to the history of Europe and North America, postcolonialism asserts the centrality of colonialism and imperialism to appreciating the intertwined histories of societies which, from a historicist perspective, are presented as separate entities differently placed on a scale of progress. So, if postcolonialism challenges a particular normative model of linear historical progress, it does so by also challenging the geographical image of distinct, self-contained societies upon which this model depends.

On the basis of these introductory remarks, the rest of this chapter will explore three dimensions to the field of postcolonialism. First, it will consider the 'origins' of this field of academic inquiry in the seminal work of Edward Said. Second, it will elaborate on what is perhaps the most significant contribution of this whole field. This is a particular model of power, one which connects ideas about discourse and textuality to more worldly issues of institutions, organizations, economies and markets. Third, the chapter will reflect on some of the broader moral and philosophical problems raised by postcolonialism, particularly as these concern issues of universalism, cultural relativism, and how to approach the task of cross-cultural understanding.

The Imaginary Geographies of Colonial Discourse

Edward Said's *Orientalism* (1978) is the single most important reference point for the emergence of postcolonial theory. In this book, Said argued that western conceptions of identity, culture, and civilization have historically been built on the projection of images of the non-West, and specifically of images of the so-called 'Orient'. These images could be negative and derogatory, or positive and romantic. In either case, the identity of the West has been defined by reference to the meanings ascribed to what is presumed to be different from the West, its non-Western 'other'. Said provided one of the most influential accounts of a more general theory of cultural politics understood as a process of 'othering', an understanding that has come to define a whole range of academic research in the social sciences and humanities. According to this understanding, identity is socially constructed in relation to other identities, in a simultaneous process of identification with certain groups and differentiation from certain other groups. At the same time, this process of construction is hidden or disavowed, so that it is common for identities to be presented as if they were natural. If identity is relationally constructed, then it works primarily by excluding some element that takes on the role of the other, an image of non-identity that confirms the identity of the self or the collective community. For geographers in particular, this theory is influential because it presents identity formation as a process of controlling boundaries and maintaining the territorial integrity of communities or selves.

One reason why Said's argument proved so influential was his use of Michel Foucault's notion of *discourse* to explain the power of cultural representations in laying the basis for colonial and imperial domination. Said provided one of the first fully worked out applications of Foucault's ideas, arguing that ideas and images were not free-standing, but were part of whole systems of institutionalized knowledge production, through which people and organizations learnt to engage with the world around them. *Orientalism* has come to act as the focal point of discussion precisely because it is a text in which the critique of colonial and imperial knowledge is brought into uneasy communication with poststructuralist theory. One way in which postcolonial theory emerged was therefore through increasingly sophisticated theoretical debates over issues of representation, identity and power. Another is through a process of empirical application of Said's original emphasis on knowledge and power. Said's analysis of orientalist discourse implied that a whole array of institutions produced different forms of knowledge through which the non-European world was discursively produced for Europe. Colonial and imperial power was inscribed in and through administrative and bureaucratic documents, maps, romantic novels, and much else besides. The critical force of Said's book was to make a strong connection between the ideals of high culture and learning – literature, theatre, science, and so on – and the world of grubby politics, power and domination. *Orientalism* provided a theoretical template through which a diverse set of institutions and representations could be given coherence as objects of analysis – as examples of colonial discourse – by being subjected to interpretive protocols loosely drawn from literary studies. All sorts of things could be understood in terms of discourse and the production of colonial subjectivity – scientific writing, historical documents, official reports, literature and poetry, the visual arts, as well as academic discourses such as anthropology, geography, or linguistics. The range and diversity of sites through which colonial subjectivities were constructed and contested is the condition for the interdisciplinary impulse of colonial discourse analysis.

In *Orientalism*, Said referred to orientalism as a form of 'imaginative geography'. His claim was that orientalist representations were really self-generating projections of Western paranoia and desire, and were not based on any detailed knowledge of different cultures and societies. As Said describes it, orientalism has two dimensions. There is a store of ideas about the Orient which have been produced over centuries through which the Orient was staged for the west. In turn, from the late eighteenth century onwards this reservoir of images and knowledges is drawn upon to direct the actual course of European territorial expansion and appropriation. Young (1990) identifies this as the central tension in Said's account. On the one hand, Said holds that the 'Orient' is essentially a misrepresentation, which reflects projections of fear and anxiety but which bears little relation to the actualities of the complex societies it purports to name and describe. Yet, on the other hand there is the suggestion that such misrepresentations become effective instruments of colonial power and administration. Said does not adequately theorize the means by which knowledge about other cultures becomes effective as an instrument in the exercise of power over those cultures. His only gesture in this direction is the distinction between 'latent' and 'manifest' orientalism, the latter presented as the means by which a static and synchronic essentialism is narrated into practical historical situations. In such a formulation, Europeans always find what they expect in the Orient, and the actualities of colonial contact and administration do not fundamentally interrupt the structures of understanding that frame any encounter with the 'real' Orient.

Said's original formulation of orientalism as a form of imaginative geography therefore bestows two theoretical dilemmas upon the analysis of colonial discourse. The first is the problem of how to account for the translation of 'knowledge' which is purely imaginative and non-empirical into knowledge that is practically useful in administering complex social systems like colonial bureaucracies, markets, and so on. The second problem is how to conceptualize anti-colonialist agency from within this understanding. The idea that colonial discourses are entirely the product of colonizers' imagination implies that there exists some pristine space, untouched by the experience of cross-cultural contact, from which authentic agency and resistance must emanate. But it is precisely this sort of 'nativist' understanding that Said has been consistently opposed to. Both of these dilemmas can be traced back to the theoretical model of colonial discourse sketched in *Orientalism*, and specifically to the unresolved problems inherent in Said's original formulation of orientalist discourse as a form of 'imaginative geography' which produces the Orient as the projection of a western will to mastery. Said argued that colonialism is discursively prefigured in the various representations through which the Orient as an imagined location is first constructed. It is this strong sense of projection and prefiguration that is most problematic, because it implies that colonial discourses were self-generating. And this tends to run counter to the strongest critical impulse of Said's work, which is the decentering of self-enclosed narratives of Western progress by showing the ways in which societies are the products of a constant traffic of cultural practices and traditions.

It is worth noting that there are, in fact, two overlapping tropological schemata through which the relationship between culture, identity and space is presented in Said's original formulation of 'imaginative geography'. The first trope one finds is the psychologistic one of the west projecting its anxieties and paranoia onto another spatial realm, through which the 'Orient' is constituted as the fully formed mirror image of western self. This suggests that the essentials of colonial knowledge are formed prior to and in the absence of the actual event of colonial contact. Invoking Gaston Bachelard to describe how distant places are invested with significance from afar by the 'poetic' ascription of meaning, orientalist discourse is presented as producing meaning from a 'here' about a 'there' in advance of actually going 'there'. In his eagerness to stress that colonial discourse involves a misrepresentation of complex realities, Said is forced to posit a core of orientalist knowledge which escapes the principle of inescapable entanglement of peoples and places. The Orient thereby emerges as the fantasy projection of an autonomous will to power.

There is, however, a second tropological schema at work in Said's original account. This presents orientalism as a discourse which stages its own performance, and through which orientalist representations were produced for a European audience. It underlines the sense that actual colonialism is prefigured at the level of culture, in such a way that the actual encounter with the 'real' Orient appears as a carefully directed and minutely orchestrated *mise-en-scène*, involving a pre-established script faithfully followed by each and every actor. Such an understanding still requires that the texts of such a 'discursive formation' be read as the expressions of a paranoid group psychology produced wholly in a metropolitan context and having no purchase on any 'real' Orient at all. The theatrical metaphor thus remains subordinated to the emphasis on poetic projection. However, perhaps we can free this second, theatrical trope from this overriding emphasis on the imaginative prefiguration of actual events. Rather than thinking of colonial practices as more or less perfect performances of already highly rehearsed scripts, we might instead read the colonial archive as made up of the traces of extensive exercises in improvisation. If the discursive production of colonial space is to be fruitfully understood on analogy with a dramatic production, then we should not think of the scenes so produced as realizations of a single autonomous *ur*-script which is the model for each of its own performances. If these performances have a script, then it is one whose existence resides nowhere other than in the contingencies of its repeated (re)enactments. Such a metaphorical flight might lead us towards new ways of reading the textual artefacts of the imperial archive, ones which do not rely on positing of a single coherent will animating each utterance, and which are able to think of colonial discourses as the products of the contingencies and contestations of the ongoing reproduction of colonial and imperial relations. This implies reading textual materials as a reflection neither of an imperial will to power, nor of the popular mood, nor of the intentions of ruling powers. Rather, it implies reading them as traces of the wider practices, institutions, and routines of which they are often the only surviving remnants. It is an understanding that directs attention away from the contents of texts, towards a concern with what they are practically used to do.

The reason for thinking of colonial discourse along these more 'performative' lines is that this answers to an important criticism of the standard model of colonial discourse derived from *Orientalism*. This is the complaint that colonial discourse has too often been theorized as a coherent product of colonializing powers. This tends to hide from view the mediations and relations through which colonialism and imperialism developed (Thomas, 1994). This criticism implies the need to shift away from a strong emphasis on irredeemable Manichaean conflict between colonizer and colonized, towards concepts which focus upon processes of cross-cultural communication. This is the task undertaken by Mary Louise Pratt's (1992) work on colonial representations. In her notion of the 'contact zone', one finds a strong empirical and theoretical argument for relocating the site of production of knowledge into an interstitial zone of colonial contact, negotiation, and contestation, which enables the constitutive role of non-western agency and knowledge in the production of such discourses to be acknowledged. Pratt's work is just one example of the shift in colonial discourse analysis and postcolonial theory towards a strong emphasis on the fully relational constitution of representations and identities. In Homi Bhabha's (1994) work, the emphasis is upon colonial subject formation as an inherently ambivalent process of emulation, mimicry, and subversive trickery, giving rise to forms of hybrid subjectivities.

This shift in the ways in which identity, geography, and power are conceptualized is also evident in Said's own work, which after *Orientalism* came to focus much more explicitly on the interconnections and entwinements of different societies and cultures (Said, 1993). Said constantly emphasizes the moral imperative of asserting that different cultures, peoples, and societies both did and could coexist in the same spaces and times, and that the critical task was to find routes to this form of non-exclusivist accommodation as a means of reckoning with the shared histories of colonialism and imperialism.[1] A crucial dimension of Said's original argument in *Orientalism* was the importance of knowledge in staking claims to territory. Colonial discourse can be understood as revolving around a three-way relationship in which relations between European or western colonizers and non-western 'native' subjects is mediated by representations of land, space, and territory. Characteristically, this relationship involved representing non-western spaces as empty, or inhabited only by ghostly subjects, or untended, in ways that legitimized colonial and imperial intervention in the name of proper stewardship of people and land. One of the strongest legacies of colonialism, Said argued, is a clear connection between ideas of exclusive possession of territory and exclusivist conceptions of cultural identity. Authentic and essentialist conceptions of identity are often associated with exclusivist claims to territory and space. In turn, this geographical imagination of identity leads to the persistent understanding of colonialism in terms of simple oppositions between colonizers and colonized. It is a consistent theme of Said's academic and political writing to contest both the connection between identity and territory, and simple notions of colonizer and colonized. The postcolonial world is, in his view, much more messy, messed up and compromised than this simple opposition suggests.

I have dwelt at length on Said's work, and in particular *Orientalism*, because it is hard to underestimate the significance of this work in the development of postcolonialism as a strand of academic interdisciplinary work. Said's work has offered an important route through which geographers have been able to engage in broader cross-disciplinary debates with historians, anthropologists, cultural theorists and others with similar interests in questions of space, territory and identity. As a central element of postcolonial theory more generally, theories of colonial discourse analysis have contributed to the process of 'decolonizing the mind' by challenging the self-image of the west as a self-determining, self-contained entity which is the unique origin of a universalizing history and culture. I now want to turn to a consideration of what is perhaps the most misunderstood issue in postcolonialism, namely the question of how the power of representations is theorized in this field.

Representation, Subjectivity and Power

Said's critique of Western representational systems raises a fundamental issue of whether and how it is possible to represent other cultures, other identities, or other communities. The answer to this question depends on two related questions. First, should cultural difference be conceptualized according to an image of discrete spatial entities? I shall address this question in the next section. Second, should practices of representation be conceptualized in zero-sum terms? I shall address this question in this section. If colonialism and imperialism involve the denial, denigration, and negation of the cultural traditions of subjugated groups, then political opposition to these processes can be characterized in part as a set of struggles for the right of communities to represent themselves. But the concept of representation has become a recurrently problematic theme in cultural theory. Social constructionist arguments depend on a particular epistemological argument about the active role of representations in constituting the realities they purport to represent. The critical force of this sort of argument – as a critique of racist stereotyping, or of patriarchal gender stereotypes, for example – actually depends on a rather unstable combination of two related arguments about representation. On the one hand, there is the *general* epistemological position that all knowledge is constructed through representations. On the other hand, there is the *specific* argument that some representations are misrepresentations, implying that certain representations are actually better than others.

Rather than getting caught up in interminable debates about whether cast-iron accurate descriptions of the world are actually possible, postcolonialism asks us to keep in mind the intimate relationship between representation in an epistemological sense and representation in a political sense, where this refers to a set of practices of delegation, substitution, and authorization. The real thrust of the critique of representation is to throw into question the modes of authority through which particular styles, forms, or voices come to be taken as representative of whole traditions, communities, or experiences. When thought of in political terms, there is an important distinction between thinking of representation as *speaking for* others and *speaking as* another. The latter notion supposes complete substitution for the other, a claim to authority on the basis of identity. In this second model, representing is understood in zero-sum terms: speaking on behalf of others is akin to usurping their own voices as one's own. The critique of representation in postcolonial cultural theory is primarily animated by a deep-reaching critique of identity thinking, and of associated norms of immediacy, authenticity, and spontaneous expression. In this respect, the former practice – speaking for others – keeps in view the contingent authority upon which such delegation depends for its legitimacy. In postcolonial theory, Gayatri Chakravorty Spivak's (1988) essay, 'Can the subaltern speak?', takes up precisely this set of arguments. It makes a clear distinction between two senses of representation: representation as *depiction*, and representation as *delegation*. Both of these senses of representation imply a process of substitution between the represented element and the representative intermediary: for example, a painting in a gallery stands in for a landscape it depicts, and a member of Parliament stands in for the constituents who elected him or her. But the second example immediately raises a set of questions about the authority of delegative representation: who voted for the MP, and to what extent do MPs faithfully represent the wishes of the voters. Representation in this sense is not a zero-sum game, but one which proliferates claims and counterclaims. Spivak's argument is that these sorts of questions also pertain to representation in a depictive sense. The argument is not that one can never have accurate depictions – of landscapes, voters' preferences, and so on – but that there is a degree of partiality involved in any representation that is not an error, but marks the point at which questions of authority and legitimacy proliferate (see Barnett, 1997).

The implication of this reconceptualization of representation is that critical attention should be focused on questions of who speaks, or to put it another way, on questions of agency. Now, agency is not simply a synonym for individual free will. It is, rather, a term that implies a set of relations of delegation and authorization; it combines a sense of self-guided activity with a sense of acting on someone else's behalf, or as their agent. Postcolonial theory's close association with the idea of discourse is often thought to be a limitation. The idea of the 'discursive construction' of subjectivity seems to imply that people's agency is wholly determined by the systems within which they are placed. Ideas of discourse are often associated with 'entrapment models' of subjectivity, in which people are seen either as wholly determined by discourses, or else as heroically resisting their placement within them. In postcolonial theory, this contrast leads to an interpretive dilemma for academics: 'You can empower discursively the native, and open yourself to charges of downplaying the epistemic (and literal) violence of colonialism; or play up the absolute nature of colonial domination, and be open to charges of negating the subjectivity and agency of the colonised, thus textually replicating the repressive operations of colonialism' (Gates, 1991: 462). This dilemma derives from different ideas of just what the purpose of academic analysis is. Some people consider the aim to be one of recovering and asserting the 'voices' of the oppressed or silenced. On these grounds, postcolonial theory is expected to offer a theory of resistance, gleaned from the evidence of colonial sources and archives. Now, this is a perfectly legitimate aim, and even a rather noble one. But it is not the only purpose that can guide analysis and interpretation. I would argue that what is most distinctive about postcolonial theory is that it is less interested in reading representations as evidence of other sorts of practice, and more concerned with the actual work that systems of textual representation do in the world.

This argument is likely to raise some eyebrows. It has become common to argue, particularly in geography, that postcolonial theory spends too much time with texts and representations, and that more attention needs to be paid to 'material practices'. Invoking figures of the 'material' world has a sort of magical cachet in the social sciences, but we should be a little wary of this sort of knock-down criticism of postcolonialism, and for two reasons. First, postcolonial theory's critique of representation should lead us to be suspicious of arguments that appeal to some sort of unmediated access to the 'material' world that does not have to pass through the hoops of particular idioms, vocabularies, and rhetorics. Second, it is an argument that fails to acknowledge that postcolonial theory's focus on textuality is neither an index of being interested in 'just texts', nor an indication of a grander argument that the 'world is like a text'. Rather, this tradition of thought is concerned with thinking through the quite specific sorts of power that can be deployed by the use of textual apparatuses like books, printing presses, reading practices and so on. In this respect, what is most distinctive about postcolonial theory is a particular conception of power. Terms such as 'representation', 'discourse', and 'textuality' all converge around a shared sense that knowledge is a critical resource in the exercise and contestation of political authority. Perhaps for disciplinary reasons, postcolonial theory has tended to focus on particular sorts of knowledge – 'soft' knowledge contained in literature and other aesthetic forms. But it is worth noting that this focus has helped to transform literary studies itself. It is hardly adequate to present it as a discipline concerned only with intensive readings of the hidden meanings of texts. It is just as likely to be concerned with the economics of publishing, the politics of education policy, or the social relations of reading. In each of these sorts of critical endeavours, the interest is with the ways in which texts get *used* to particular effects in a broader web of social relations – used to make friends, to train experts, to convert people, and so on. The 'work' that texts, or discourses, or representations do is not, from a postcolonial perspective, imaginary or ideological; it is not about making people think certain things, believe in certain values, or identify with certain subject positions. It is practical: above all, it is about uneven access to literacy, and by extension, to vocabularies of self-definition, practices of comportment, and rituals of distinction. It is concerned with how people are made and make themselves into subjects and agents who can act in the world. Embedded in wider practices, texts enable certain sorts of agency, in the double sense described above, by providing a mediated source of knowledge through which people can act as subjects of their own actions. In this focus on the power of textually mediated subject formation, postcolonial theory therefore acknowledges the density of representations and the durability of texts; it does not look through them to another reality or inside them for layers of meaning, but takes seriously the weight that they carry in the world.

Geographies of Understanding

In concluding, I want to consider the other question that was raised at the start of the previous section – the question of how to conceptualize cultural difference with the aim of fostering cross-cultural understanding, or what David Slater (1992) has called 'learning from other regions'. Understood as a version of social constructionism, postcolonial criticism leaves us with a dilemma: in so far as its critical edge comes from arguing that representations of non-Western societies are just that – representations – then the question arises whether it is possible to ever accurately describe unfamiliar cultures and societies. A strong social constructionist would appear to deny this possibility, in so far as all description is held to be context- and culture-specific. But the question of cross-cultural interpretation remains central to the postcolonial project. The strong impulse of Said's work is to affirm the value of robust empirical knowledge as a basic premise of interpretation and evaluation. Similarly, Spivak has consistently asserted the importance of empirical knowledge to the work of interpretation, and has gone so far as to affirm the importance of a revivified area studies.

This call for robust area-based knowledge is interesting precisely because geography is one of the disciplines associated with the production of area-specific knowledge of regions, cultures and societies. But what is notable about the encounter between postcolonialism and geography is the extent to which the critique of colonialist paradigms and legacies, when made through epistemological arguments about the construction of truth claims, has reinforced an interpretive turn in the discipline that promotes a general aversion towards values of objectivity, empirical validity, and explanation. This interpretive turn, marked by a set of scruples about representing other cultures and societies, is in danger of jettisoning one of geography's most enduring popular legacies, which is a sense of worldly curiosity: 'any sense of Western scholars claiming to represent, claiming to know, "other societies" has become dangerous territory' (Bonnett, 2003: 60). The problem with this seemingly impeccable respect for the particularities of other traditions is that it threatens to install an oddly indifferent attitude of tolerance towards other perspectives. By supposing that any judgement as to the validity of knowledge claims is itself suspect, the common-or-garden variety of social constructionism invests specific persons, styles, or practices from other places with the status of being representative of whole cultures. It therefore promotes a style of cultural relativism that, in its suspension of judgement, makes cross-cultural learning impossible by presenting any and all forms of geographical curiosity as morally suspect (see Mohanty, 1995).

My argument is that this style of tolerant indifference or cultural relativism manages to miss the real challenge of postcolonialism. If one of the ways of 'postcolonializing geography' is to address a set of embedded institutional practices of teaching, writing, and publishing (see Robinson, 2003), then another is to follow through on the implications of the postcolonial critique of historicism for the ways in which we imagine the geographies of cultural difference. In particular, postcolonialism should not be understood as a simple, all-encompassing dismissal of the universalistic aspirations of modern humanistic culture. In large part, writers like Said and

Spivak criticize Western traditions for their failure to be adequately attuned to the forms of communication through which a genuine pluralistic universalism might develop; these forms involve developing an ear for other ways of apprehending the world, opening up to other ways of knowing.

There are two points worth making in respect to this challenge of reconstructing a pluralist universalism, an attitude which would be less focused on the scruples of representing other cultures, and more open to the styles of sharing that come from a reworked style of geographical curiosity. First, it is worth reminding ourselves that cultures or societies are not arranged as if they are tight, concentric circles (Connolly, 2000). Postcolonialism teaches us that coming from one place, belonging to a particular culture, or sharing a specific language does not enclose us inside a territory. Rather, it implies being placed along multiple routes and trajectories, and being exposed to all sorts of movements and exchanges. The tendency to conflate the affirmation of cultural pluralism with an assertion of incommensurable values in fact misses the real force of postcolonial criticism, which takes as its target ways of thinking about difference in territorialized ways – in terms of them and us, inside and outside, here and there. The master tropes of postcolonial theory – of hybridity, syncretism, diaspora, exile, and so on – are not only all geographical metaphors. They are, more specifically, all metaphors of impurity and mixing. They therefore retain a strong sense of the importance of thinking about the geography of identity, but do so without modelling this geography of identity on an image of clear-cut and indivisible demarcations of belonging. Difference is not a barrier to relating and understanding, but is their very condition.

Second, one of the key insights of postcolonial criticism is that 'the West' is not a self-enclosed entity, but is made 'from the outside in'. This is one of Fanon's key arguments, but taken to its logical conclusion by postcolonial theorists, it implies that supposedly 'Western' forms (democracy, or rationality, or individualism) are not straightforwardly Western at all. Rather, they have multiple origins and pathways, and are formed out of the amalgamation of various practices and strands of thought. This is a fundamental issue, because it indicates the way in which postcolonial criticism takes as its target not just Western paradigms, but also the dominant critical paradigms of modern anti-colonialist nationalism, which still often appeal to images of authentic culture, and thereby reproduce forms of 'nativism' that can be deployed by authoritarian regimes to justify the authoritarian usurpation of power.

The relativist interpretation of postcolonial theory promoted by some of its champions as well as criticized by many of its detractors therefore needs to be contrasted to a reading that is at once more radical and more liberal in its implications. This alternative reading starts from the observation that postcolonial theory has engaged in a sustained criticism of a dominant imagination of space, one which renders cultures and societies as enclosed, territorialized entities with clear and tight boundaries. It is from this image of space that all the dilemmas, scruples, and reassurances of cultural relativism arise. It is no accident that an alternative imagination of space – in terms of movement, mobility, translation, and porosity – should have arisen out of a field of work that is prevalently populated by literary scholars. As we have already seen, postcolonial theory is often taken to task for being *too* textual. I have already suggested that this criticism might be missing an important point about how power works through institutionalized practices of subject formation. But another reason why we need not accept this criticism at face value cuts to the heart of geography's favoured subject matter – conceptualizations of space, place, and scale. Rather than presuming that postcolonial theory needs to be supplemented by geography's robust materialism, we might acknowledge that we geographers have something to learn from literary theory precisely because a concern with the *material* things that literary scholars are traditionally occupied with – books, the printed word, the formal qualities of textuality – opens to view a set of spatialities that are much more fluid, mobile, tactile, and differentiated than the ones that social scientists often favour.

Here then, in conclusion, are three reasons why postcolonial theory is not relevant only to geographers, but important precisely because of the fact that it is predominantly a variety of literary theory concerned with issues of textuality. First, it teaches us important lessons about the ways in which power operates in the modern world, through the mediated production of subjectivities. Second, by problematizing seemingly neutral practices like reading, writing and interpreting, it broaches questions about the ways in which cross-cultural understanding depends not on the mastery of meaning but on openness to difference, to developing an ear for the other, and on relations of translation. And finally, in the focus upon practices through which textual meaning is produced and enforced, postcolonialism opens up an alternative conceptualization of spatiality that is not 'metaphorical', and therefore not in need of being beefed up by some added 'materiality', but emerges from a careful attention to the *textures* of symbolic communication itself.

NOTE

1 This is one of the themes which connects Said's cultural theorizing with the other facet of work for which he is best known, namely his strong advocacy of the cause of Palestinian independence (see Gregory, 1995; 2004).

References

Barnett, C. (1997) 'Sing along with the common people: politics, postcolonialism and other figures', *Environment and Planning D: Society and Space*, 15: 137–54.

Bhabha, H. (1994) *The Location of Culture*. London: Routledge.

Blaut, J. (1993) *The Colonizer's Model of the World: Geographical Diffusionism and Eurocentric History*. New York: Guilford.

Bonnett, A. (2003) 'Geography as the world discipline: connecting popular and academic geographical imaginations', *Area*, 35: 55–63.

Connolly, W. (2000) 'Speed, concentric cultures, and cosmopolitanism', *Political Theory*, 29: 596–618.

Fanon, F. (1967) *The Wretched of the Earth*. Harmondsworth: Penguin.

Fanon, F. (1991) *Black Skins, White Masks*. New York: Grove.

Gates, H.L. (1991) 'Critical Fanonism', *Critical Inquiry*, 17: 457–70.

Gregory, D. (1995) 'Imaginative geographies', *Progress in Human Geography*, 19: 447–85.

Gregory, D. (2004) *The Colonial Present*. Oxford: Blackwell.

James, C.L.R. (1963) *Beyond a Boundary*. London: Hutchinson.

Mohanty, S.P. (1995) 'Colonial legacies, multicultural futures: relativism, objectivity, the challenge of otherness', *Publications of the Modern Languages Association*, 111(1): 108–18.

Ngûgî wa Thiong'o (1986) *Decolonising the Mind: The Politics of Language in African Literature*. London: Heinemann.

Pratt, M.L. (1992) *Imperial Eyes: Travel Writing and Transculturation*. London: Routledge.
Robinson, J. (2003) 'Postcolonialising geography: tactics and pitfalls', *Singapore Journal of Tropical Geography*, 24: 273–89.
Said, E. (1978) *Orientalism*. Harmondsworth: Penguin.
Said, E. (1993) *Culture and Imperialism*. London: Chatto and Windus.
Sidaway, J. (2000) 'Postcolonial geographies: an exploratory essay', *Progress in Human Geography*, 24: 591–612.
Slater, D. (1992) 'On the borders of social theory: learning from other regions', *Environment and Planning D: Society and Space*, 10: 307–27.
Spivak, G.C. (1988) 'Can the subaltern speak?', in C. Nelson and L. Grossberg (eds), *Marxism and the Interpretation of Culture*. London: Macmillan, pp. 271–313.
Thomas, N. (1994) *Colonialism's Culture: Anthropology, Travel, and Government*. Princeton, NJ: Princeton University Press.
Young, R.J.C. (1990) *White Mythologies: Writing History and the West*. London: Routledge.
Young, R.J.C. (2001) *Postcolonialism: An Historical Introduction*. London: Routledge.

Editors' Passnotes

PASSNOTES – POSITIVISM

What is it about?: Using scientific principles and methods to study social issues.

Who does it?: Gerard Rushton, Art Getis.

Where did it come from?: 1950s – recognition that geography was too descriptive and needed to be more scientific in its method to explain spatial patterns.

Where's it going?: GIS (though not the critical kind)!

Who started it?: Auguste Comte and the Vienna Circle.

How do you do it?: Through identifying laws, spatial modelling and hypothesis testing.

What's the evidence?: Quantitative data.

Tell me something I don't know about it: Not many people claim to be positivists even though others label their work in this way.

Where would/did you find these geographers?: The University of Iowa and Wisconsin (US); Bristol and Cambridge (UK).

PASSNOTES – HUMANISM

What is it?: A people-centred philosophy concerned with meanings, values and understanding our taken for granted experiences of 'being-in-the-world'.

Who does it?: Robert Sack (1997) provides a humanist conception of place making.

Where did it come from?: A discontent with, and critique of, positivistic human geography.

Where's it going?: Its legacy is work on moral geography and ethics, and the study of landscapes and everyday life in cultural geography.

Who started it?: In Geography?: Yi Fu Tuan, Anne Buttimer, Edward Relph and David Ley.

How do you do it?: In-depth engagement with individuals; exploring meanings and the human imagination through texts, language, sound, art, etc.

What's the evidence?: Interpretive accounts of human experience.

Tell me something I don't know about it: There are close connections between humanism and historical perspectives in Geography.

Where would/did you find these geographers?: It's about individuals remember!

PASSNOTES – FEMINISM

What is it?: A philosophy that theorizes gender relations.

Who does it?: Collectives (e.g. Women and Geography Study Group of IBG/RGS; and Geographical Perspectives on Women speciality group of the AAG)

Where did it come from?: A concern with understanding women's place in the world and a critique of the masculinist nature of geography.

Where's it going?: Troubling the category 'gender'.

Who started it?: A defining moment was Jan Monk and Susan Hanson's paper in *The Professional Geographer* – 'On not excluding half of the human in human geography'.

How do you do it?: Close attention to, and politicization of, the research process (from a focus on power; acknowledging your positionality; giving something back to research participants).

What's the evidence?: Anything you like (quantitative, qualitative, etc.), it's not what you do, it's the way that you do it.

Tell me something I don't know about it: Men can do it too.

Where would/did you find these geographers?: In the pages of *Gender, Place and Culture.*

PASSNOTES – MARXISM

What is it?: An uptake and moving beyond of Marx's ideas so as to foster a wider realm of human freedom and an attitude of non-conquest toward the non-human.

Who does it?: Versions have seeped into much of critical human geography.

Where did it come from?: A critique of 'establishment' geography. Influenced by political uprisings in Europe, Australia, Africa, Asia and the Americas.

Where's it going?: Marxism rejects categories so it is always going outside the box!

Who started it?: Some say Marx (others say Hegel, Spinoza or capitalism itself).

How do you do it?: Reading Marx is one way but don't stop there.

What's the evidence?: The ubiquity of capitalist social formations.

Tell me something I don't know about it: Marx was a starting point for 'post' philosophers like Foucault, Derrida and Spivak. Marx disavowed Marx-ism.

Where would/did you find these geographers?: Wherever you find Geography.

PASSNOTES – BEHAVIOURIALISM

What is it about?: Understanding the psychology behind individual spatial behaviour.

Who does it?: Reg Golledge, time-geographers.

Where did it come from?: Psychology.

Where's it going?: Its legacy may be in psychoanalytic geographies.

Who started it?: Lowenthal's paper on geographical experience and his typification of Environment, Perception and Behaviour research were influential.

How do you do it?: Measurement e.g. structured questionnaires, cognitive maps, etc.

What's the evidence?: Mental maps.

Tell me something I don't know about it: Behavioural research has contributed to sub-disciplinary fields such as cartography and disability geography.

Where would/did you find these geographers?: In the pages of *Environment and Behaviour.*

PASSNOTES – STRUCTURATION THEORY

What is it?: An approach to social theory that explains intersection of human agency and the wider social structures within which we operate.

Who does it?: Time-geographers; some health geographers.

Where did it come from?: Giddens' critique of historical materialism.

Where's it going?: It's being eclipsed by Actor-Network Theory.

Who started it?: The sociologist Anthony Giddens.

How do you do it?: By understanding context and the contingencies.

What's the evidence?: Models of time–space relations.

Tell me something I don't know about it: Anthony Giddens' friendship with Derek Gregory at Cambridge facilitated the flow of his ideas into geography.

Where would/did you find these geographers?: Health geography, sociology.

PASSNOTES – REALISM

What is it about?: A philosophy that seeks to identify causal mechanisms.

Who does it?: Andrew Sayer.

Where did it come from?: Bhaskar's realist theory of science.

Where's it going?: Continues to be widely supported.

Who started it?: Sayer was the key figure in introducing realism within human geography.

How do you do it?: Process of abstraction.

What's the evidence?: Things, relations, actions, discourse …

Tell me something I don't know about it: Physical geography also engaged with it in the 1990s.

Where would/did you find these geographers?: Scattered across British and Scandinavian geography and social science departments.

PASSNOTES – POSTMODERNISM

What is it?: A loss of faith in modern myths (e.g. 'progress' via 'reason' towards a 'final' 'glorious' 'order').

Who does it?: Ed Soja, Michael Dear (in geography). Jean-François Lyotard (philosopher) and Zygmunt Bauman (sociologist) are amongst the most reliable guides.

Where did it come from?: We're talking a genuine *Zeitgeist* here – a reflection of the spirit of an age, the colour of the times, the mood of contemporary culture …

Where's it going?: The term's already outmoded, but the ideas persist in various guises.

Who started it?: Lyotard did much to popularize the term.

How do you do it?: Ironically.

What's the evidence?: How modern of you to ask! (See? Irony).

Tell me something I don't know about it: Lyotard isn't named after a gym garment; the garment was named after a French trapeze artist, Jules Léotard (1830–1870).

Where would/did you find these geographers?: Los Angeles, cyberspace, *elsewhere* …

PASSNOTES – POSTSTRUCTURALISM

What is it?: It's a loosely grouped set of texts and philosophies that are radically anti-essentialist.

Who does it?: Most geographers under 35.

Where did it come from?: Its genealogy can be traced back 200 years+ through continental philosophy.

Where's it going?: Some argue it's fragmenting into other connective forms of theorising such as Actor-Network Theory.

Who started it?: It dates from Kant.

How do you do it?: Deconstruction; unsettle categories; challenge binaries.

What's the evidence?: Discourse and new cartographies.

Tell me something I don't know about it: It's inspiring geographers to find new ways of writing.

Where would/did you find these geographers?: In libraries.

PASSNOTES – ACTOR-NETWORK THEORY

What is it?: It's a theory that understands the world as multiplicity of different connections in which non-humans also have agency.

Who does it?: Latour and Law are the big cheeses; in Geography, Thrift, Whatmore, Murdoch.

Where did it come from?: Sociology of Science.

Where's it going?: Towards theories of chaos and complexity.

Who started it?: Michel Serres and Bruno Latour.

How do you do it?: Theorizing networks of association.

What's the evidence?: Whatmore's *Hybrid Geographies.*

Tell me something I don't know about it: It uses the term actant rather than actor.

Where would/did you find these geographers?: Open University (UK).

PASSNOTES – POSTCOLONIALISM

What is it?: A critique of colonialism and western-centric views of the world; and a reconceptualization of representation.

Who does it?: Bhabha, Spivak, Pratt; and in Geography: Alison Blunt, Jane Jacobs, Derek Gregory, David Slater.

Where did it come from?: Intellectuals who wrote about anti-colonial struggles and challenged western versions of culture. But its academic roots are in literary and cultural studies.

Where's it going?: The pressure is on to show more concern with material practices rather than representation.

Who started it?: Edward Said's book *Orientalism* has been most influential in its development.

How do you do it?: Analysis of colonial archives.

What's the evidence?: Discourses, representations.

Tell me something I don't know about it: It has involved close engagement with poststructuralists like Derrida.

Where would/did you find these geographies?: In cultural studies.

Part 2
People

The first part of the book dealt primarily with arguments. Leading scholars in the discipline made a case for their particular ways of knowing. In this part we turn to autobiographies to help elaborate more fully some of the personal factors that influence how geographers come to know their world. Philosophies as ways of knowing are derived from day-to-day living and, as such, they are intimately entwined with the lives of their practitioners. The previous part focused on some major philosophies that engage geographers. In this part, some well-known figures in the field talk about how they came to engage particular ways of knowing and doing.

Within contemporary human geography there has been a recent emphasis on situated or contextual knowledges (see Chapter 12); personal writing is seen as an important strategy to challenge the disembodied and dispassionate nature of much academic writing. This second part focuses on how the work of a number of influential contemporary geographers has been shaped by their academic context, place, and personal experiences. In the process, the part highlights the way that some approaches to the discipline may remain constant through a career, may evolve gradually or may undergo a radical shift in the face of disciplinary or societal movements.

The part is made up of autobiographical accounts of how ideas and work are shaped by philosophies, personal experiences, place and time. These different testimonies highlight the contradictions, ambiguities, stabilities and flux in individuals' writing and practice. Journeys through and between places are central to the stories in these testimonies. Some talk about a profound love of place, and the construction of a geographical imagination that began in youth and continues in a connected way to where they live now. Others focus on experiencing different people and places and how that experience changes the way they think about the world; while still others focus on solving problems evident from observing landscapes and change. Some of the commentators talk about the process of journeying and mobility and how those processes weave connections and the contexts of their work. There are those who see change in the way they move from institution to institution, while others see the importance of change in the institutions that they remain at throughout their careers. All are indebted in huge ways to past teachers and influential colleagues. These personal connections are important, making these authors who they are as individuals as well as respected leaders in geography.

Academic careers are equally about entertaining interesting and unsought after opportunities as they are about stalwartly following convictions. To some, the journey and the road less travelled is the most important thing. For David Harvey, for example, articles, books, and prestigious appointments pale in significance compared with a continuous lived process of learning and exploration. The love of this process, this way of life, is shared in different ways by each of the authors in this part. Linda McDowell writes about the excitement of being part of the women's movement in the UK and shares with Harvey roots that enable a clear appreciation of class struggles. For Harvey and David Ley, the process is always embedded in a singular love of places. Ley notes the ways that people and places intertwine to promote powerful opportunities. Both Ley and Gerard Rushton talk about what can happen when new ideas are coupled with the dynamics of young students in one place at one moment in time (e.g. Pennsylvania or Iowa in the 1960s), while Richa Nagar highlights the importance of inspiring mentors (particularly Susan Geiger). For Ley, the important focus was on humanism and social geography. Coming from a different philosophical tradition, for Rushton the challenge was to disentangle and solve spatial problems in the landscape.

While most of the writers focus on the connections with and pleasures of place, Larry Knopp draws on his experience as a white gay man to reflect on the sense of being simultaneously in and out of place, and the often unacknowledged pleasures of movement, displacement and placelessness.

Both Richa Nagar and Vera Chouinard identify the importance of the relationship between radical theories of power encountered in the geography classroom and their understandings of their own personal experiences from childhood onwards. Here, Nagar describes how her active negotiations of her own identity as a Tanzanian Asian woman of colour became interwoven in her PhD dissertation where she explored the complexities of gender, race and community. Chouinard's essay focuses on the process of being marked out as different within the discipline of geography . While other essays stress the joys of the pursuit of geographical knowledge, Chouinard describes her struggles within the academy and the toll they took on her health, and the wellbeing of her family. In doing so she shows how becoming disabled in turn shaped her own geographical agenda and reflects on connecting the personal and the political with the philosophical and the theoretical.

Mohandas Ghandhi once noted that almost anything we do is insignificant, and yet it is nonetheless very important that we do it. Each day, each idea, each opportunity, each word is small but takes us in a direction. These directions may in time look like Janice Monk's braided streams or Larry Knopp's fluid mobility, and they are always of consequence. The notion of doing provides an important conduit to Part 3, which outlines some of the day-to-day tensions between philosophical purity, theoretical cohesion and the need for pragmatism in how we do research. As some of the chapters in Part 2 suggest, conflict between ways of knowing and ways of doing often emerges in the course of research projects because of the practicalities of everyday contexts and constraints such as time, resources, and funding.

13 INSTITUTIONS AND CULTURES

Gerard Rushton

Growing up in a small industrial town in Northwest England during WWII whetted my interest in Geography at an early age. Every evening my mother would lay a large map of Europe on the floor and we would listen to the nine o' clock news and then find the locations of the battles and bombings on the map. I vividly remember climbing a hill near home to see the red night sky as a bombed Manchester burned 30 miles away. Five uncles were in His Majesty's Service: two in Europe, one in India and the other in Kenya. My father was also in service, but luckily was selected, by drawing lots, to stay in Britain when his ship was not able to transport the whole regiment. That ship was torpedoed and all perished.

It was in high school that my interest in economic geography began. In my home town of Nelson, Lancashire, there were many cotton weaving mills which closed down when in the 1930s Japan dominated the world market for cotton cloth. In 1945, the looms were brought out of storage and were soon working around the clock. Britain then depended on the industry for a large part of its export earnings, but by 1952, the mills began to close again. Some mills relocated to Tasmania. My auntie Francis agreed to relocate if she could have the same 12 looms she worked in Nelson accompany her to Tasmania. One of the remaining larger mills in Nelson hired an Urdu speaker to solicit workers from Pakistan and when I returned a few years ago, all five movie theaters were playing South Asian films.

The University of Wales

When I was an undergraduate at the University College of Wales, Aberystwyth, our economic geography textbook by William Smith had maps of Burnley, the neighboring town to Nelson, showing the mills that closed down in the 1930s. Smith argued that most of the closed mills were located on the canal and that, as a means of transportation, the canal was little used by the 1930s. Therefore, the mills had lost their locational advantage.

Knowing how these mills operated, I was unimpressed with this argument. Yet the pattern of canal-located mills closing and most others remaining open was clearly there. No point pattern analysis was needed. By bicycling around and noting the dates on the cornerstones of the mills, I established that the pattern of closing in the 1930s correlated with the age of the mill. I also found that those built before 1880 had invariably woven coarse cotton cloth, whereas most established after that date had Jacquard looms.[1] This became the model for computer software code. So my interpretation of the pattern of closures was connected to the kind of cotton weaving each mill did: those in the business of the coarse cloth had a tendency to close; those doing the fine weaving mostly remained open. This became the basis of my undergraduate thesis, so change in geographical patterns of industrial activities became my special area of study.

In 1959 the Chairman of Geography at 'Aber' called me into his office and offered a "Demonstrator" position teaching the field work components of surveying and geomorphology. The British Government had just announced its intention to end the military draft in two years. My friends were being assigned to Britain's trouble spots of the day: Cyprus, Northern Ireland, Aden, Singapore and Hong Kong. Doing geography seemed a far better way to spend the next two years.

My masters thesis was on patterns of change in industrial activities in Lancashire. When I interviewed people at 'The Board of Trade' in Manchester about industrial change in Lancashire, I soon realized that their goal was to attract American capital and American branch plants. I began to see my small area of England in a wider context.

The people I interviewed in new industrial plants in the area stressed their commitment to finding optimal locations for their activities. They often told me that it was the subsidies the UK Government offered them that attracted them to consider such peripheral places as Northern England. This was the age of rationality and new methods were being developed to solve the problem of finding optimal locations.

My adviser in Aber, Peter Mounfield, had just received his doctorate from Cambridge and he talked about the new industrial geography there. Developing cost surfaces at alternative locations was the approach, but, easier said than done. He advised me to read American economic geography and then suggested a year of study in the US. He helped me make a list of American economic geographers whose work we liked. I wrote to them to enquire about admission and support for one year of graduate study. I applied to Chicago, Wisconsin and Iowa, and also, because Mounfield had a personal friend, David Simonett, at Kansas, I wrote there too. Simonett immediately replied: 'Iowa is the place for you.'

The University of Iowa

Iowa geography in 1961 was like nothing I had ever encountered. The small faculty (six) and graduate student group (about 30) were clearly on a crusade. They saw a new geography founded in philosophical positivism and implemented with quantitative methods (cf. McCarty, 1954). This period at Iowa has been described by McCarty (1979) and by King (1979). McCarty, the Chair, had us read Sten de Geer (1923) as his inspiration on the nature of laws in geography. According to de Geer 'Geography is the science of the present-day distribution phenomena on the surface of the earth. It aims at a comparative and explanatory description of the characteristic complexes of important distribution phenomena – geographical provinces and regions – which occur on the earth's surface' (1923; 10). This was Iowa's definition of geography in the 1960s and the tools were field observations, statistics and the computer.

Iowa Professor Jim Lindberg introduced me to central place theory and to a new English translation by Baskin of Christaller's book. I was fascinated and I contrasted my view of it with the emphasis on geometric patterns of urban centers in the literature of that time. Influenced by the experience of flying across Iowa several times as a guest on a small plane, I was convinced I could predict the locations of the next towns and their relative sizes. It was far from a neat hexagonal pattern, but always it seemed to make sense as Christaller and Lösch theorized. Both these scholars were interested in the orderly behavior of people. They thought of people, both consumers and producers, as rational in a spatial context. This became my mantra in my third and final year at Iowa.

That year I decided I needed to finish this post-graduate education and move on. I had married a girl from northern Iowa – an undergraduate in English whom I'd met on a blind date. I soon saw the need to finish my dissertation quickly. I was employed during that third year as a research assistant by the Bureau of Business and Economic Research. They had a multidisciplinary project to study the future role of Iowa's small towns. As part of their study, the survey research center at Iowa State University in Ames had conducted a random sample of the rural population of Iowa. They asked each of 803 households where they shopped for a few dozen commodities. The results were on 81,000 punched cards and analysis of these was stalled. Not much thought had gone into designing an analysis plan. It was a good lesson for a young scholar to learn.

We made slow progress until I proposed to the Director that I be hired for extra hours, specifically to prove that I could replicate the results from their retail trade-area analysis surveys from a model based on this sample. It took some persuasion, but the point that finally convinced the Director was my promise to write it up for an article in the Iowa Business Digest, my first real publication, in 1964.

The project almost came to a halt when I petitioned the University computer allocation committee for a block of eight hours of computing time. They wondered why a geographer needed that much time. I had to appear before the committee to explain my model. I calibrated two large T-squares in miles and taped a large transportation map of Iowa to a table. The location of each of 1144 towns in Iowa were measured in terms of miles east and miles north of an arbitrary location in southern Nebraska. My computer program placed a four mile grid on the state and, sequentially, from each grid point, computed distances to all the neighboring towns. It then placed each of the six closest towns on a utility function I had calibrated from the survey data and it then predicted the likelihood of a person at the grid location patronizing each of the towns around.

The computer committee not only granted my request, but assigned a doctoral student in computer science as my assistant to optimize the model. The model ran early one Saturday morning. I received a telephone call from my assistant that the program had crashed after three hours. He was baffled. I raced to the computer center, stopping at a gas station to pick up a map of Iowa. I figured out where in Iowa the model was working when it crashed. I saw the problem immediately. Our program found the towns within 25 miles of each grid location. But we had not programmed what to do when there were not six towns within 25 miles. That was the problem.

We changed that key parameter and re-started the model and within the eight hours it successfully completed its work. Later, I tested the results from the model against one of the Bureau's field trade area studies and it performed very well. This was my first model of an aspect of a spatial economy in a geographic information system.

In my final summer at the University of Iowa, I worked with Reg Golledge and Bill Clark. Reg was a fellow Iowa student and Bill had just received his doctorate from the University of Illinois. Ron Boyce, the new urban geographer at Iowa, hired us to work on the continuation of the Bureau project. With little formal guidance, we crafted our own objectives and designed three papers we would write using the same survey data described above. Each of us would be principal author of one of them, but all would actively contribute. The theme of each was recovering or using the rules people used in making their spatial choices of where to shop in the context of the real choices available to them. These papers were published soon after we left Iowa.

McMaster University

After receiving my PhD I was hired by McMaster University in Ontario, Canada.

McMaster was a different culture. It was a traditional geography department, yet it had aspirations to change. I received lots of encouragement from the Chair and other faculty to develop my ideas. They encouraged me to recruit students, although they already had many talented students at all levels. Soon there was a very fine group of graduate students, among them Michael Goodchild, Bryan Massam, John Mercer and Tim Oke. They worked in several areas under different advisers, but all were dedicated to scientific geography and used quantitative models where appropriate.

After three years at 'Mac,' I received a call from Michigan State University. They had a position, half-time in geography and half-time in the Computer Institute for Social Science Research. Julian Wolpert had vacated the position to go to Princeton and he was their model geographer. The interview in CISSR was extremely demanding and the sheer stimulation of it made me want to be there. I taught introductory economic geography and a seminar in central place theory at MSU. I had plenty of time for my research. From McMaster I had brought my unfinished business which was to construct a behavioral model of spatial choice that was consistent with other models of choice in social science. Part of my positivistic philosophy was that a theory and model in geography should not be exceptional with respect to the basics of social science. I felt that my pre-MSU writing was too exceptional in this regard. CISSR was a good place to voice these doubts and I found a sympathetic group.

I became attracted to psychological measurement theory and soon found a group of scholars who were heavily engaged in choice theory and associated measurement theories at nearby University of Michigan. A psychologist colleague at CISSR showed me some non-metric multidimensional scaling (MDS) work from there and I saw a connection to the problem I had grappled with for five years: how, from the individual choices of people in different spatial contexts, could a scale be determined that showed how all such possible choices would be evaluated by them? I wrote: 'In the study of spatial behavior we are interested in finding the rules for spatial choice which, when applied to any unique distribution of spatial opportunities, are capable of generating spatial behavior patterns similar to those observed' (Rushton, 1969: 391). I introduced the concept of revealed space preferences and later found that MDS was the technical means of implementing it. I wrote several papers which supported this claim.

Back in Iowa

After two years at MSU a call came from Iowa. Professor McCarty had just retired, the department was growing in size, and they were about to make three new appointments to add to the two they had just made. All were to be in the scientific and quantitative tradition. Given my interests, the opportunity seemed too good to be true. At Iowa my interests in spatial preferences broadened to consider the problem of eliciting preferences for spatial situations that could not currently be observed. This brought the transition from methods of revealed space preference to information integration theory where soon, Jordan Louviere, a graduate student I was advising, was making impressive contributions. My interests began to switch to the other set of actors in central place theory: those who chose locations in which to put activities or those who chose activities for the locations they occupied. In other words, what set of locations maximized preferences? At MSU, I had constructed a model that showed that the same preferences, when applied in a classical central place situation, led to different groups of functions in places at the same level of the urban hierarchy. The model itself had many sections lifted from the Iowa Bureau model of 1964.

Instead of working over the actual map of Iowa, it used a grid over a classic central place hexagonal pattern (Rushton, 1971).

In 1970, The Ford Foundation asked if I would review a project in India that was designed to facilitate the provision of services to villages in areas where the green revolution was being promoted. The project was implemented by the Government of India and the Ford Foundation provided technical support.

The project organizers in India stated that their approach was based on central place theory. I criticized their plan roundly, mainly on the grounds that it appeared to be trying to lay hexagons over the regions of interest to promote the development of villages close to the theoretical nodes. In an invited visit to New York, I argued at the Foundation that the principles of central place theory should be used in this experiment but that the geometric results should not be applied. I made several suggestions about how this could be accomplished.

They asked if I would be willing to visit the project for a few months. For three months in the spring of 1971 I spent time in Hyderabad and New Delhi. Working for the Foundation was unlike academic work and living in their guest house and meeting the people coming and going from some of the best universities in the US and elsewhere was an interesting experience. I was asked to join the staff of the project but for many reasons declined. Instead, they gave me an open ticket to come and go whenever I chose, providing I would spend at least three weeks there on each visit. I made five visits in all between 1971 and 1974.

I had a reputation for being too theoretical and too impractical, and, to convince me of this, they encouraged me to visit several of their field sites. I had many opportunities to see the effects of poorly made decisions to locate services and became convinced that, without a methodology to judge their efficacy, locating services would continue to be the stuff of politics. Geography needed to develop a methodology for evaluating the location of services and for judging the effectiveness of alternative location decisions.

Allen Scott introduced location-allocation models to geography in the first volume of *Geographical Analysis* in 1969. As soon as I saw this article I knew this was the track to follow. In the summer of 1973, I received a grant from NSF to conduct a three week workshop for college teachers on location-allocation models and their role in a new location theory. 'Camp Algorithm,' as the participants called it, served to advance the adoption of location models in geography. What would August Lösch have said if he had seen such models and could see that in fact they could be implemented? I want, Lösch said, 'a spatial economic science that, more like architecture than the history of architecture, creates rather than describes.' Here was a major development to that end.

Location-allocation models advanced rapidly in the 1970s, particularly in the Operations Research community. Methods were implemented, however, mainly on small problems rather than on real applications. The data models were ill-developed to optimize the task of spatial searching which, after all, was the real purpose of the models.

For fifteen years I was consumed with making these models work for real situations. I advised several students who wrote dissertations on the subject, notably Ed Hillsman, Steve Nichols and Paul Densham. We added the centroids of Iowa townships to my geographic information system of Iowa towns and succeeded in applying the then best heuristic location-allocation algorithm, the Teitz and Bart algorithm, to this. At the time, this was considered to be an extremely large data set of about 3,000 nodes.

Two NSF grants allowed us to continue working in India and Nigeria on problems of locating services in developing countries. My colleagues were Mike McNulty in Iowa,

Vinod Tewari in Bangalore and Bola Ayeni in Ibadan. In a summary of this work, I wrote: 'Location-allocation analysis systems provide an explicit framework for diagnosing service accessibility problems, measuring the efficiency of recent locational decisions and the current levels of settlement efficiency, and generating viable alternatives for action by decision makers' (Rushton, 1988: 97).

My student, Paul Densham, and I coined the phrase 'spatial decision support systems' to recognize that location theory was no longer the retrospective study of past location decisions, but was now a tool for assisting people to make better decisions. The new location theory had to recognize how people understood and used the tools. I became puzzled at that time that geography, while embracing the rapid developments in geographic information systems, was retreating from the use of many of the models developed in the 1970s.

GIS was the new spatial decision support system of the 1990s. The problem was to link the information processing functions of GIS with the analytic functions of the models. Since a good education in GIS must include knowledge of the spatial models, I now teach a course 'Location models and spatial decision support systems'.

In 1989 San Diego State University was searching for a senior person who would compliment their very strong program in geographic information systems, the lack of which at Iowa at that time was affecting my work. With the assistance of UC Santa Barbara, they were commencing a doctoral program. Although it was wrenching to leave Iowa, which had contributed so much to my professional development and to my life, I enjoyed working with colleagues in SDSU and enjoyed getting to know Southern California and San Diego. I was able to complete papers describing new implementation methods for location-allocation models in a GIS environment. Two years later I returned to Iowa, where I sensed the beginnings of change and, eventually, a return to the department's traditional strength in geographic information science, spatial analytic models, behavior and the environment.

I had often collaborated with faculty in the health sciences at Iowa. For two years in the mid-1970s I had been a full-time director of the Center for Health Services Research. In 1993, I received a call from a Professor of Pediatrics asking if I would meet with a group that had been asked by the Iowa Department of Public Health to assist with a study of infant mortality in Des Moines. Was it true, they asked, that computers can now locate the addresses of births and deaths from vital statistics records and make maps of rates?

Tiger line files from the 1990 Census were being released that year. I agreed to make such maps by address-matching the 20,000 or so birth records and the 200 or so infant deaths. A student, Panos Lolonis, had worked with me to make projections of students in small areas for the Iowa City School District and I saw how, with a little modification, we could make a continuous distribution of infant mortality rates. I was impressed by how much this map was appreciated by the Health Department in Des Moines and I saw how much superior such maps are over the traditional maps of census tracts or zip code areas.

I became interested in the spatial change in the error structures of such maps and wondered why so few people were interested in the subject. Why did the whole area of public health make so little use of spatial analytic methods and of GIS?

I saw a lot of use of spatial statistics in this area, but virtually no use of spatial analytic methods. I sensed an interesting area to study. With colleagues and students, I wrote two papers published in Statistics in Medicine and was invited to give lectures to groups in the Centers for Disease Control. I was appointed to several peer-review panels in NIH where the term GIS was beginning to appear in research proposals sent to them. I was also appointed Chair of the GIS oversight committee for the National Cancer Institute's Long Island Breast Cancer Project. Currently, I have grants from CDC and NIH to research the development of more effective spatial analysis methods for measuring the cancer burden in local populations. The principles of spatial search are still a major feature of these efforts.

What is my current philosophy of geography? I remain dedicated to a scientific approach: searching for theories, laws and models that can predict human spatial behavior; developing theory that links spatial behavior to the spatial structures it generates; and maintaining, developing and validating methods of spatial analysis to support decision-making in significant arenas of human endeavor. If I am labeled a stalwart, unreconstructed positivist, I gladly accept that charge. For me and for my students, I firmly believe the best is yet to come.

NOTE

1 Jacquard looms are known today by many computer science students for their use of a sequence of punched cards to instruct a machine to weave complex color patterns in cloth.

References

De Geer, S. (1923) 'On the definition, method and classification of geography', *Geografiska Annaler,* 5: 1–37.

King, L.J. (1979) 'Areal associations and regressions', *Annals, Association of American Geographers,* 69: 124–8.

McCarty, H.H. (1954) 'An approach to a theory of economic geography', *Economic Geography,* 30: 95–101.

McCarty, H.H. (1979) 'Geography at Iowa', *Annals, Association of American Geographers,* 69: 121–4.

Rushton, G. (1969) 'Analysis of spatial behavior by revealed space preference', *Annals, Association of American Geographers,* 59: 391–400.

Rushton, G. (1971) 'Postulates of central place theory and the properties of central place systems', *Geographical Analysis*, 3: 140–56.

Rushton, G. (1988) 'Location theory, location-allocation models, and service development planning in the third world', *Economic Geography,* 64: 97–120.

14 PLACES AND CONTEXTS

David Ley

Intellectual knowledge, like other social phenomena, emerges from the intersection of imagination, practice, and context. In this short essay I shall outline the places and (some of the) people who have shaped my own academic development. This autobiographical account exemplifies my research bent of tracing the relations between place and identity, extending it to my own formation as a social and cultural geographer.

Early Influences: from Swansea to Windsor

Where does biology end and social context begin? Was it the promptings of DNA or some tribal conformity which prescribed that in my extended family five of the seven university entrants up to my generation studied geography? And was childhood stamp-collecting – as others have noted, an early indicator of a geographical predisposition – a response to some inherited patterning, or to opportunity (my father's occasional gifts from foreign sea captains in the ports of South Wales)? And how did an adolescence preoccupied with team sports intersect with recognition of the frequent coincidence between university athletes and geographical studies? In these taken-for-granted collisions and collusions a broader configuration took shape, prompting certain paths to be taken rather than others, and cumulatively orchestrating more and less likely trajectories.

From Windsor to Oxford: the Power of Mentoring

By adolescence certain influences assumed greater clarity. It has been my good fortune to have been instructed by superb teacher/mentors, including Roy Yabsley and Colin Brock at Windsor Grammar School, Paul Paget at Jesus College, Oxford, and Peter Gould at Pennsylvania State University. Without their formative and imaginative direction, outcomes would have been different. Colin's charismatic influence deflected me away from history and to geography as a university subject; Paul ignited the interest in social geography for me as he did for so many others (Clarke, 1984); while an unexpected letter from Peter effectively ended a potential career as an urban planner in Britain and led to a focus on inner city research in first the United States and then Canada.

These mentors brought specific contributions. At secondary school, field studies were an essential ingredient of knowledge, and the forced marches in the milieux of the Thames Basin – across the chalk downs, through the clay vales, and up to the sandy heathlands – created a compelling pedagogic lesson of spatial association, of an ordered world to be discovered. This education continued at university, with field trips among the diverse and rapidly changing landscapes of the Oxford region, and longer forays to the dramatic Celtic ends of Europe, the landscapes of Connemara and the Scottish Highlands that stretched the concept of *ecumene* to its limit, as striking beauty obscured a miserly endowment for human settlement even without the historical intervention of oppressive landowners. The culmination of an Oxford degree in those days was a regional interpretation of a relatively small area, and my selection, scouted while on site at a field hockey tournament, was a region of the westernmost Weald of Sussex in the south of England. In this rural area of villages and small towns the unfolding of a rapid progression of geological outcrops was accompanied by a sympathetic human geography of parish boundaries, land use and settlement, a vernacular landscape that had consolidated through the centuries, had largely leapfrogged the interventions of industrial urbanization, and was being ossified by retirees and long-range commuters to postindustrial London who, as rural gentrifiers, sought the apparent authenticity of historic preservation.

It was in undertaking this regional study that I became, in an enveloping visceral sense, smitten by geography. Cycling down country lanes in early summer, from village to village and from small town to small town, observing the lie of the land, engaging in casual and semi-directed conversations, was a sensuous reception of sounds, sights and smells, of palpable geographical presence, as well as an intellectual challenge of problem-solving. The work (if such it was) not only sedimented a conviction about the importance and rewards of field study, but also established a founding principle of the centrality of empirical work in establishing geographical regularities and conceptual development, a statement that would seem banal were it not for tendencies in human geography in the past 40 years to privilege theoretical abstraction that has freed itself from the inconvenience of empirical accountability. For – as I have frequently discovered – the empirical world has no shortage of surprises, as agricultural labourers' cottages in Sussex may be occupied by London bankers, just as millionaire mansions in Vancouver might be the home of immigrant households with Canadian incomes below the poverty line (Ley, 2003). Theorists who forsake empirical work commonly reproduce their own presuppositions as results and thereby lose significant opportunities for learning. A second principle emerged. An interpretive study is integrative, and the task of synthesis encourages an intellectual vocabulary of nuance, of 'more or less' and 'most of the time'. It allows for contingency and exception. In contrast a more analytical methodology, notably of the type that C. Wright Mills (1959) chastised as abstracted empiricism, might be more likely to press harder with lines of causality.

Of course the tradition I am describing endured serious criticism during the quantitative revolution. In part those criticisms were well placed, for regional interpretation could easily become formulaic and conceptually lazy. At Oxford in the late 1960s there was great excitement as students passed around dog-eared copies of a truly radical message in the pages of early issues of the *Journal of Regional Science* (Barnes, 2004), reinforcing word of more conceptually innovative geography emanating from departments in Cambridge and Bristol and in more distant parts like Lund, Chicago, Pennsylvania, and Washington. So when in spring of 1968 an unanticipated invitation to undertake a graduate degree in Pennsylvania arrived, I flew like a moth towards a bright light.

Pennsylvania: The Shock of the New

The Penn State department in 1968, in fact America in 1968, were exceptional places of hyperstimulation. A civil war was under way in the classrooms of Penn State geography, and the four young Turks – Ron Abler, John Adams, Peter Gould and Tony Williams – had both the zeal of youth and the scent of victory. Their own momentum was enhanced by the stimulating presence of David Harvey and Julian Wolpert as visiting professors, by the appointment of Roger Downs as a superb instructor, and by the visits of Gunnar Olsson and Les King, other regional members of the 'Michigan Inter-University Community of Mathematical Geographers', who showed up periodically for energetic conversation at the departmental watering hole on College Avenue. Peter Gould was the animator of these events and an extraordinary intellectual presence (Haggett, 2003).

But the United States too was on the verge of civil war in 1968, with rebellion in the inner cities and riots on university campuses. Perhaps most remarkable was the manner in which these turbulent arenas of scholarship and the nation scarcely intersected. Among the faculty at Penn State, it was a member of the old guard, Wilbur Zelinsky, who was the most public opponent of the Vietnam War and the domestic injustices in American society. Meanwhile courses on simulation, Markov chains (topics that students requested David Harvey to teach!), linear programming, canonical correlation, and so on, had this in common. They were preoccupied with an internal world of theory, with technique and logic, rather than an external world of people and place. Indeed to my disappointment (and initial incredulity) I discovered that creative course assignments set to 'learn the techniques' consisted of imaginary data sets that were contrived to show off the power of the method. This privileging of 'theory' as the greatest good seems to me to be fundamentally misguided, and no less so with structuralism, social theory and cultural theory, some of the frameworks that followed the logical positivism at Penn State. The issue of course is not the need for theory, which is self-evident, but the privileging of theory which can make scholarship overly introverted, at its worst a type of intellectual makework programme. A better balance is represented, for example, by my colleague Gerry Pratt's new book, *Working Feminism* (2004), with its concern to work with, indeed to struggle with, theory in face of the rude surprises of evidence – though I suspect my own predilection might be to see still less explicit theory and a fuller empirical account.

It was at Penn State that I discovered the city, through courses with John Adams and especially Julian Wolpert, a visitor from the Regional Science Department at the University of Pennsylvania, who invited me to his home in suburban Philadelphia for a weekend, when we drove through North Philadelphia and witnessed the extensive territory of what in those days we uncritically called the black ghetto. Through a summer job in a high school equivalency programme in the American South, I had recently entered the astonishing world of American race relations, and Wolpert's encouragement fuelled my own religious ethics of social justice, and led to the ethnographic study of a section of North Philadelphia that became my doctoral dissertation.

I have reflected on theoretical, methodological, and ethical issues associated with that research before (Ley, 1988; Ley and Mountz, 2000) and will not do so again. Two points, however, may bear repeating. Like all human works, my study was in part a product of the intellectual and social environment of its time, an integration of authorial agency and encompassing context. To my reading of it today, what has aged in the account is the busy conceptual scaffolding around the study, while what has survived is the interpretation of the practice of everyday life in an environment of chronic constraints and stressors. My conclusion is that we should hold onto our conceptual apparatus lightly. Second, though an ethnography, the study included some modest quantitative analysis of a neighbourhood questionnaire. While ethnography has become a popular methodology among human geographers in the past decade, invariably a form of methodological purity has prevailed, with little or no pursuit of anything but qualitative data. This predilection rules out of the scholarly record valuable large data sets that can throw light on general relationships by engaging much larger samples than are possible in ethnographies. My preference is for a messier strategy of triangulation, building off several methodological positions, rather than a purer model that favours a single methodological base point.

Vancouver: Reflections *in situ*

Relocation to Vancouver (like the earlier move to Pennsylvania) is testimony to the strength of weak ties, for just as a letter out of the blue despatched me to the United States, so a chance corridor meeting made me reconsider the unthinkable condition of not returning to the United Kingdom following graduate study. I arrived in Vancouver in 1972 to a very different institutional, national and urban environment than I had come to know on the East Coast. The department at the University of British Columbia (UBC) had strengths in physical, cultural and historical geography, and saw itself as the intellectual other to the quantitative innovations in Seattle, Pennsylvania, Toronto and elsewhere. Appointed at UBC in the same year was Marwyn Samuels, who like me had illustrated the pluralism of a putatively quantitative department by writing a dissertation with Anne Buttimer at the University of Washington on geography and existentialism. UBC wanted to show off its new appointments and asked Marwyn and me to devise a title for a double header that would introduce us to geographers in the region. We came up with the title 'The Meaning of Space' and presented our papers to a local meeting of the Canadian Association of Geographers in November 1972. From that session, and the fruitful conversations that developed from it, was conceived the edited volume *Humanistic Geography* which appeared six years later.

In many respects the position identified in my essay in that volume is one that I have followed in most of my research since that time, including forays to such theoretical fields as human agency, postmodernism, and globalization that tend to work from the same presuppositions (e.g. Ley, 2004). Somewhat different, and a clear legacy from graduate learning, are projects that have used simple statistics to examine relationships in larger databases (e.g. Ley and Tutchener, 2001). Aside from the database analysis, the abiding message has been a concern with the intersubjective projects of everyday life among groups occupying urban places – actions contextualized by the influence of other groups (agents of the state, the market, or other collectivities in civil society) and also by a much broader set of contextual processes, comprising relations that may be summarized in such concepts as postindustrialism, multiculturalism or neoliberalism. Such a position is methodologically and theoretically eclectic, informed by theory but not confined by it, and allowing for surprises in the world to reshape conceptual agendas. Consideration of positionality, the taken-for-grantedness of the scholarly pose, is a basic building block in work that observes the social and cultural embeddedness (but not determination) of all of life.

After some 20 years of research on the social and cultural geography of Canadian cities, examining such issues as neighbourhood organizations, urban politics, changing labour and housing markets (especially gentrification), in the mid 1990s a very different opportunity presented itself: to establish an interdisciplinary network of researchers in the field of immigration and urbanization, an initiative known as the Metropolis Project. I had just completed a book that brought together my gentrification work (Ley, 1996), and was at a career stage where a significant reorientation was welcome. A consortium of federal government departments established four research centres across Canada, with the provision of six years of assured funding, later renewed for a further five years. A keyword of Metropolis was partnership, involving relations with different levels of government, with NGOs and community representatives, and with an international network of close to 20 countries. As one of the two Directors of the Vancouver Centre (RIIM) from 1996 to 2003, life as a more or less solitary researcher came to an abrupt end. While general policy themes are identified by government sponsors, funds are channelled through the Social Science and Humanities Research Council, and scholars have considerable freedom in defining specific topics (see the Vancouver Centre's website, www.riim.metropolis.net). The broader story of Metropolis has yet to be told, but it has proven an interesting exemplar of policy relevant research, of interdisciplinary and international enrichment, and of engagement with government and the community. Metropolis qualifies as an example of the current call for geographers to be more active in public policy research (Dicken, 2004).

Not the least of the opportunities has been the stable funding horizon that has provided significant professional development resources for graduate research. Students have benefited not only from research funds but also from access to a network of personnel, data and field opportunities, and from the presence of each other. Reflecting the emphases of the project, a significant harvest of student publications has ensued (including Ley and Smith, 2000; Rose, 2001; Waters, 2002; Mountz, 2003; Teo, 2003) and with it a new generation of social geographers examining race, ethnicity and immigration in the city.

Conclusion: Continuity and Change

To spend an entire working career in a single department may seem to be a failure of geographical imagination. However, not only are there advantages in establishing continuity in a place, but also the places we occupy both undergo change and emerge as multifaceted, for through the life course one moves from the sites of young adulthood – inner city, apartment based, with sustained travel to leisure and learning sites across the metropolitan region – to different urban settings as a parent and citizen involved in children's activities and institutional programming. Moreover, cities (and departments) change around us. In 1972 I was advised that Vancouver was a quasi-British city, that included Brits, cricket, and temperate weather, and indeed in the early 1970s the British flag flew from all public buildings. Some time later that decade the practice ended, and no one today would make such an ethnic designation in a city where 80 per cent of a large immigrant cohort originates in Asia, and where the Chinese-origin population alone nudges 400,000. To a considerable degree I have hybridized with the city, leading to a new, but never final, reconfiguration of identity and place.

References

Barnes, T. (2004) 'The rise (and decline) of American regional science: lessons for the new economic geography?', *Journal of Economic Geography*, 4: 107–29.

Clarke, C. (1984) 'Paul Paget; an appreciation' , in C. Clarke, D. Ley and C. Peach (eds), *Geography and Ethnic Pluralism*. London: Allen and Unwin, pp. xiv–xvii.

Dicken, P. (2004) 'Geographers and "globalization": (yet) another missed boat?', *Transactions of the Institute of British Geographers,* n.s. 29: 5–26.

Haggett, P. (2003) 'Peter Robin Gould, 1932–2000', *Annals of the Association of American Geographers,* 93: 925–34.

Ley, D. (1988) 'Interpretive social research in the inner city', in J. Eyles (ed.), *Research in Human Geography*. Oxford: Blackwell, pp. 121–38.

Ley, D. (1996) *The New Middle Class and the Remaking of the Central City*. Oxford: Oxford University Press.

Ley, D. (2003) 'Seeking *Homo economicus*: the Canadian state and the strange story of the Business Immigration Program', *Annals of the Association of American Geographers,* 93: 426–41.

Ley, D. (2004) 'Transnational spaces and everyday lives', *Transactions of the Institute of British Geographers,* n.s. 29: 151–64.

Ley, D. and Mountz, A. (2000) 'Interpretation, representation, positionality: issues in field research in human geography', in M. Limb and C. Dwyer (eds), *Qualitative Methods for Geographers*. London: Arnold, pp. 234–50.

Ley, D. and Smith, H. (2000) 'Relations between deprivation and immigrant groups in large Canadian cities', *Urban Studies*, 37: 37–62.

Ley, D. and Tutchener, J. (2001) 'Immigration, globalisation and house prices in Canada's gateway cities', *Housing Studies*, 16: 199–223.

Mills, C. Wright (1959) *The Sociological Imagination*. New York: Oxford University Press.

Mountz, A. (2003) 'Human smuggling, the transnational imaginary, and everyday geographies of the nation-state', *Antipode*, 35: 622–44.

Pratt, G. (2004) *Working Feminism*. Edinburgh: Edinburgh University Press.

Rose, J. (2001) 'Contexts of interpretation: assessing urban immigrant reception in Richmond, BC', *The Canadian Geographer,* 45: 474–93.

Teo, S.-Y. (2003) 'Dreaming inside a walled city: imagination, gender and the roots of immigration', *Asia and Pacific Migration Journal,* 12: 411–38.

Waters, J. (2002) 'Flexible families? "Astronaut" households and the experiences of lone mothers in Vancouver, BC', *Social and Cultural Geography,* 3: 117–34.

15 MEMORIES AND DESIRES

David Harvey

When I was around 12 years old, I had my first lesson on the geography of North America. We drew a map of the eastern seaboard and marked on it something called the 'fall line.' It stretched from New England to Georgia and recorded where the rolling foothills of the Appalachians abutted onto the flat alluvial coastal plain. Its name derived from the numerous waterfalls to be found there. These waterfalls had social significance because they provided water power for innumerable mills that spawned towns and villages and eventually large cities. Today, the fall line is roughly marked by interstate 95, which connects a whole string of cities up and down the eastern seaboard of the United States.

As I studied that map in the dark and gloomy days of postwar Britain, I dreamed that one day I might visit North America and explore its wonders. The idea seemed hopeless then, even though I had relatives in the United States (they sent us food parcels in the Second World War). We were too poor, and it was all too far away. Little did I then imagine that I would spend more than half my life living on that fall line.

If you had told a 12-year-old boy living in the despairing days of immediate postwar Britain that this was the future in store for him, every muscle in his body would have been aquiver with excitement. That boy had to rummage in his imagination (TV was yet to come, and he got to the cinema as a treat only once every six months) to construct flights of fancy that landed him in Rio, Rangoon, San Francisco, or Benares. And more than once he decided to run away from home and explore the world only to find that if it was sunny in the morning, it was raining in the afternoon (an elementary fact of British meteorology), and that sharing a hollowed-out tree in the rain with an assortment of insects was not anywhere near as comfortable as bathing in the maternal warmth of home. And so it was that his interest in what I now call the dialectics of space and place (the way experiences in place always mesh with broader spatial relations) began.

Undergraduate Studies at Cambridge and an Empire's Decline

The fantasy of escape and exploration was nourished by the knowledge that the world was open to be explored. Those maps with so many parts of the world colored red as somehow 'belonging' to Britain indicated wide-ranging choices of territories available for inspection. But British power was declining. Many of my teachers at Cambridge had experience in the military or colonial service. They seemed to regret the loss of Empire, while accepting that it should evolve, though only in 'sensible' ways. It is easy in retrospect to criticize their imperial vision, their paternalism, and their colonial thinking. But what stays with me more positively is the incredible love they evinced for the countries and the peoples they worked among and studied. Through innumerable anecdotes, I learned of the curious conflicts among the colonizers and the colonized, as well as the more obvious symbiotic conflict between the two over what to do and where to do it. The everyday struggles around the use of the land, over power and social relations, over resources and meanings, came alive in those anecdotes and remain fundamental to my geographic education.

The catalytic moment for me occurred when Britain, France, and Israel colluded to try to take back the Suez Canal from Egyptian control. Even my father, who never expressed any overt political opinion but gave off the aura of a respectable working-class patriot who accepted that the aristocracy had been born to rule benevolently over the nation and the Empire, expressed disgust. I was just 21 and in my final year as an undergraduate. I abandoned my studies for a whole term to argue vehemently about politics and turned resolutely anti-imperialist thereafter.

Local/Global Dialectics

I did my undergraduate thesis on fruit cultivation in the nineteenth century in my local area (I picked fruit to earn extra money – a pittance – every summer from age 14 onward) and continued such studies through to the end of my doctoral dissertation, 'Aspects of agricultural and rural change in Kent, 1815–1900'. Exploring and excavating the deep roots of my own landscape, my own locality, became a central obsession. Intimate sensual contact with the land went hand in hand with study of it. I became deeply immersed in that world.

As part of my dissertation research I read the local newspapers from 1815 to 1880. It took me a whole summer to do it, and it was an incredibly rewarding experience. Anecdote upon anecdote all added up to an intricate picture that saw personal lives articulated with abstract social forces, making for those glacial changes that in the end have incredibly deep consequences for the landscape and social life. The newspapers changed their format and their social content as the century wore on. When I sat back and reflected, I recognized that I had witnessed the rise and fall of a certain kind of regional consciousness, an upheaval that depended on changing means of transport and communications, as well as on more general economic, technological, and social changes. The speed and spatial range of communications was clearly a dynamic shaping force in historical geography.

Yet there was another theme writ large in that experience. When I looked at the data on the hop industry, the cycles in plantings, output, and spatial spread and contraction correlated almost exactly with business cycles in the British economy. Agricultural distress or affluence in mid Kent was a function of changing discount rates in the London financial markets, which depended on trade conditions more generally. Finance capital and geographical forms were, as I now would put it, intimately and dynamically connected.

I record all this because I now think of that summer as one of the most formative experiences of my intellectual life. The reading occurred, of course, against the dual background of the local intimacy that I had long cultivated and the fantasy of escape to a wider world that I had long harbored. But the experience gave me insights and resources that I have drawn upon ever since; it plainly underlies much of what I myself write about the circulation of capital and the spatial and temporal dynamics of global and local relations. The local and the global, as we would now put it, are two faces of the same coin.

A Love of Place and the Arrogance of Class

So my preference is to go places and just hang out. I did that on many trips to Sweden in the 1960s and in Paris for several summers in the 1970s when I was collecting materials and writing up on the historical geography of the Second Empire. Long Greyhound bus trips in the United States in the 1960s, with stopovers to visit with people such as Bill Bunge in Detroit (a very formative influence), or down into the depths of Mexico (where I designed the format of *Explanation in Geography*), were typical of my wanderings.

The formal world of academic life too often seems removed from such tactile experiences of the world. There is always a connection between what I write and what I feel, and what I feel depends on where I place myself and how I react to people and situation. To walk the streets of Baltimore or to talk with workers in a Burger King, for example, is to experience outrage at the waste of lives and opportunities, the patent injustices and stupid inefficiencies, the gross neglect that demands rectification. Experiences like that impel me to write. They fuel my academic rage.

Call it class envy, prejudice, or war, but Cambridge taught me about class in a way that I had not earlier experienced, for I had been raised in a fairly humdrum town where the main class distinctions were between professionals (mainly military) and respectable and dissolute working classes. My father's mother was the last of an aristocratic line that had fallen on hard times. She disgraced herself by marrying a professional naval man, but the aristocratic heritage was there. Like my father, I was prepared to accept some notions of aristocratic privilege until I got to Cambridge and felt its abusive qualities firsthand.

My mother's father, on the other hand, was of 'the aristocracy of labor'. Scottish by origin, he was a skilled worker in the Amalgamated Engineering Union. He said little and was quiet and dutiful – which was just as well because his wife was extremely determined, strong, and opinionated. She was the ambitious and pushy daughter of an agricultural laborer. She was also a strongly outspoken socialist who would shop only at the co-op. I recall her standing, in the middle of the Second World War, in the co-op denouncing Winston Churchill (our much revered war leader) as a 'rotten bugger' who cared nothing for the working classes. She responded to the surprised looks by admitting that Hitler was an even 'rottener bugger' and that it probably 'took one rotten bugger to get rid of another rotten bugger'. I evidently inherited some of her political rage. If you look at my class heritage as a whole, of course, only one significant class is missing: that of capitalist. I sometimes think I inherited anti-capitalism in my DNA.

Graduate Years and Changing Gears

I had acquired a Leverhulme scholarship to study in Sweden for a year, mainly on the grounds that the Swedes had better population data going back into the eighteenth century. When I arrived in Uppsala in 1960, I was unceremoniously dumped into a room alongside some strange bear of a figure named Gunnar Olsson – an event that both of us freely acknowledge was one of those fortuitous accidents that have long-standing consequences. Along with many others (Chorley, Haggett, Ullman, Garrison, Berry, Morrill, and Hägerstrand in leading roles), we helped to bend the structure of formal geography, against considerable opposition, to our collective will. The immediate effect was that I dropped the research project on Swedish demographics and just hung out all over Sweden, learning a tremendous amount about what it meant to live in a strange and foreign land, while retooling myself with all sorts of ideas and prospects for undertaking new kinds of research armed with different philosophical foundations and methods. This was the project that was to preoccupy me throughout my subsequent years at Bristol University, with a talented faculty that included people such as Michael Chisholm (in his sensible years), Barry Garner (a wonderful drinking companion), Peter Haggett, Allan Frey, and Mike Morgan. This project (aided and abetted by a year teaching in Penn State with Peter Gould as mentor) was to culminate in the writing of my first major book, *Explanation in Geography*, in which I sought to explore the rational and scientific basis for geographical knowledge by way of philosophy of science.

But I confess to being inwardly torn throughout much of this period. On the one hand, the political, intellectual, and hence professional project pointed toward the unity of all forms of knowledge under the umbrella of positivism and toward the rational application of such knowledge to the general task of social betterment. On the other hand, I still had that lust to wander and diverge, to challenge authority, to get off the beaten path of knowledge into something different, to explore the wild recesses of the imagination as well as of the world.

I turned in the manuscript in the summer of 1968 with near revolutions going on in Paris, Berlin, Mexico City, Bangkok, Chicago, and San Francisco. I had hardly noticed what was happening. I felt sort of idiotic. It seemed absurd to be writing when the world was collapsing in chaos around me and cities were going up in flames. The balance between active engagement and academic work is always tough to negotiate, and the whole issue still bothers me immensely. I try to retain an activist connection by attachment to social movements and some level of participation where I can. Such participation always remains an important source of inspiration, and I hope I can translate some of that inspiration into the world of academia.

Johns Hopkins and Baltimore

In any case, I felt a crying need to retool myself again, to take up those moral and ethical questions that I had left open in *Explanation* and try to bring them closer to the ground of everyday political life. In 1969, I was hired at Johns Hopkins in Baltimore to work in an interdisciplinary program dealing with geography and environmental engineering.

The internal attraction of Johns Hopkins was its interdisciplinarity. I thrived on it, much as I had as a graduate student. Figures such as Vicente Navarro, Rich Pfeffer, Nancy Hartsock, Donna Haraway, Emily Martin, Katherine Verdery, Ashraf Ghani, Alejandro Portes, and Neil Hertz (to name a few) became part of my intellectual firmament. The other attraction was my location in Baltimore, a city deeply troubled by social unrest and impoverishment, one of those cities that had gone up in flames the year before I arrived. I wanted to put my skills to work to try to deal with urban issues and to do so in a directly reformist and engaged way. My new department already had some research under way on inner city housing. I immediately became engaged in that work and, together with my first graduate student at Johns Hopkins, Lata Chatterjee, did some very detailed studies on housing finance and government policy in the city. This formed the empirical background to my thinking about urban geography in a new way.

All of this set the stage for the publication of *Social Justice and the City*, a book that contrasts what I called 'liberal' with 'socialist' formulations of urban issues. I began to read Marx seriously around 1970. The journal *Antipode* had been launched a bit earlier, and socialist, Marxist, anti-imperialist, and anarchist thought had an organized focus within the Association of American Geographers (AAG). Ben Wisner, Jim Blaut, David Stea, and Dick Peet at Clark University, along with many of their graduate students, formed the nexus, with the awesome but difficult figure of Bill Bunge always lurking in the background.

These were heady days of discovery for me, both in Baltimore and beyond. For once in my life it seemed that professional, personal, and political life merged into one turbulent stream of continuous innovation backed by revolutionary fervor and cultural power. Our research was dismissed as unreliable and irrelevant. The publisher's readers of *Social Justice and the City* called it incoherent and unreliable, and recommended its rejection. A key essay I did, 'Population, resources, and the ideology of science', was rejected out of hand by the *Annals of the Association of American Geographers* on the grounds that it had nothing to do with geography (it came out in *Economic Geography* instead).

By the time the battle was over, the opposition had conceded (though usually with multiple caveats) that the field of Marxist geography might be intellectually coherent, even empirically relevant, but refused to engage with it for reasons of politics.

Limits and Paris

I had discovered earlier that Henri Lefebvre had written on the urban question (I cited him briefly in *Social Justice*), and it seemed that the urban question was taken seriously in France. Manuel Castells had published *La Question urbaine* in French in 1972. I met and listened to him with great interest several times in the mid 1970s. He encouraged me to come to France and created an attachment in Paris. I got a Guggenheim to go there in 1976–7.

I came out of the French experience wanting to convert my attempt to construct a better urban political economy into nothing less than a project to overhaul all of Marxian theory in order to encompass historical and geographical questions more competently. I conceived and wrote *Limits to Capital* to remedy the problem.

If I began with heady arrogance, the writing of that book was terribly humbling. Crucial support came from Dick Walker (by then at Berkeley) and from Neil Smith (whose key thesis work, which later became the book *Uneven Development*, intertwined in all sorts of ways with my own project) and Beatriz Nofal (both then graduate students), who shared some of the agony. The more I worked at it the more complicated it became. I was desperate. *Limits* tested my own limits. It humbled me to write it. I knew then how much I could never know. Yet it was also a serious achievement. To my surprise and disappointment, *Limits* was neither widely read nor, as far as I could tell, influential with anyone very much apart from those specifically interested in geographical and urban questions. I then discovered the limits that disciplinary tribalism placed on free exchange. Economists would not take geographers seriously, and the sociologists had their world system theory, and so on.

Writing a book like that takes its toll. Fortunately, friends and colleagues dragged me into political activities. Neil Smith was particularly insistent in those years, frequently pointing out the pitfalls of lapsing into theory with no political praxis (I lost count of the number of picket lines he had me walking on)! I also became involved with solidarity work in Central America and spent time with a former colleague and his wife, Chuck Schnell and Flor Torres (she later became a special assistant to Daniel Ortega), doing support work for the Sandinistas (including occasional journalism with my then partner, Barbara Koeppel), operating out of Costa Rica. I witnessed there the devastating effects of US imperialism firsthand.

Ignorance of real geographical information (as opposed to competency in the techniques of geographical information systems) increasingly appears to me as a deliberate means for the prosecution of narrow and self-serving US imperial interests. No wonder geography is so marginalized and so badly taught in the US educational system! It allows a privileged elite to have its way with the world without any serious protest.

Oxford and Postmodernism

I was offered – to my considerable surprise – the Halford Mackinder Professorship of Geography at Oxford. For all sorts of odd reasons, I went there in January 1987, even though I was reluctant to leave Baltimore (by then my adopted hometown) and my salary was cut by half.

Above all, I think Oxford reminded me of my origins. The question of England and its relation to what had been Empire was still all over the place. I was forcefully reminded of what I had long ago rebelled against. The smugness of much of Oxford repelled me, and I could not understand how the left in Britain, within geography and without, was taking such wishy-washy positions in relation to Thatcherism. I wrote a few tendentious articles to that effect and got everyone annoyed. And then I launched in and wrote *The Condition of Postmodernity*.

Condition was the easiest book I ever wrote. It just poured out lickety-split without a moment of angst or hesitation (perhaps that is why it is so readable). I wanted to prove that Marxism was not as dead as some proclaimed and that it could offer some very cogent explanations of the dynamics then occurring (much as it had in the case of the Baltimore housing market in my earlier work). The argument worked well enough to provoke considerable and occasionally irate opposition among many postmodernists (particularly of a strongly feminist persuasion). Because I had been deliberately provocative, what should I expect? As time went on, it became clear that *Condition* was a forceful account to be reckoned with and that it had helped many people put a perspective on events that had hitherto been lacking. It became a best seller (widely translated into foreign languages).

But there was an interesting sidekick in all this because the kind of Marxism that was working here was quite different from that which had dominated in the early 1970s. Marxists as a whole had never taken very seriously questions of urbanization, of geography, of spatiotemporality, of place and culture, of environment and ecological change, and of uneven geographical development. *Condition* worked precisely because it took these geographical matters as central rather than peripheral to Marxian thinking.

Hope and Dreams

By 1993, I was back on the fall line in Baltimore. It was not an easy move. I went from a position of power in a large geography department to being a minor and marginalized figure in a department dominated now by engineers in a very corporatist engineering school that cared only about grants and sponsored research. I returned to Johns Hopkins under unfavorable conditions, though a position of sorts was found for Haydee, my wife (an oceanographer with strong training in fluid dynamics), and I had support from many individuals in other departments.

Though I did not know it, my own permanence was seriously threatened by lack of blood flow to the heart. I sometimes think that my articulation of dialectics as a relation between flows and permanences subconsciously reflected that condition! The five heart bypasses came just after I had done the index to *Justice, Nature, and the Geography of Difference*. For all of its lapses, I regard *Justice* as one of my most profound geographical works. The book does for me in a geographical way what *Limits* did for me from the standpoint of political economy.

Trying to keep the connective tissue of experience, of thought, of writing, and of just being in the world, all together is what life is all about for me. So I wrote a book called *Spaces of Hope*, which focused on possibilities and conversations about alternatives (the missing chapters of *Justice*). The book was inspired in part by the 'living-wage campaign' in Baltimore and Johns Hopkins (with some of my students heavily involved), and it used Baltimore as both foreground and backdrop for the exploration of utopian ideals. But it turned out to be my last Baltimore hurrah. Things just got too dismal for words in the department and the university, and when an attractive offer came to join the Anthropology Program at the Graduate Center of the City University of New York (with Neil Smith and Cindi Katz as colleagues), I jumped at the chance. It proved a shot in the arm and an opportunity to engage on an even broader canvas of thought and activity. The outcome has been a curious kind of closure of the circle, for I have now written a book on *The New Imperialism* which takes me from the moment of a fading British imperialism of my early youth to the sudden surge into an overt American militaristic imperialism of today. The book draws upon 30 years of teaching Marx's *Capital*, as well as upon many of my other previous writings, but also adds much that is new. But if it harks back to a sense of dynamics acquired by a graduate student in a long summer reading local press reports, and if it turns a certain imperial gaze acquired in the waning days of the British Empire upon the continents of knowledge as well as upon the contemporary world ... well, that is just how books get written!

I hope, because I have both the memory and the desire to change the world into a far, far better place than that which now exists. Life without hope is the death of desire. But in the world at large, it is things that count. The processes so fundamental to their production fade into nothing. We can hope, of course, that the things capture something about the joys and frustrations, the irritations and sublime moments, entailed in their production. But I regard my books as essentially dead things crystallized out of a continuous lived process of learning and exploration. Now I must let the text go to stand in the world as a fixed, static, and unchangeable document. But the dialectic of living does not stop here, neither for you nor for me. Dreams can come true, if only for that small segment of space–time in which we are able to sustain them.

16 EXPERIENCES AND EMOTIONS

Robin A. Kearns

The Lure of the Far Away

During my high school years, I had a prized poster pinned on my wall. While others had pinups of motorcycles or film stars, the poster I gazed at depicted American national parks. Despite spending a childhood in close connection to the spectacular coast and hills of the Northland region of New Zealand, I dreamt of visiting faraway places like Yellowstone and Jackson's Hole. This fascination for places with exotic names came from my mother, Heather. She loved the sight of planes overhead, joked that she was born with jet fuel in her blood and found joy in visiting far-flung places like Jasper, Alberta (knowing it from a John Denver song).

My parents had come to New Zealand as migrants from Britain in search of new experiences. The opportunities for my father, a veterinarian, were more challenging in remote Northland than in the manicured landscapes of Suffolk. They never returned to live there, although my mother quietly yearned for the glens of Scotland. That sea journey at age four perhaps sowed seeds of the geographical imagination for me, with stops in Curaçao, Panama and Tahiti.

Growing up in the regional city of Whangarei, we were never wholly at home. Bereft of soulmates my parents looked to the outdoors for connection, and my curiosity for landscape was roused. My father, a keen ornithologist, took me on avian census-taking exercises in out-of-the-way places, and, once my sister was a walker, we'd regularly go tramping in the rugged hills beyond the city. By teenage years I could identify most trees, rocks, birds and insects and, from my mother's influence, developed a love of the written word. While the science of physical geography turned me off in later years, those formative times gave me a love of the land and empathy for the natural world.

Getting There

The dream of exploring North America was partly fulfilled in 1977 when, having completed high school, I left for the US as a Rotary Exchange Student. I took flight with considerable relief having failed all but two of my final year public examinations. My highest grade was in English; the 39 per cent in geography reflected the absence of anything that interested me in the curriculum. Industrial Europe seemed distant and dull, and the only field trip that year was to our teacher's pig farm. There was more to the world than this and my protest was leaving the exam room early.

Once in the tiny town of Sardis, Mississippi, the world paradoxically became larger. I learnt a great deal about race, history and national identity. I attended a school that was 95 per cent African American. On the first day someone asked if New Zealand was 'up near Wyoming'. It rapidly became apparent that while school was a great cultural experience, travel would be a preferable educational option. I took up every opportunity. As my parents had hosted exchange students, I knew more about the programme than my local Rotary hosts. My local counsellor began to ask 'Where are you off to now?' whenever I phoned him. I reached 37 states that year along with a few of the national parks featured on that poster.

The wildest trip was accompanying a trucker who was driving to New York delivering hospital beds to Pilgrim State Hospital on Long Island, then one of the largest psychiatric institutions in the US. Seeing the huge hospital populated by distressed patients, and having hardly slept for three days, contributed to a lasting impression that may, at some level, have fostered a later interest in places of mental health care.

While America had attracted me at age 17, I would never have particularly chosen to spend a year in the Deep South. But, in retrospect, choosing the country but not the region was fortuitous, for that year on the edge of the Mississippi delta left a deep imprint.

Finding Geography

I returned from America and enrolled at the University of Auckland. My hopes to be a psychologist were dashed by finding that year one psychology was intensely tedious and technical. It had nothing to do with studying community dynamics as I had imagined. Feeding rats and being introduced to Van Morrison music by my lab partner were the only relief. I passed with C grades. But in geography courses, to my surprise, I gained A passes. In these, there seemed to be room to piece together my knowledge and experience of the world. In my second year I took English and began to regret not having taken it at the outset. By way of compromise, I returned for an MA and persuaded the Dean to allow me to do both geography and NZ literature. While English was first love, geography ultimately offered many more career opportunities.

Auckland had, and still has, a strong masters programme. For a thesis, my connections to rural Northland and the attraction of fieldwork on familiar turf led me to research the politics and land use change associated with a community irrigation proposal. The potential of the rich volcanic soils close to Whangarei for kiwifruit production was being recognized. The social as well as the spatial order was changing. It was a grand summer visiting farms on a borrowed Yamaha 50, enjoying my first taste of interviewing amid the volcanic landscapes of the Maungatapere district. That experience left me a legacy of interest in rural communities and a fascination for the conflicting interests of individuals, communities and institutions.

Towards the end of 1982, my thesis supervisor, Warren Moran, encouraged me to apply for a scholarship for doctoral study. Anne Buttimer visited and was similarly encouraging. I began to realize there was a world of geography beyond the thesis! Canada sparked my curiosity through having a geologist uncle in Toronto. The Commonwealth Scholarship application asked that six destination universities be listed in preferred order. However, Brent Hall, a kiwi who'd returned from doing his own PhD in Canada, said there was only one university worth going to: McMaster. I took his advice and then five months later received a telegram offering the scholarship. Having listed just one university, I can only imagine the adjudicators thought I knew exactly what I wanted. The opposite was the case. I knew next to nothing about McMaster, but couldn't turn down a paid passage to Canada.

Ontario in the Eighties

After I recovered from the shock of the mid August humidity, the next adjustment was to geography at McMaster. Gone was the land-based focus of Auckland in the 1980s. Instead I found myself within a hotbed of debate on urban and economic aspects of social life. I felt as if my carefully filled bucket of assumptions about the world was overturned. I struggled to find classes that were of interest. For the first semester, I took Ruth Fincher's urban political economy course and, for the first time, grappled with the deeper structures that underlie society.

These were days of feeling profoundly disconnected from known territory. I grieved proximity to the sea. Each Great Lake seemed like a great emptiness. Yet I was also deeply excited at being within a thoroughly international cohort of grad students. The degree of sociability between academics and grad students was particularly enjoyable. Grappling with structuration theory with Derek Gregory in Michael Dear's living room seemed a great deal more doable than reading the journals. Yet, I was glad to create networks unconnected to geography. Engaging with the Catholic left gave me new friendships (including my partner Pat who was beginning medical school) and insights into a variety of struggles within and beyond Canada. However, none of the struggles seemed to touch the concerns of aboriginal Canadians. So when a chance encounter led me to Catherine Verrall, a self-effacing Quaker who convened the local chapter of the Canadian Alliance in Solidarity with the Native Peoples, my graduate years were changed profoundly. A fascination with First Nations' spirituality and land struggles was nourished within an organization comprising of (at the time, radically) native and non-native people working together. It was a privilege to spend weekends on reserves learning of traditions and in dialogue with artists.

It all left me at a loss for a doctoral research topic. I fleetingly considered transferring universities, quitting even. The enthusiasm for matters Marxist and quantitative at McMaster left me as cold as the looming winter. The opportunity that gave me hope was Michael Dear and Martin Taylor's recently completed work on community attitudes to mental health care facilities. Martin, with whom I was taking a course in environmental perception, suggested I might examine community mental health care from the perspective of psychiatric inpatients resident in inner city boarding houses. Thus began a fascinating and challenging opportunity to enter, at least partly, the world of mental health care.

Influences beyond, as well as within, McMaster shaped my thinking during those years (1983–7). I found my way to a Centre for Ecology and Spirituality on Lake Erie and for four summers participated in colloquia facilitated by Thomas Berry, a sage ecophilosopher who described himself as a 'geologian'. At last I found *rapprochement* between the earth and deeper questions of meaning. Poetry also loomed large. I wrote and read as part of the Hamilton Poetry Centre and we hosted various Canadian poets passing through. The Toronto Harbourfront readings were a regular haunt and an opportunity to meet writers ranging from Janet Frame to Lawrence Ferlinghetti. All this, in combination, shaped my passions for good writing and the good of the earth. At times, my colleagues and supervisors at McMaster were puzzled. But for me geography would always be a matter of the heart as well as the mind.

As my thesis research unfolded, I spent time each week volunteering at a drop-in centre and would return with clothes smelling of smoke and my mind spinning. I experienced frequent attempts by droppers-in to convert me, or convince me of wild and wonderful things. Others just rocked in their seats. The university campus and the care centre felt like worlds apart. After the harrowing experience of comprehensive examinations (two eight-hour exams followed by an oral on the questions I chose not to answer), I was free to work on the thesis research.

Qualitative approaches were unheard of at McMaster at the time. I was flying by the proverbial seat of my pants, collecting narratives from respondents who surely thought my questionnaires were akin to yet another psychological test. Fellow graduate students provided important perspectives on my thesis work. Glenda Laws' knowledge of welfare restructuring and Susan Elliott's panache for statistical analysis were particularly helpful.

Another strong influence was Norman White, an intriguing psychiatrist whose socio-ecological ideas on health continued to inform my research for many years. Norman revelled in engaging with geographers who could think outside the medical model. Another influence was, and is, Michael Hayes who had returned to geography after a masters in epidemiology. He rapidly became a soulmate whose catch-cry of 'remember the gradient' served as an enduring reminder of the deeply structured opportunities for wealth, health and power in society. Michael's commitment to social justice, tempered by his irrepressible good humour and enthusiasm for good music, contributed to the enjoyment of my later Ontario years. As my PhD oral examination loomed, my supervisor posed the question: 'Do you see yourself becoming a "career academic"?' I recoiled from the affirmative: the 'publish or perish' adage filled me with ambivalence.

Back to the City of Sails

It was a weightless feeling that accompanied me back to New Zealand, suddenly with 'Dr' in front of my name. I'd received a Medical Research Council postdoctoral fellowship to undertake community mental health research in Auckland. Arriving back in my old department was cosy yet unsettling. Former teachers were now colleagues but none seemed to share common research interests. I set about trying to repeat my doctoral project but it felt a bit stale. I also seemed to be collecting rejection letters from my attempts to publish from my PhD.

People working in the mental health system were obliging and curious at a geographer's questions. The most surreal experience during that period was being put in touch with a mental health client who wished to be interviewed at work rather than the clinic. The address was on K Road, in Auckland's red light district. Feeling vulnerable at the prospect of going alone, I persuaded Pat to come along. Once within, and having run the gauntlet of the proprietor, we entered a room decked out in S&M paraphernalia. Our awkwardness must have been matched by that of the interviewee, given the unusual sight of an anxious young couple entering her room looking like health inspectors. That experience was one of a series leading me to an enduring interest in the ethics of social research.

During the time of my postdoc, Christopher Smith of SUNY Albany came to New Zealand on sabbatical and recruited my involvement in grant applications seeking to research relationships between housing and mental health. I warmed to Chris's humour, humility and working-class sympathies. Shortly before he left, the funding was approved, leaving me the daunting prospect of running a huge grant and employing a team of interviewers in two cities. It was a case of leaping into the deep end. The advice of my architect friend Tony Watkins, who had earlier encouraged me to go to Canada, rattled in my head: 'Bite off more than you can chew, then chew like hell.' It became a career motto.

Within our first year together in New Zealand, Pat had the opportunity of a medical locum in the Hokianga district, an area with which I had long-standing connections. In those glorious postdoctoral days of minimal teaching and no administration, I tagged along and concocted research on the meaning of the health system. To gauge the social significance of the community clinics I 'hung out' for a morning at each, noting the conversations around me, while pretending to complete the crossword puzzle in the daily newspaper. This made-up method, which might now struggle to pass an institutional ethics review, served to generate stark evidence of the place of these clinics in the social fabric of community. The importance of their form (as medical clinics) was clearly rivalled by their function (as *de facto* community centres). This impromptu project became a foundational research experience and the results, published in *Social Science and Medicine,* went on to assist the local community in resisting amalgamation and asserting the importance of its system to central bureaucrats. It served as an early and unexpected career satisfaction.

Placing Myself

Towards the end of the postdoc, I was offered a lectureship at Auckland and began to diversify my research. The key construct that began to inform my thinking was place as a recursive relationship connecting tangible location with experience and identity. Drawing loosely on comments in John Eyles' *Senses of Place* book, this idea subsequently helped to inform my understanding of a range of community studies and the associated narratives. Around that time, I was approached by colleague Steve Britton to contribute to a book on restructuring in New Zealand. My initial reaction was that I had little to contribute. Restructuring sounded too economic for me. Eventually I was persuaded to write sections on Maori housing initiatives and the Hokianga health care system. The resulting book, *Changing Places,* influenced my thinking and became a collective expression of New Zealand geographers' understandings of the radical shifts in policy sweeping the country in the 1980s. Some of the sadness of Steve's premature death in 1991 was allayed by a growing collaboration with friend Alun Joseph of Guelph with whom I have explored the links between rural health care services and broader restructuring processes.

After writing papers on home birth and rural health clinics, I was drawn to reflect on method and meaning in medical geography at the 1990 New Zealand Geography conference. Frustrated at the prospect of a long wait for publication in small-run conference proceedings, I took a chance and sent the paper to the *Professional Geographer.* To my surprise it was accepted. But reaction to my proposal for a reformed medical geography was quick. Two of the established names in the field, Jonathan Mayer and Melinda Meade, delivered a scorching retort to which I was offered the chance of rebuttal. I recall a senior colleague telling me, 'You've bitten off more than you can chew there, mate.' Throwing caution to the wind, I introduced more reasons why geography should escape the shadow of medicine and engage with a more progressive discourse of health and wellbeing. Postmodernism appealed to me for its legitimation of varied identities, its opportunity for eclecticism, and its challenge to the certainty of science. We needed to 'make space for difference', as I entitled a later essay in *Progress in Human Geography.*

This quest for difference followed me into research methods. I last analysed data using multivariate statistics in the mid 1990s and have since seen myself as a qualitative methods person. Creativity of method, for me, has led to an expansion of theoretical explanation. A key target for method has been talk. Place encourages, is formed by, and incorporates (literally and metaphorically) talk. A transdisciplinary conference in Auckland in 1996 on narrative and metaphor sent me back to old case materials, seeing new possibilities for analysis. At last my love of language could flourish!

Having been introduced to cultural geography when an undergraduate student as something exotic and largely found elsewhere, I found the reinvention of this field in the 1990s exciting. The local and political were suddenly in fashion. Getting to northern hemisphere conferences allowed me to make acquaintances with those whose work I admired. Subsequent visits by Sue Smith, Peter Jackson, Isabel Dyck and Graham Moon widened geographical horizons at Auckland, and in some cases led to new collaborations. A key influence in the 1990s was Lawrence Berg who was completing his PhD at Waikato and led me to think critically about 'race', place and power. My personal commitment to supporting Maori aspirations in health and housing was finding new theoretical foundations that could translate into teaching and research. While I have resisted self-identification as a 'critical' geographer (for me it risks being perceived as holier-than-thou), new cultural geography infused and enthused my thinking. This was also occurring for Wil Gesler of UNC Chapel Hill. Although we spent very little time together (I visited Chapel Hill once for two days) we ended up publishing two books together which, in combination, have assisted in widening the horizons of health geography through bringing together ideas about culture, place and health.

Research among Family and Friends

My research interests have frequently been associated with developments close to home. While some prefer to keep home and work life separate, I have never been able to create strong demarcations. A critical perspective on the world is simply a way of seeing, and social theory assists the clarity of vision. When our son was born, I built on earlier work on homebirth and received a grant to explore mental health issues before and after childbirth. Later when Liam was admitted to the local children's hospital, his linking of metaphor and place ('Dad take me to the Starship where the robot is') led to a series of investigations of the spaces of corporate health care with Ross Barnett of the University of Canterbury. As another long-term collaborator, Ross's knack of sketching the 'big picture' of politics and capital has complemented my eye for detail and understanding of the nature of place.

Latterly, with two school-age children, the role of primary schools as the heart of neighbourhoods has become acutely apparent. So has the way that traffic can bring congestion and threaten the health of that educational heart. At our local school, a small project questioning the transport preferences of children and adults in 1999 led to an ongoing collaboration with Damian Collins. The introduction of Auckland's first 'walking school bus' (a scheduled walk to and from school staffed by adult volunteers) soon followed. Within four years, over 100 such routes are operating across Auckland. There is great satisfaction in seeing research gain legs and political traction.

In a small country, networks matter a great deal. As in graduate days, I have been keen to maintain connections from beyond, as well as within, geography. A key involvement beyond geography over the last 16 years has been the NZ Public Health Association. Invariably the sole geographer participating in annual conferences, I've become known and can now take students to a region of the country, locate a member of the PHA and very soon have a dialogue going on local health and community concerns. Getting away to be refreshed by the buzz of the bigger world offshore is another key element to being a geographer in a small country. For maintaining wider networks, the two-yearly international symposia in medical geography have been crucial and far more satisfying then the grand scale of general geography conferences.

My involvement in multiple networks helps to offset my discomfort at the prospect of being regarded as a specialist. The generalist in me wants to keep moving on. I am currently being stretched by new research on primary care reform, tuberculosis, and sustainable settlement. In all three projects, I am the sole geographer. While I continue to write with, and enjoy the friendship of, a dispersed network of geographical colleagues, my local research is increasingly transdisciplinary. This move has also occurred within teaching. While my enthusiasm for teaching cultural and health geography is undiminished, the opportunity to teach within public health programmes brings the satisfaction of seeing non-geography students realize the important of space and place.

As I await a bus home from the university, I often see a sociology professor who tells me that geographers spread themselves too thinly. To an extent, he's right. There's a lot of ground to cover. In the course of my work, I repeatedly defy expectations of what geographers do. From early years, I was driven to look out on, yet be involved in, the world. There are few limits to the scope of our research commitment. As I've acknowledged, we are a magpie discipline – picking up on ideas, opportunities and attractive theories. But for me, this breadth of scope is geography's attraction. In biting off more than I can chew, I've learnt to rely on, and learn from, friends and colleagues to help digest the data and complete the picture. The heart of place can be appreciated, but not understood, alone.

17 PERSONAL AND POLITICAL

Vera Chouinard

How does one end up doing geographic research from one philosophical and theoretical position rather than another? How does what happens in one's life in particular places and times alter one's approach to understanding geographic phenomena such as the development of particular cities or urban areas?

In this chapter I outline some of the intellectual, personal and political forces that have shaped my philosophical and theoretical approaches to doing geographic research. In doing so, I hope to show that becoming a particular type of geographer involves much more than learning a particular approach from books. Rather it is an often messy, sometimes confusing and certainly contested journey through learning which is as personal and political as it is intellectual.

Early Days: Learning at Daddy's 'Knee'

I have always had a fascination with power: who has it, who doesn't, and what differences this makes in people's lives. This is one of the earliest signs that I might eventually become radical and feminist in my thinking about the world and people's places in it. I am not sure where this came from but, looking back, I am sure that it was nurtured in places of childhood in which I grew up with a father who enjoyed almost absolute power and authority in our home. Challenging his authority was a rare and risky business, resulting in corporal and psychological punishment. Add to this the fact that my father was a large, imposing man and it is no wonder that I grew up in fear of him. Life with father was especially stormy during my teenage years as I began to rebel against adult authority. One particularly vivid set of memories is when I would come home from school eager to discuss issues such as pollution and poverty – trying to express new-found opinions about difficult issues of the day – as we sat around my father's dinner table. As I sat down and began to speak, my father would angrily insist that the only opinions I was 'allowed' to have on such matters were the same as his own. If I disagreed, which I almost always did given how different our politics were, then down his fist would come upon the dinner table sending the peas and usually my entire dinner plate hurtling off the table while he shouted, 'I'm right and you're wrong!' While my mother rushed to pick up broken dishes and pacify my father I would lapse into a temporary angry silence, only to come back stubbornly another day to try to say what I thought despite my father's anger and intimidation.

What I learned at my father's 'knee' and supper table was that men had the power to oppress women – determining how they lived, particularly in places under their control such as the home, and even what they could and could not say and think. My fascination with the causes of such inequalities in power grew although it would be a long time yet before my studies and life more generally began to give me answers to why such inequalities persisted.

Learning Geography Positivist Style

As a young woman who had been regularly silenced and intimidated in her home, I was looking forward to the freedom of inquiry and debate that university life promised. Little did I imagine then that universities are also places where knowledge is deeply and sometimes violently contested; that knowledge is itself a potent form of power and something that can be used in oppressive as well as liberating ways.

It is probably fair to say that, for most of my undergraduate days, I remained largely unaware that social scientists disagreed about the philosophical and theoretical perspectives one could use to help develop explanations of phenomena. As a psychology major, I discovered that we had theories about how perception worked or about the inner psyche which we then tested deductively against empirical data. When a course in Third World development persuaded me to switch to major in geography (unlike my other geography courses, here we looked at inequalities in power, for example between First and Third World nations), I learned to explain geographic phenomena in similarly positivistic ways. I was in a geography department in which positivism and micro theories of urban and regional development held sway (i.e. theories which attribute phenomena such as suburbanization or gentrification to micro-level causes such as individual economic utility maximization) but were seldom if ever compared to other rival philosophies of science and theories of sociospatial change. Still, these perspectives informed everything we did. No wonder, then, that when I was asked to read David Harvey's *Explanation in Geography,* an extended account of positivistic explanations in geography, for a geographic thought class, I read it not as a discussion of one possible philosophical approach to geography, but simply as *the* way scientific explanation was done. Nor is it any wonder that, while learning micro theories of urban and regional change, on some level I kept feeling that issues of social power were being left out.

Radical Beginnings

It was at the masters level that I began to question, at least implicitly, positivistic approaches to geography and to realize that geographers disagreed, sometimes vehemently, about the theories they should use to help explain phenomena such as inner city or regional decline. My long-standing interest in issues of social power was leaving me dissatisfied with the micro-economic and behavioural approaches to explanation popular with many instructors at the University of Toronto. Yet, like most students, I was uncertain what to do about this. Although one or two human geographers located in this department 'dabbled' in radical theory and debates, such as Manuel Castells' writings on cities as sites of consumption in advanced capitalism, by and large geography was a conservative department. For the most part, then, the increasingly influential subdiscipline of radical geography was pretty much passing this department by.

As a result, faculty in the department did not encourage students to explore more radical approaches to geography. The almost entirely male faculty also had a strong sense of hierarchy and of privilege when it came to telling students how we could and couldn't do 'good' geographic research. I discovered this, for example, when dealing with one professor who insisted I would like approaching my work from a micro-economic modelling perspective if only I'd give it a serious try. I went along far enough to read important works such as Rawls' theory of justice. I also worked on my multivariate statistical skills but found myself more fascinated with the many problematic assumptions on which such ways of manipulating and explaining data are based than I was by any apparent correlations, spatial or otherwise, that I found. Perhaps micro individual-level theories of urban and regional change, and multivariate statistical analysis, were just not for me!

As my frustration with these approaches to geographic research grew, so too did my interest in finding out what 'radical' geography was all about. I had overheard a PhD student discussing his work with a radical planning professor with another student in the hallway one day and decided to approach him. I remember being shy and having a sense that this subject was 'taboo' when I stopped him and asked him what this was about. For whatever reasons, he was reluctant to say much about it: a word or two and he was on his way. So I decided to find out for myself. Since I didn't yet know many geographers working in this area I began by reading a key source, the writings of Karl Marx. It was a fascinating if lonely intellectual journey. But the more I read the more I understood that Marxist theory argued that it is macro-level differences in power in society, notably differences in people's locations within the class structure of society, which are the primary reason that some people and places thrive while others struggle to barely survive. It was, for example, sudden and rapid shifts in capital investment into and out of places such as urban neighbourhoods or one-industry towns, I realized, which determined who lived in these places and how (e.g. wealthy or poor, homed or homeless, employed or unemployed). I began reading more recent Marxist work, much of it inspired by French structuralist thinkers, about questions such as why urban environments were being redeveloped for more affluent citizens while the poor were losing their homes. Gradually, and despite the sometimes impenetrable language, the reasons cities and regions were changing in ways that gave rise to problems such as homelessness, deindustrialization, and exploitation of Third World workers began to make more sense in a macro process (cf. individual causation or statistical correlation) sense.

Even at this relatively early stage in discovering more radical ways of understanding society and space, in hindsight my new philosophical and theoretical directions were changing my approach to understanding the world as well as my life and politics. Consistent with the historical materialist philosophical tradition, I was becoming more interested in developing a historical understanding of processes of urban and regional change – one which attended to changing conditions of material life in society and space. My life and politics were also changing. I was becoming much more alert, for example, to the ways in which academic environments were shaped by disparities in power. Not only was power at stake in struggles over what did and didn't count as geographic knowledge, but our working lives were shaped by our different locations in wider social structures of power. As graduate students working as teaching assistants, for example, we lacked job security, benefits, and the power to resist exploitative working conditions. As I began to make these connections between radical theories of power and my own life, I started to become more active on issues affecting relatively marginalized groups (something that characterizes my work as a geographer to this day). So, for example, I agreed to serve as a vice-president of my teaching assistants union and learned about other aspects of struggles over power at universities: that we couldn't talk openly on the office phones because they were monitored by RCMP officers keeping tabs on suspected 'radical students', and that universities were not collegial communities when it came to bargaining and labour matters – especially with relatively powerless groups such as graduate students. Of course, we were privileged in the wider society in which it was difficult to afford university education, and those who could were more likely than others to become members of what some scholars termed an elite core of the labour force. I was awakening to the political realities of power and oppression and the contradictory ways that I and others found ourselves situated within relationships between the more and less powerful in society and space.

It had also become clear that the Toronto geography department was not a place in which my emerging interests in radical geography could flourish. Fortunately, nearby McMaster University had attracted a small group of radical geography faculty (Michael Dear, Ruth Fincher, Michael Webber) and in September of 1981 I went to work on my PhD with them. It turned out to be a wonderful place to study more radical approaches to the discipline – with lively seminars and interesting graduate students to share ideas with. The early 1980s was a time when Marxist approaches to urban and regional change were flourishing in geography but also coming under attack by proponents of other philosophical and theoretical approaches. I recall being infuriated, in particular, by humanistic critiques which alleged that Marxist approaches to geography were necessarily structurally deterministic and hence neglected the role of human agency and struggle in social change. And I wrote a reply to one such paper (with Ruth Fincher) pointing out that more causally complex Marxist explanations of social and spatial change treated social structures as limiting but not determining outcomes which were also shaped by human agency (Chouinard and Fincher, 1983).

What I didn't realize at this time was how hotly Marxist and other radical approaches to geography were being contested within geography departments such as ours. This is because it was the professors who were training me who were on the 'front lines' of conservative backlashes against approaches challenging more conventional, positivistic approaches to the discipline. Nonetheless, this backlash also found expression in my work. In 1984 I published, with Ruth Fincher and Michael Webber, a paper entitled 'Explanation in scientific human geography' in which we explained how a postpositivist realist philosophy of science provided the basis for the rigorous, scientific testing of hypotheses in Marxist explanations of social and spatial change (Chouinard et al., 1984). At the time, I attributed my interest in writing this piece to my long-standing interest in issues in philosophy of science. I realized later, however, that it was also a rebuttal to faculty in our department who continued to insist that there was only one scientific way to do human geography and that Marxist geography was by definition unscientific – and an attempt to encourage others to widen their horizons and admit that one could do rigorous and important work from diverse philosophical and theoretical vantage points.

The topic I chose for my PhD research was the struggle for cooperative housing for low-income people in Canada. I was interested in the role of the capitalist state in regulating struggles for alternative housing (i.e. decommodified) and the extent to which grassroots groups were able to achieve progressive ends within the limits of state policies and procedures. In accord with a complex causal approach to using Marxist theory (see Edel, 1981), the state was conceptualized not as a structure which determined outcomes in cities and regions (e.g. particular types of social housing projects) but as a 'terrain of conflict' over the power to determine how low- and moderate-income people would be housed (see Chouinard and Fincher, 1987). Finally, I was able to use a theoretical approach which allowed me to ask questions about how struggles over power were played out and who 'won' and 'lost' as a consequence!

Among the many memorable moments from the days of working on my dissertation, one stands out in terms of influencing my perspective on the politics of doing geographic research. This was my first meeting with a leading activist in the Canadian cooperative housing movement. We were having lunch on Queen Street in Toronto and I was hoping to persuade him to help me network with the activists and policy-makers who would help make my project a reality. Our conversation turned to the question of why he should help me get this research project started. I remember saying something about how it would help document not only policy changes over time but also the history of struggles for cooperative housing in places such as Toronto; thus adding to our knowledge of the movement and its outcomes. 'But,' he cut in with his fork stabbing the air for emphasis, 'We already know all that'. As I sat there, ego quickly deflating, I realized he was absolutely right. As a researcher I might at best pull pieces of the co-op housing story together but it was, after all, *their* story; that is, the story of the people who had been fighting at the front lines for a non-inflationary housing alternative. What can and should researchers contribute beyond a simple retelling of the stories and knowledge of others? How can we do radical research in ways that empower the marginalized groups we work with? While I did not then nor do I now have all the answers to these difficult questions, I do know that it is important that we continue to ask them of any geographic research we do. For they remind us that there are crucial gulfs and power imbalances between the world of academic research and the real-world struggles for social change with which we seek to connect. And that to be a radical geographer is in part to be caught up in the persistent contradictions of activism in and through the academy.

Out of the Frying Pan and into the Fire? Life as a Junior, Radical, Female Professor

My PhD studies were drawing to a close and I would soon find myself facing the challenges of being a radical member of the academy. When I was hired into a tenure-track position in the same department where I completed my PhD (those who had trained me had left) I faced a situation in which I was the only radical, Marxist professor in the department and the only woman. Unfortunately, these differences would become bases upon which colleagues marginalized myself, my work and my students. Being marked out as a negatively 'other' or different member of a department (see Kobayashi, 1997 on processes of differencing) is a gradual process – one that creeps up on you through mundane everyday experiences. I recall being initially puzzled and frustrated, for instance, as student after student ended up in my office telling me that although they were interested in doing geographic research from radical perspectives, other professors were advising them against this because 'it wasn't real research' or 'it wouldn't get them a job' and so on. I would counter such narrow-minded advice by reminding students that universities were, at least in principle, places where diverse perspectives and approaches to research can and did thrive; after all it was debate across a range of perspectives that advanced knowledge, and not everyone doing and thinking the exact same thing!

The difference that being a woman made also became clearer and clearer. Whereas at the graduate level, I had naively assumed that we women were 'equal', as a junior faculty member it was very clear that we were anything but. At a personal level, the signs ranged across everything from male colleagues reminding me that if my career 'didn't work out' I could always 'stay home with the kids', to inappropriate comments on my appearance and child-bearing intentions, to being ignored when speaking in faculty meetings (unless a male faculty member reinforced what I'd said as important and deserving a response), and to being professionally harassed. The latter was especially painful and distressing – damaging my health and my ability to do my job.

Other events made it clear that the kinds of personal negative differencing I was experiencing were part and parcel of systemic discrimination against women in geography. I recall quite vividly for example the first faculty hiring process I was privy to after becoming a junior professor. What was striking about the process was the way in which female candidates were evaluated by very different and inappropriate criteria in comparison to their male counterparts. The latter were assessed by the usual standards of academic merit (e.g. number of publications, grants, special awards) while the former were assessed, quite oddly it seemed to me at the time, by personal circumstances which 'might' affect their working lives (notably whether or not a woman had a male partner in another city and would therefore commute to her job and potentially spend less time on campus). The gendered inequities in such evaluations are clear: candidates ought to be assessed by the same criteria based on ability to do the job.

Through experiences such as these I found my feminist consciousness gradually awakening as was my sensitivity to the role of differences of gender in geographic change. Feminist colleagues and students also nudged my thinking along and encouraged me to incorporate struggles over gender relations and roles into my work on changing geographies of the local state (e.g. Chouinard, 1996).

Looking back on my early days as a lone radical, female geography professor in a largely conservative department, I think it is fair to say that I inadvertently became a lightning rod for whatever misimpressions about radical, Marxist and feminist geography, and about women in academia, persisted in this place. It was a hostile, lonely and difficult place to work for myself and the associated stress took its toll on my health and on the wellbeing of my family.

Disability and Differencing

In 1990 I was diagnosed with rheumatoid arthritis, an incurable immune-system-based illness characterized, in my case, by severe unremitting inflammation in all the joints in the body. This illness was a devastating personal blow to someone in her early thirties with a new career and a young family. It was also, as it turned out, a devastating professional blow because it marked the beginning of a long and gruelling struggle for accommodation of my needs as a disabled professor. Under Canadian human rights laws, employers such as universities are required to accommodate disabled workers' needs to the point of undue hardship (usually defined in terms of prohibitive financial costs). Too often, however, as in my case, legal rights enjoyed in principle are difficult to win in practice. Further, my disability and struggles for accommodation also helped to mark me out as even more negatively different than I'd become as a radical and feminist professor.

My struggles for accommodation covered virtually every aspect of my work as a disabled professor – from the need for an accessible parking spot adjacent to the building I worked in (an eight month fight) through to the need for ground floor and elevator access to the office in which I worked (over two years), to the fight against discriminatory ways of evaluating my job performance and paying for the work I did (over a decade). These and other struggles also helped to open my eyes to the significance of diversity in radical and feminist geography, to the need to open up categories such as 'woman', 'child' and 'man' to diversity in where and how people live – as disabled or non-disabled, heterosexual or homosexual, for instance (Chouinard and Grant, 1995).

They also made me aware of the many social and spatial barriers that disabled women, children and men face in the fight for inclusion. Personal experiences, such as having to take legal action against my employer and fight for over 12 years for accommodation at work, have taken on political meaning as I've come to realize, through disability research and activism, how many disabled women, children and men fight, day in and day out, to have opportunities that the non-disabled often take for granted. Personal moments, such as the time I was forced to stand in a conference session in excruciating pain because no one would give up a seat for me even though I begged them to do so, spring to mind as I watch physically impaired women struggle to get into a room without an automated door or a facially disfigured child retreat from the taunts of other children into isolated play. The personal is deeply political – connecting us in our marginalization, our sense of outrage and our diversity.

Geography Lessons: Connecting the Personal and Political with the Philosophical and Theoretical

Contemporary geography has come to encompass a rich diversity of philosophical and theoretical perspectives. Here I've talked about becoming a radical, feminist and disabled geographer and some of the things I've learned along the way. Among these are how impossible it is to disentangle who we are and how we do geography from where we've been and how we've been caught up in processes of differencing. This is perhaps easiest to see from the margins – from such vantage points as being the only woman faculty in a department or the only visibly disabled professor for example. It is harder to see from vantage points of privilege because of the normality and taken-for-grantedness that attaches to, for instance, being white in places of whiteness. Yet, it is vital that we learn to make such connections between how and why we do geography the way we do – not only so that we can respect and learn from diversity in philosophical and theoretical ideas but also so that we are aware of people and perspectives who remain 'outside the project' of geography. By this I mean groups such as the disabled whose vantage points and expertise promise to teach us a great deal about how the world works and for whom, and about what needs to change if we are to work together towards a more inclusive world. By pushing the borders and boundaries of the geographic project in such ways, we open our discipline and ourselves to new and exciting ways of understanding and changing the world we share.

References

Chouinard, V. (1996) 'Gender and Class Identities in Process and in Place: The local state as a site of gender and class formation?', *Environment and Planning A*, 28: 1485–506.

Chouinard, V. and Fincher, R. (1983) 'A Critique of Structural Marxism and Human Geography', *Annals of the Association of American Geographers*, 137–45.

Chouinard, V. and Fincher, R. (1987) 'State Formation in Capitalism: A Conjunctural Approach to Analysis', *Antipode*, December: 329–53.

Chouinard, V. and Grant, A. (1995) 'On Being Not Even Anywhere Near "the Project": Ways of Putting Ourselves in the Picture', *Antipode*, 27(2): 137–66.

Chouinard, V., Fincher, R. and Webber, M. (1984) 'Empirical Research in Scientific Human Geography', *Progress in Human Geography*, 8(3): 347–80.

Edel, M. (1981) in Scott, A.J. and Dear, M.J. (eds), *Urbanization and Urban Planning in Capitalist Society*. London: Routledge.

Kobayashi, A. (1997) 'The Paradox of Difference and Diversity (or, Why the Thresholds Keep Moving)', in John Paul Jones III, Heidi J. Nast and Susan M. Roberts (eds), *Thresholds in Feminist Geography*. Lanham, MD: Rowman and Littlefield Publishers. pp. 3–9.

18 DIFFERENCE AND PLACE

Linda McDowell

The opportunity to reflect – in print – on one's own academic life, research interests and preoccupations is both flattering and intimidating. I realized as I sat down to think through this comment that I had been continuously employed in academic life for just about 30 years and so I am entering the last quarter of my career. I'm a representative of that fortunate generation of British women, born in the postwar years, helped into a good and free education through the grammar school system that was in existence in England and Wales in the late 1960s, and then supported through university by what now seems the almost unimaginable largesse of the state, committed to the financial support of students in higher education – but of course only an elite few. When I went to university in 1968, less than 8 per cent of women in my age group had the same opportunity. It is a significant mark of progress that now over 40 per cent of all the relevant age cohort in Britain enter higher education, as many if not more young women than young men, although the transfer of costs to individuals and their families is regrettable. But these women of my age, although relatively few in number, have proved to be a significant lot – at least as far as the development of the second wave women's movement and feminist theory and scholarship is concerned. And this is the focus of my paper, because, above all, my work has been influenced by and set within the huge and exciting flowering of feminist-inspired work within and outside geography. I began my academic life as an urban geographer and slowly metamorphosed into an economic geographer, although as I argue below, and as feminist theorists have long insisted, the separation of work from home, the urban from the economic, or daily life from working life is analytically unsatisfactory, challenged by the evident connections that are held in place, in the main, by women's domestic labour. A focus on the connections between what were in general defined as separate spheres by geographers is the connecting thread in my published work over the decades.

Before I explore these connections and my changing emphases, I want first to emphasize that academic work is always the product of collaboration. Even those papers and book attributed to a single author are located in academic and policy debates between near and distant scholars, theorists and practitioners, greatly facilitated now by email and the explosion of new journals on the web. I have been fortunate to work with and to be influenced by conversations and collaborations with a significant group of feminist geographers and others working within a largely leftist critical social theory approach from the late 1970s onwards. In the immediate circle of geographers, in at the beginning as it were of feminist work in our discipline, were Jo Foord, Jane Lewis, Jackie Tivers, Eleanore Kofman and Sophie Bowlby in the UK and Damaris Rose and the late and much missed Suzanne MacKenzie in Canada. These conversations spread in part through the establishment of a women and geography working party in the IBG (later the Women and Geography Group; see Women and Geography Study Group, 1984; 1997), to include, among others, Melissa Gilbert, Susan Halford, Michelle Lowe and Gill Valentine. All these scholars enriched my understanding of urban issues, as did a wider group of feminists, including Mary Evans, Nanneke Redclift and Clare Ungerson, then at the University of Kent at Canterbury which was where I began my academic career. Indeed, one of the first explicitly feminist papers I produced (McDowell, 1983) had its origins in a conference paper given at UKC. I remember it well because I had just discovered that I was pregnant. And pregnancy, birth and childcare are the other events irretrievably bound up for me with my writing and teaching, influencing not only the time available but my understanding of the constraining influence of the built environment and the spatial segregation of different types of urban activities as well as gendered differentiation in the workplace.

The issues about negotiating the built urban fabric and patching together services separated in time and space that I addressed in an academic sense in that 1983 paper took on a new meaning and materiality for me as I changed status from a mobile young professional to an encumbered mother of two, or rather as, again with the help of numerous others, I tried to combine the two, struggling against both the tyranny of the built environment and the diurnal patterns of university timetables. I reread Hägerstrand's work on the constraints of time and space with a new understanding. I was not alone in combining employment and motherhood as numerous feminist friends were involved in the same combination of what was still then, even in the 1980s, referred to as 'women's dual roles'. And, as has often been noted about the writings of the second wave feminists, the changing concerns of that discourse reflected our/their own life course, as emphases shifted from the domestic labour debate, through childbirth and childrearing to 'the change' and ageing. My most recent papers (McDowell, 2003a; 2004) are based on fieldwork I have been undertaking with elderly women, born about the same time as my own mother. In the years between I have undertaken a study of middle-class women and men working in a pressured environment – merchant banking (McDowell, 1997), not universities, but there are parallels (a point Doreen Massey, 1995 has also made in her work on high-tech industries) – and more recently on the social construction of masculine identities among young men affected by deindustrialization and facing working lives in feminizing service sector occupations (McDowell, 2003b). In both pieces of work, there are clear connections with and reflections of my own career as a professional worker and as a mother of a son.

One of the great joys, of course, of being an academic feminist theorist is being able to draw on personal experiences while analysing the sociospatial construction of gendered identities. This very strength is also a weakness, however, as the bitter debate in the mid 1980s about the white middle-class heterosexist emphasis in both the theoretical and the political demands of the women's movement were challenged by women of colour and lesbian separatism. Throughout that decade, as the diversity rather than the commonalities of women's lives demanded both theoretical and empirical attention, the authority and right of academic feminists, largely white and middle class, to speak for others was challenged. In part as a response to this recognition of diversity but also connected to broader theoretical shifts, in the 20 years or so since I have been publishing feminist work, both feminist and urban theory have changed, expanded and developed in numerous ways (see for example Fincher and Jacobs, 1998). In the early 1980s, for example, urban theory was beginning to emerge from what now seems an unnecessarily restrictive straitjacket of searching for its own 'theoretical object'. Influenced by Althusserian Marxism as interpreted by Manuel Castells (1978), many geographers and sociologists had defined their object of analysis as the provision of the goods and services of collective consumption, regarded then, before full-blown Thatcherism and Reaganism showed how wrong we were, as essential for the maintenance and reproduction of cities.

Feminist theorists, in similar theoretical blinkers, were at the same time asserting the necessity of domestic labour in the maintenance and reproduction of men, children, everyday life and the capitalist system, before the advent of fast food, massage parlours, shirt pressing services and singles bars put paid to this argument too (Ehrenreich, 1984; McDowell, 1991). Further, the increasing mobility of international capital demonstrated that the reproduction of particular working-class families in particular places was barely of any concern at all. And yet at the time this joint focus on collective and household maintenance services was productive. Even though most of the renowned urban analysts of the era tended not to have noticed, it was women's role as key providers of both types of service – both as employees of central and local government services and as housewives and mothers in millions of individual homes – that kept cities running. Throughout those years, however, feminist arguments about interconnections and complexity were generally ignored within our discipline. But it might also be argued that the other social sciences, in their turn, ignored the significance of space and place in understanding the nature and distribution of economic and social life. Neither of these assertions stands up today. There has been an exciting explosion of work across the social sciences focusing on the significance of spatial difference and its constitution within and across a range of different spatial scales: a point I shall return to in a moment. First though, I want to pick up again my point about the significance of diversity by returning to that 1983 paper about the gender division of urban space.

One of the most evident signs of the time when that paper was written lies in its singular focus on 'women'. Here both feminist theory and urban studies have changed. More recent work on the rich diversity of urban life and the experiences of women from different class backgrounds, ethnic and national origins, sexual identities, abilities, ages, and household circumstances are submerged in this singular term. The focus was on 'women', I think, not from ignorance – as a rich strand of sociological and historical work by feminists, as well as, of course, that long geographical tradition about residential segregation and the city, made it abundantly clear that the city was an arena of great diversity – but because the theoretical tools to link this recognition to an explicitly feminist geographical analysis had not yet been developed. As I wrote in the early 1980s a singular discourse about women's oppression and the necessity of equal treatment informed our work. Neither the 'equality/difference' debate (Phillips, 1987), now subsumed within a wider debate about the significance of redistribution and recognition in claims for equality (Fraser, 1997), nor the shift in emphasis towards diversity in feminism and the particularity of place in geography, was yet well developed, as older theoretical traditions emphasizing regularities, even a search for spatial laws, or at least explicitly spatial processes, were still evident. Thus, my 1983 paper included a swipe at locational analysis, as well as, as I have already suggested, its positioning within a materialist approach. Nevertheless, the arguments about the familial and gendered assumptions made concrete in urban form were important and influenced a rich set of empirical studies analysing, for example, the connections between the growing numbers of single women, sole mothers and dual-career households and gentrification: not just a 'rent gap' nor a lifestyle choice but a response to the changing position of women in the contemporary cities (see for example Bondi, 1991). A parallel set of analyses about the connections between urban form and sexuality have also drawn on and extended the earlier analyses of urban structure (see for example Valentine, 1993).

Perhaps one of the most important shifts in feminist work about the connections between gender and urban form, as well as the move to analyse the role of state institutions as elements in the oppression of women, has been the much greater recognition of ambivalence and contradiction and the multiplicity of women's lives and geographies (Women and Geography Study Group, 1997; McDowell, 1999). The early second wave women's movement was, in retrospect, perhaps rather dour and serious in its initial theorizing about, for example, the labour theory of value and its connection to domestic labour or its insistence on the oppressive nature of marriage and the nuclear family. Thus, what was absent from my paper was any awareness of the contradictory role of the home and the family in many women's lives, as a place of hard work and inequality undoubtedly, but also as a place of love and joy, the location of rest and recreation as well as respect and pleasure. While feminists of my generation wrote rather dismissively of women 'dusting their lives away' or of their false consciousness if they promised to obey on marriage and to honour a man by working unpaid in his home, we forgot that the home is also a private site of intimacy as well as, as bell hooks (1991) reminded us, a haven for some of the most exploited women, whether working class or members of a minority population, and a place from which political protest and resistance to the norms of what hooks termed the white heteropatriarchy might be organized. Thus Jane Humphries (1977), in an early and controversial paper about the domestic lives of working-class women in the nineteenth century that received a rather hostile reception, insisted on the rationality and the mutual support that lay behind a conventional gender division of labour in working-class families, as well as on its basis in female exploitation. Her analysis has worn well. The general assertions in my paper now seem more particular: the demands of white middle-class women for equality in the workplace and in the home loom large but unexplored in their background.

My paper also reflects the emphasis on material conditions then dominating feminist theorizing. Two decades on, as I wrote a new paper on the city for a book on feminist theory (McDowell, 2003c), I was struck by my current focus on issues of meaning and representation as well as much greater attention to the significance of different spatial scales in the social construction of meaning. In recent work on cities and on feminist theorizations of urban space, the focus moves both up and down spatial scales as well as, of course, emphasizing their interconnections. We now recognize that place, the city, is a locus in a network of interconnected sociospatial process in which the 'local' and the 'global', and any scale in between, are mutually constituted, as Doreen Massey (1984) insisted. I was fortunate to work with her at the Open University for a decade and her analysis of spatial divisions of labour and the constitution of space, place and scale has had a lasting impact not just on my own work but on the whole discipline. Thus, current feminist work in and on the city, including my own, might, for example, discuss the meaning of the 'home' at the smallest spatial scale of room layouts within a dwelling, as well as address the connections between the 'homeland' and the nation and idealized constructions of a particular version of nationalist femininity, reflected both in national myths and in everyday lives in the home. In this recent work, in which alternative versions of the home are imagined, constructed, and connected to place and to territory in different ways, as well as struggled against and resisted, a far wider range of sources than was common 20 years ago is drawn into analyses.

In common with the wider 'cultural turn' in the social sciences, feminist geographers now place greater emphasis on questions of meaning and representation as well as draw more readily on artistic and literary sources in explorations of the connections between femininity and urbanity, on the gendered meanings of urban space as well as on women's lived experiences, in unpicking the links between material inequalities and the cultural meaning of difference in understanding still persistent gender divisions. In this work there has been a productive coincidence of effort by feminist geographers and by social philosophers. Here I think of the recent influence of the US theorists Iris Young (1990) and Nancy Fraser (1997) on my own and others' analyses of the changing nature of inequality and difference, as well as Judith Butler's (1990; 1993) influential analysis of gender as a performance, and more recently perhaps Martha Nussbaum's (2000) extension of Amartya Sen's concept of capacities and capabilities as a way through the complex discussion about social justice and equality in a postmodern theoretical landscape. The last few years have been a demanding but extremely productive period in the development of critical feminist thought.

It is also beginning to be an extremely productive period in urban theorizing. It seems to me that the late 1980s and early 1990s were perhaps somewhat an era of marking time in urban theorizing in general and feminist geographical work on the city, despite the undoubtedly significant work that was undertaken in mapping and measuring the extent of the differences in women's opportunities and lives from men's in particular cities, in the main I think because the vexed question of how to define 'the urban' continued to haunt many urban geographers. What has been so interesting about recent work – and surely it is significant that most theorists now talk about 'cities' rather than the urban – is the emphasis on global interconnections, on cities as nodes, networks, spaces of flows, in which diasporic populations forge new lives but also maintain ties between and connect in new ways places in different parts of the world. As the sociologist Stuart Hall (1990) insisted and feminist theorist Chandra Talpade Mohanty (1991; 2003) demonstrated for women migrants, the Third World is now in the First World, transforming understandings of spatial divisions and connections, as well as individual lives. Newer work on identity, on women's waged and unwaged work, on formal and informal employment, on commodified domestic labour, on artistic and other cultural movements in global cities, on political resistance in cities is now under way, moving beyond what once was regarded as 'urban' processes and patterns, blurring the distinctions between urban, economic, social and cultural geography, and infusing work on cities with a new rationale and a great surge of enthusiasm. To be part of this new and exciting wave is a pleasure in the same way that to have been part of the early flowering of feminist geographical work was both a pleasure and a privilege. It seems to me that feminist theorizing, in all its rich multiplicity and variety, has come in from the cold of a separate category or subdiscipline 'outside the project' (Christopherson, 1989) and is now a key part of mainstream geographical analyses.

References

Bondi, L. (1991) 'Gender divisions and gentrification: a critique', *Transactions of the Institute of British Geographers*, 16: 190–8.

Butler, J. (1990) *Gender Trouble*. London: Routledge.

Butler, J. (1993) *Bodies that Matter*. London: Routledge.

Castells, M. (1978) *The Urban Question*. London: Arnold.

Christopherson, S. (1989) 'On being outside "the project"', *Antipode*, 21: 83–9.

Ehrenreich, B. (1984) 'Life without father: reconsidering socialist-feminist theory', *Socialist Review*, 73: 48–57.

Fincher, R. and Jacobs, J. (eds) (1998) *Cities of Difference*. New York: Guilford.

Fraser, N. (1997) *Justice Interruptus: Critical Reflections on the 'Postsocialist' Condition*. New York: Routledge.

Hall, S. (1990) 'Cultural identity and diaspora', in J. Rutherford (ed.), *Identity: Culture, Community, Difference*. London: Lawrence and Wishart, pp. 222–37.

hooks, b. (1991) 'Homeplace: a site of resistance', in *Yearning: Race, Gender and Cultural Politics*. London: Turnaround, pp. 41–9.

Humphries, J. (1997) 'Class struggle and the persistence of the working class family', *Cambridge Journal of Economics*, 1: 241–58.

Massey, D. (1984) *Spatial Divisions of Labour*. Basingstoke: Macmillan.

Massey, D. (1995) 'Masculinity, dualism and high technology', *Transactions of the Institute of British Geographers*, 20: 487–99.

McDowell, L. (1983) 'Towards an understanding of the gender division of urban space', *Environment and Planning D: Society and Space*, 1: 59–72.

McDowell, L. (1991) 'Life without father and Ford: the new gender order of post-Fordism', *Transactions of the Institute of British Geographers*, 16: 400–19.

McDowell, L. (1997) *Capital Culture: Gender at Work in the City*. Oxford: Blackwell.

McDowell, L. (1999) *Gender, Identity and Place: Understanding Feminist Geographies*. Cambridge: Polity.

McDowell, L. (2003a) 'The particularities of place: geographies of gendered moral responsibilities among Latvian migrant workers in 1950s Britain', *Transactions of the Institute of British Geographers*, 28: 19–34.

McDowell, L. (2003b) *Redundant Masculinities? Employment Change and White Working Class Youth*. Oxford: Blackwell.

McDowell, L. (2003c) 'Space, place and home', in M. Eagleton (ed.), *A Concise Companion to Feminist Theory*. Oxford: Blackwell, pp.1–11.

McDowell, L. (2004) 'Workers, migrants, aliens or citizens? State constructions and discourses among post-war European labour migrants in Britain', *Political Geography*, 22: 863–86.

Mohanty, C.T. (1991) 'Cartographies of struggle: third world women and the politics of feminism', in C.T. Mohanty, A. Russo and L. Torres (eds), *Third World Women and the Politics of Feminism*. Bloomington, IN: University of Indiana Press, pp. 1–47.

Mohanty, C.T. (2003) '"Under western eyes" revisited: feminist solidarity through anti-capitalist studies', *Signs: Journal of Women in Culture and Society*, 28: 499–535.

Nussbaum, M. (2000) *Women and human development: the capabilities approach*. Cambridge: Cambridge University Press.

Phillips, A. (ed.) (1987) *Feminism and Equality*. Oxford: Blackwell.

Valentine, G. (1993) 'Negotiating and managing multiple sexual identities: lesbian space–time strategies', *Transactions of the Institute of British Geographers*, 18: 237–48.

Women and Geography Study Group (1984) *Geography and Gender: An Introduction to Feminist Geography*. London: Heinemann.

Women and Geography Study Group (1997) *Feminist Geographies: Explorations in Diversity and Difference*. London: Longman.

Young, I.M. (1990) *Justice and the Politics of Difference*. Princeton, NJ: Princeton University Press.

19 LOCAL AND GLOBAL

Richa Nagar

Rumor has it that Kothi Sahji, the grand eighteenth-century house where I grew up in the old city of Lucknow – and where the warrior Begum Hazrat Mahal took shelter at the time of the Indian revolt against the British – was constructed from the building materials stolen from the Asaf-ud-daula Imambara. My grandfather, who was to become an eminent Hindi novelist by the time of my birth, started renting this *kothi* located in the historic neighborhood of Chowk for 100 rupees in 1958. It was in this *kothi*, in the narrow, bustling lanes surrounding the *kothi*, and in the manner in which the rest of the world related to those lanes and the *kothi*, that my first and most deeply felt encounters with geographies of difference, inequity, and social injustices happened.

My childhood memories are filled with times spent with cousins, neighbors, domestic workers and their children in the big and small courtyards; in the winding, narrow stairwells; and in the 'secret' little doorways and tunnels that interlinked many of the old houses and tightly compressed lanes of Chowk. Chowk, by the way, was considered as the only real tourist attraction of Lucknow, because it was here that the past glory of Shi'i nawabs nurtured not only the Lakhnavi Urdu and culture of modesty, finesse and hospitality, but also the craftsmanship of the local Sunni and Hindu artisans, and the practices of Khattry business families who hired (and exploited) them.

Inside the spaces of the *kothi* – at once imposing, stifling and nurturing – I came to admire my grandfather's genius and his popularity among people of all classes. And in the same spaces, I watched my mother being shunned by the family because of her parents' extreme poverty. I learned how child labor got transformed into a lifetime of bonded labor through the stories of Baba who raised me and my sister. I was taught what my sociospatial and behavioral limits were as the oldest girl in the extended household. And I saw my young and ambitious father gradually becoming a prisoner of his own body as he battled with an aggressive muscular dystrophy. Immediately outside the *kothi*, I met *bhangi*[1] women and men who inhabited the other side of our residential lane, and who came with their baskets every day to collect the filth from our homes and non-flush latrines. In the covered alley beside the *kothi*, I knew girls of my own age who cooked, ate and slept with their families with only an eighteenth-century arch over their heads. These were girls who never got a chance to go to school or to use a 'real' toilet; whose growing, barely clothed bodies filled their mothers' hearts with fears; and who were married off and had babies by the time I reached college.

When I was seven, my mother – who had then become an assistant teacher of Hindi in a primary school – rebelled in a startling way. In a household that prided itself in serving the Hindustani literature and people's theater, and in a community where 'English schools' were considered both elitist and beyond financial reach, she demanded that her daughters be sent to an *Angrezi* school and announced that she would spend her earnings to help with the fees. Her victory resulted in my admission in 1976 (followed by my sister's in 1979) to La Martiniere, a school founded by a French general which is known as much for its high-quality education as for the historical role its boys played in fighting against Indians and helping restore British power during the revolt of 1857. In retrospect, the journey to La Martiniere – barely two miles away from Kothi Sahji – was at once the most traumatic and enabling journey that changed the course of my future life.

From Chowk to La Martiniere

When I arrived in La Martiniere, I became silent. The people, sounds and sensations that throbbed in the veins of Chowk were very far removed from this world. I was surrounded by Anglo-Indian teachers, administrators and boarders; and by daughters of bureaucrats, professionals, military officers, local legislators and business families, who were raised in the modern 'residential colonies' of Lucknow in nuclear families, who often spoke English comfortably, and who chatted about travels, films, novels and parties that I had never heard of. To many of them, Chowk was a 'backward Muslim interior' where everyone wore *chikan* fabrics, chewed betel leaves, flew kites or visited the courtesans. I tried hard not to feel embarrassed of belonging to Chowk or of coming from a joint family that did not own a car or a house, and the most sophisticated members of which could only speak broken English. I searched for words and points of connection as I traversed back and forth on cycle rickshaws between Chowk and La Martiniere. For the next nine years, I struggled to muster the tools to translate the pieces from one world of my childhood and early teenage to another.

My latter years in La Martiniere were not as difficult as the initial ones, however – partly because my sister and I had co-devised several survival strategies; partly because I started finding solace in Hindi creative writing; and partly because I had earned a reputation as a so-called *pundit* of things non-English. It was toward the end of the La Martiniere days that I also discovered Ms McClure. Raised in Burma, Ms McClure disapproved of two things: her husband's cigarettes, and girls who 'stitched like *mochis*'[2] (shoemakers) in her sewing class. But she loved stories by Rudyard Kipling and geography textbooks by Goh Cheng Leong. Even with her weakness for monsoon-Asia-type regional geography, Ms McClure effectively communicated to us that everything in our world happens in space and place, and one cannot ever escape geography! Although Ms McClure never said it in so many words, somehow her adoration of geography convinced me that it was possible for geographers to move between many worlds without compromising their passion for any of them. I think it was while listening to one of Ms McClure's lectures in 1983 that I decided to become a geographer. Three years later, this decision was to become a second battle point in my household ...

The Aborted Journey to Allahabad

The desire for geography led me to pursue my BA at Avadh College where I studied anthropology, geography and English literature; but at the masters level, geography was non-existent in Lucknow. No one in my entire clan had ever heard of sending a daughter away to study something as inconspicuous as geography! A hundred relatives interrogated my father: 'Where will you find the means to put her in a hostel?' If I had been selected for a medical or an engineering program, it would have been worthwhile to beg or borrow, but that was not the case: 'What wonders will she accomplish with an MA in geography? Work for the Geological Survey of India?' And then, there was the bigger question lurking behind these minor anxieties: 'What if she does something that disgraces the family?'

In some ways, perhaps, my father shared bits of all of these fears, but deep inside him he also had faith in his children. He decided that since the Government of India had agreed to give me a merit scholarship of 150 rupees per month to pursue higher studies, I should be allowed to go Allahabad, four hours by train from Lucknow, to get my masters degree in geography.

But a year of intense student activism led to the academic year 1986–7 being declared a 'zero session' at the University of Allahabad. As I devoured Bengali and Russian classics (in translation) and waited in vain for classes to start, my father convinced me that this was a good time to develop my creative faculties. I worked for the Educational Television, received lessons in writing and directing children's plays, transcribed life-history interviews with theater activists, and reviewed Kathak dance performances for the local dailies. Hindi literary writing continued to pull me as I began to publish short stories and poems in magazines such as *Dharmyug* and *Sarika*. I also became the unofficial personal assistant of my grandfather, who was then fighting glaucoma and diabetes. I took dictation as he narrated his last novel, answered his mail, accompanied him to seminars in Lucknow and Delhi, and escorted him to Bombay when the well-known film maker Shyam Benegal invited him to discuss one of his novels for a film project.

The one-month trip to Bombay with my grandfather impacted me deeply. It exposed me to a vibrant political and artistic atmosphere and to the excitement of being in a big city in a more 'advanced' part of the country; and it triggered in me a desire to move beyond Uttar Pradesh. I decided against going to Allahabad when studies resumed there and applied instead to the Universities of Bombay and Poona. With student political activity delaying the start of the academic year in Bombay this time, I found myself starting a new life in the Savitribai Phule Hostel in Pune in August 1987.

From Pune to Minneapolis

The University of Poona was a subsidized state university and boasted one of India's best geography departments. Of the large number of students who came to pursue geography at Poona, almost 80 per cent were Marathi-speaking men from middle- to lower-class farming families in the adjoining districts. Of the small minority who was educated in the English medium, four students in 1987–9 were from the Pune metropolitan region, and two (including me) were from the capital cities of Manipur and Uttar Pradesh. The composition of our regional, class and educational backgrounds made the linguistic medium of instruction an interesting challenge for instructors and students alike. Nevertheless, the department managed to give all its students a two-year immersion in all key subfields: geomorphology, climatology, human geography, economic geography, cartography and research methods. Although everything we covered in these areas was dominated by the work of British, US and German geographers, the department did a good job of introducing us to geographers working in Maharashtra – Dikshit, Diddee, Sawant, and Arunachalam.

Ironically, however, all this geography remained untouched by larger political issues that had captured the imagination of students on campus: the furor over Rushdie's *Satanic Verses* and the murder of *sati* Roop Kanwar in Rajasthan. Terms such as Marxism, feminism, political economy, imperialism or even colonialism never became part of our classroom discussions. Our training remained faithfully entrenched within positivist, Malthusian, and neoclassical paradigms no matter what we chose to specialize in during our last semester in the program. For me, all these sociopolitical influences remained confined to the extra-curricular realm, and I did not imagine that they could ever become part of geography – until I came to the University of Minnesota.

Minnesota was the most unbelievable accident of my life. Three factors facilitated this accident. First, Jayamala Diddee urged me to contact Joseph Schwartzberg because he had authored *An Historical Atlas of South Asia*. Second, my resourceful roommate in the hostel decided to take the TOEFL and GRE so she could apply to electronic engineering programs in California. Third, I did not look for Minnesota on the world map until after I got a MacArthur fellowship to study geography there!

From Minneapolis to Dar es Salaam

The trip from Delhi to Minneapolis on 31 August 1989 was the most difficult trip I have ever taken anywhere. The joint household had split; my family had been in the grip of some serious illnesses and economic hardships; and my presence was needed in Lucknow. Although I received nothing but complete support for my decision to go to the US, the circumstances in which I left made me feel guilty and fearful.

But exciting things were in store for me! The MacArthur program had just begun to generate tremendous opportunities as a community of international and US students came together with a dynamic group of left-leaning faculty at Minnesota to create new interdisciplinary agendas in a post-1989 world. I was particularly drawn to the conversations happening among the African studies scholars in the MacArthur program, as well as the energetic discussions on oral histories, personal narratives and popular memory that had animated the work of a large group of feminist scholars at Minnesota. At the same time, postcolonial approaches had started stirring exciting critical conversations about feminisms and the projects of ethnography.

The energy created by the MacArthur program was nourished by the Geography Department which encouraged me to grow theoretically and methodologically in the directions that were drawing me, and I decided to focus my doctoral research on the South Asian communities in postcolonial Tanzania, under the support and advice of six inspiring mentors. Philip Porter and Susan Geiger, my co-advisers, taught me the importance of telling stories in academia – without losing a sense of responsibility and commitment either to the people I was studying, or to the issues I wished to confront and struggle for. Eric Sheppard and Helga Leitner exposed me to the most exciting ideas in social and economic geography and cultivated the spaces where their students could come together to expand their own – as well as geography's – horizons. Ron Aminzade and Prabhakara Jha made me attentive to temporality and postcoloniality, and showed me what interdisciplinarity was all about.

After having immersed in the race politics of the US and developing a strong identity as a woman of color, confronting the racialized realities of East Africa in a physical way was jarring. As a woman from India who had arrived in Dar es Salaam (Dar) via the USA and who didn't easily fit the stereotypical category of a local *muhindi* ('Asian'), I was sometimes treated as an honorary *mzungu* ('European'). But soon I found myself negotiating and actively exploring other layers of politics as well – class, caste, religion, language, neighborhoods, as well as those of sexual practices and privileges – in a Tanzania that was shifting from being Nyerere's dream to a thoroughly liberalized multiparty democracy. The worlds that I had moved between – from Chowk and La Martiniere to Poona and Minneapolis, as well as my ancestral links to the Gujarati language – gave me the tools and passion to analyze the complexities of gender, race and community in the everyday spaces and identities of South Asian immigrants in Dar. And all this happened right as the Babri Masjid was being razed in Ayodhya near to my hometown, and the effects of BJP's rise in India could be felt as strongly among the upper caste Hindus in Dar as Ayatollah Khomeini's preachings could be heard in the Khoja Shia Ithnasheri Jamaat. All these interwoven processes became the subject matter of my dissertation, as well as a string of nine articles and book chapters that followed between 1995 and 2000.

From Geography to Women's Studies

In 1995, I started my first teaching job at the Department of Geography in Boulder, Colorado. To put it in Minnesota English, Boulder was different! And I do not mean simply its physical geography. Both interdisciplinarity and engagement with transnational politics were difficult to carry out at Colorado, especially in the face of the overt hostility that was frequently expressed against faculty who happened to have a combination of specific traits (relatively young, radical women of color who mentioned US imperialism in their undergraduate lectures, for example). However, I did find wonderful colleagues to learn from and grow with. Don Mitchell, Lynn Staeheli, Tony Bebbington and Tom Perreault in geography and Michiko Hase, Kamala Kempadoo and Alison Jaggar in women's studies, in particular, gave sustenance to mind and soul. David Barsamian of Alternative Radio became a source of political nourishment while Amy Goodman did some of that work through the radio waves every morning.

But the massive shift in institutional culture and context that Boulder brought into my life sparked questions that went beyond Boulder. As I moved from being an adopted 'daughter' of Dar to becoming an assistant professor at Colorado, I found myself caught between intellectual, political and personal commitments I had made to communities in three continents. In strategic terms, I learned to respond to the administration's message of 'publish or perish'. But I was deeply troubled by the realization that the only things that counted were those that could be discussed or consumed within western academic circles. There was hardly any institutional space to act on my sense of accountability to the people and issues I had studied in Dar. Any efforts to make my work 'travel' beyond the Anglophone academy in ways that could become meaningful to people that mattered in Dar or Delhi were deemed extracurricular – in the same way that politics surrounding Roop Kanwar and Rushdie became extracurricular in Poona.

My work in Tanzania also made me aware of other difficulties pertaining to the question of relevance in scholarly knowledge production. In challenging the dominant image of all Tanzanian Asians as exploitative male traders, I highlighted the narratives of people from varied caste, class, religious, sectarian and linguistic locations. In highlighting the relationality of identity, space and power, I focused as much on the lives of cab drivers, sex workers, and 'racially mixed' people (who were both accepted and shunned by the 'pure' Asians) as on the prosperous merchants, professionals, and community leaders. However, the sociopolitical power wielded by affluent Asians in Tanzania – combined with my position as a non-Tanzanian – meant that I could not share the critiques of communal organizations and leaders articulated by Asians who were relegated to the margins, without risking the latter's social lives or livelihoods. While my work could have proved helpful for those interested in fostering progressive interracial alliances in Tanzania, this fear of backlash by community leaders prevented me from publishing a book on my research. An ethnographic focus on the practices of the elite, however exciting theoretically or empirically, seriously limited the spaces available to me for producing knowledges that could advance progressive politics in Tanzania 'on the ground'.

Struggling with these questions made it necessary for me to ask why I wanted to be in the US academy, and what kind of academic work I wanted to do. Conversations with Saraswati Raju and Satish Kumar in Delhi, as well as a project with David Faust that focused on discursive and material divides produced by English medium education in postcolonial India, added new layers of complexity to this struggle. As I tried to work through these layers, I became frustrated with the limitations posed by narrow conceptualizations of reflexivity in critical scholarship, that rarely addressed how to generate conversations (and produce knowledges) that could move across the borders of the academy, classes and continents. Incidentally, Susan Geiger, who was finishing her book *TANU Women*, was also becoming disillusioned with popular approaches to reflexivity. Together, our mounting dissatisfaction created fertile ground to sow the seeds of a collaborative project, entitled 'Reflexivity, positionality and identity in feminist fieldwork: beyond the impasse'.

And somewhere in the middle of all these searches, I decided to shift my institutional home to women's studies – a 'field' where I felt I could blend commitments, genres, and theories in a more undisciplined way. Through a mix of exciting developments, I found myself returning to Minnesota in fall 1997 to make a new beginning in women's studies.

From *Mujhe Jawab Do* to *Playing with Fire*

Susan Geiger's untimely death in 2001 cruelly interrupted our collaboration. But the leukemia that destroyed Susan's body could not kill the quest that our collaboration had inspired: a quest to create new forms of accountability in feminist knowledge production, not only through a self-reflexivity about how 'researchers' are always inserted in politics of identities and categories, but also through a serious interrogation of how our institutional and geopolitical positions contribute to rendering our work relevant – or irrelevant – across the boundaries of the northern academy (wherever that north might be geographically located). For me, this quest – combined with prior associations with feminist activists in India – translated into a process of imagining new collaborations with non-governmental organization (NGO) workers and activists in Uttar Pradesh.

The process began with *Mujhe Jawab Do*, a study of a rural women's street theater campaign against domestic violence in Chitrakoot District. This work shared the ongoing commitment of postcolonial feminists to destabilize ethnographic practices that perpetuate the idea that it is only 'women' who live in the Third World – not the institutions or subjects of feminism. But as I faced the reality of how NGOs and donor-driven visions of empowerment were deradicalizing grassroots feminisms, it became clear to me that any effective intervention in transnational politics of knowledge production would have to be accompanied by a reshaping of dominant intellectual practices. It would require – among other things – collaborative agendas created with grassroots activists, to concretely grapple with the forms and languages in which new knowledges ought to be produced, and the ways in which those knowledges can be shared, critiqued, used and revised across multiple sociopolitical, institutional and geographical borders. These concerns found expression in an ongoing journey with eight NGO activists in Sitapur District, that resulted first in the creation of a book in Hindi called *Sangtin Yatra*, and then its English version, *Playing with Fire*.

The Hindi book has faced a warm welcome as well as a backlash, and both have advanced the authors' struggle against depoliticization propagated by donor-driven programs that seek to 'empower' rural women in the global south. As we await readers' responses to *Playing with Fire*, my collaborators and I remain convinced that it is only through more collective journeys across borders that we can create new intellectual and political possibilities to grow and flourish on our own terms, in our own spaces, and in our own languages.

NOTES

1 On the one hand, the terms *bhangi* and *mochi* reinforce categories created by practices of untouchability. On the other hand, these terms specify (rather than hide) the continuing degradation and dehumanization to which members of these caste groups are subjected. I use these labels here to mark this specificity.

2 See note 1.

20 MOVEMENT AND ENCOUNTER

Lawrence Knopp

Introduction

My ideas and work have been shaped by a variety of forces and circumstances, including various personal experiences, engagements with the ideas and works of others, and travels through space and time. Of course these are not separable, and I have discussed elsewhere their unity in terms of a wide variety of themes, crosscurrents, tensions, and transformations (Knopp, 2000). These have characterized and continue to characterize my ongoing process of *becoming* as a geographer, a gay man, a radical, and an activist (among other things).

For purposes of this chapter I focus more narrowly on the significance of one particular lens through which these processes can be considered: the experience of *movement*. Being in motion is of course part of what it means to be embodied and human. But it is constructed and experienced differently by different bodies and consciousnesses. For many moderns, especially queer moderns, movements of stunningly diverse types, quantities, and scales are particularly acute and common experiences. Our bodies, consciousnesses, and creations circulate through space as agents and artifacts of production, consumption and meaning-making.

This experience of hypermobility has certainly shaped my own thinking and work as a geographer, and as an increasingly important feature of (post?)modernity it deserves scrutiny in terms of its broader impact on geographic thought[1]. In this chapter, then, I discuss both of these through a consideration of how my own and other queers' movements have impacted geographers' thinking about spatial ontologies, most notably ontologies of place, placelessness, and movement.

I locate my discussion within the broader context of contemporary crises in social science and geography, which I believe to be part and parcel of broader crises of contemporary society[2]. I focus on spatial *ontologies* (rather than, say, epistemologies or methodologies) because I believe that epistemological and methodological issues have generally received more attention in this current crisis than ontological ones (the main exception being debates about essentialism, particularly in the context of questions of identity vs subjectivity[3]).

I focus as well on a particular kind of movement – what I call quests for identity – that I regard as being crucial to my own process of becoming as well as, I suspect, that of many other queer (post)moderns (and, indeed, many moderns generally, queer or otherwise). By 'quests for identity' I mean personal journeys through space and time – material, psychic, and at a variety of scales – that are constructed internally as being about the search for an integrated wholeness as individual humans living in community (if not society). What this amounts to, in my mind, is a search for emotional and ontological security. It is an effort to create order out of the chaos that is fractured identity combined with oppressive structures of power. While there may not be a lot that is new about this phenomenon in general terms, I do believe that in the current historical moment (in western individualist cultures, anyway) it takes a particular form. This is the creation and transformation of what sociologists call 'communities of limited liability'[4] (Janowitz, 1952) into collective identities that demand a place at the table in some kind of a liberal imagination. For gay men, as for other oppressed groups, this means seeking people, places, relationships, and ways of being that provide the physical and emotional security, the wholeness as individuals and as collectivities, and the solidarity that are denied us in a heterosexist world.

Gay Men's Journeys: Quixotic Quests?

My personal story, like that of many gay men, features very material geographical quests for identity in the sense that I have described. Indeed, the basic contours of this experience are common to many sex/gender 'outlaws' and indeed to members of any group who struggle with stigmatization based on some kind of very strong psychosocial experience or desire that is culturally pathologized (e.g. gender 'dysphoria').

One of the most obvious manifestations of this was my distancing from family and community of origin in order to 'come out'. That this is extremely common for gay men is well documented in the gay studies literature (Leap, 1995; Miller, 1989; White, 1982). Story after story features people who are either rejected by or voluntarily disavow their roots and then *move* in order to 'find themselves'. Frequently this entails traveling great physical distances over long periods of time, and it is common not only for gay men from unsupportive families and communities but, interestingly, for those from supportive ones as well. This was certainly the case for me. While I had come out in a formal sense to very supportive family and friends several years before moving away from them, and while several key formative experiences as a gay man took place in the burgeoning and generally well-supported gay community of my hometown urban environment in Seattle, Washington, I did not feel free of what nonetheless felt like the oppressive gaze of my family and hometown until I actually left Seattle. The unlikely environs of a small college town in Iowa afforded me the freedoms and opportunities to explore what being a gay man would mean to *me*. Ironically, this milieu – quintessentially middle American and conservative in a general sense, but mediated by an atypical liberal academic culture – showed me, in a way that the protective cocoon of family and community of origin could not, that the price of being openly gay in US society did *not* have to include the loss of certain privileges associated with social status, gender and class. Indeed it taught me that some of these privileges – most notably male privilege – can actually trump homophobia and heterosexism in some situations, and challenged me with disturbing ethical dilemmas in the process (see Knopp, 1999 for a more detailed discussion).

That and subsequent experiences in a variety of places have convinced me that for many – perhaps even most – gay men, coming out is about much more than just finding and/or creating new families, new relationships, new communities, and new *places* wherein certain counterhegemonic norms prevail, and self-actualization is perceived as possible. It is also about testing, exploring, and experimenting with alternative ways of being in contexts that are unencumbered by the expectations of tight-knit family, kinship, or community relationships – no matter how 'accepting' these might be perceived to be. In New Orleans, for example, where I lived and conducted fieldwork for my PhD dissertation when I was 30 years old, I was challenged by membership and participation in a gay community whose racism was generally quite unabashed, and that worked automatically to my advantage as a white man. In the process, my eyes were opened to the more subtle ways in which I had benefited from white skin privilege in both the gay and broader communities of supposedly more 'liberal' Seattle and Iowa City. This helped me to recognize similar processes – having to do not just with race but also with class – when I moved to another self-consciously 'liberal', predominantly white, and predominantly working-class place (Duluth, Minnesota), where I have lived and worked since.

It should come as no surprise, given the drive for many gay men to flee oppressive families and communities of origin, that rural to urban migrations of gay men are common (Grebinoski, 1993; Fellows, 1996). But so, my experience suggests, are migrations *down* an urban hierarchy (my own is a case in point), as well as international, interregional and intra-urban migrations and movements, and a general embracing of cosmopolitanism (Bech, 1997). This is because movement and changes of environment in and of themselves can be every bit as important for gay men (and others) engaged in identity quests as the particular characteristics of their origins and destinations.

Clearly these sorts of mobility practices are common for many people in contemporary individualistic societies and cultures, especially those with the means to be physically mobile, such as those with class, race, and/or gender privilege. Such privilege has certainly been a feature of my own story. But these practices' connections to a rather urgent perceived need to reinvent the self shine through particularly strongly in many, many accounts of gay men's lives, not just those of the privileged (see for example Northwest Lesbian and Gay History Museum Project, 2002).

At the same time, my experience and that of many others reveals gay men's rather fraught relationships to the new places they encounter and create, and a nostalgia, at times, for the places left behind. A persistent disappointment with every new environment I have encountered (and/or attempted to create) has led me to re-engage on new terms with past places from which I became alienated. This reflects the entirely predictable failure of a plainly utopian imagination that is at the heart of any quest of this nature. While life may be better as a result of these movements, it sometimes is not, and in any event there is no escape from the dominant culture of sex and gender to which we all belong and from which we almost all spring. Will Fellows (1996) illustrates this ambivalence particularly well in his compilation of first-person stories by mostly expatriate 'farm boys': the vast majority of his narrators no longer live on farms or in rural areas but express a sense of loss about this, along with a sense of at least partial alienation from their (chosen) new environments. Frank Browning (1996) captures a similar fraught relationship to place in his very personal account of his own journeys, as well as those of others, in his book *A Queer Geography*. In my own case, a midlife nostalgia for my urban gay origins, and a continuing ambivalent relationship to my adopted 'home' in Minnesota, have led to the construction of a 'two home' solution in which travel between them is every bit as crucial to my sense of self as simply being in one or the other (see Knopp and Brown, 2003, for a fuller explanation).

It is this attachment to movement that I find particularly interesting, along with a corresponding ambivalent relationship to both placement and identity. I and many gay men find the quest itself to be a source of considerable pleasure (Bech, 1997). For many of us (certainly for me), it becomes a source of ontological and emotional security as well. And while this too has its disappointments, the idea of movement, flux, and flows as important ontological sites in and of themselves, for both gay men and geographic thought, has been underappreciated and underdeveloped in the literatures on both. The fact is that being simultaneously in and out of place, and seeking comfort as well as pleasure in movement, displacement, and place*less*ness, are commonly sought after experiences for *many* people.

Ontologies of Place, Placelessness, and Movement

So how have these experiences shaped my ideas and work, and what are the implications for geographic thought more broadly? I focus here on three issues that have occupied varying amounts of my and other geographers' attention: *place*, *placelessness*, and *movement* (in particular, diffusion). I have worked with the first primarily in the context of place-based gay and lesbian politics and social movements (though I have also explored issues of place as they relate to nationalisms and nation-building projects; cf. Knopp, 1990; 1995a; and Knopp, 1997; 1998). The third I have begun to explore in the context of how gay male bodies, cultures, and politics circulate between metropolitan and non-metropolitan environments (Knopp and Brown, 2003). But the second I have only recently begun to think about. All three, however, have been conceived in multiple ways by different kinds of scholars representing various philosophical, epistemological and ontological commitments.

From my perspective, the identity quest experience makes very problematic almost all existing ontologies of place, placelessness, and movement, but particularly those that have their origins in some kind of modernist and/or structuralist thought. Such thought is almost by definition grounded in all sorts of binaries and essentialisms (e.g. 'gay', 'straight', 'man', woman', 'public', 'private', 'inside', 'outside'). Yet my and others' identity quest experiences demonstrate quite clearly that human subjectivities are multiple, fluid, and fractured. Like Heisenberg's subatomic particles, they refuse to be pinned down, much less to have order imposed upon them. A new and radical anti-identity politics of hybridity and fluidity, variously termed diasporic, postcolonial, postfeminist, and 'queer' (among other appellations), speaks against these modernist and structuralist essentialisms in a way I find quite compelling. But as I have also expressed elsewhere, denying the materiality and material consequences of essentialisms also has its dangers (Knopp, 1995b). So for me some kind of postmodern or poststructuralist ontological perspective with clear ethical and political groundings offers the best hope for accommodating this experience. But what might this be? Clearly diasporic notions of identity such as Paul Gilroy's *Black Atlantic* (1993) can help, but these kinds of postcolonial efforts seem much more interested in the ontological significance of movement as it pertains to identity than in the ontology of movement itself. Nigel Thrift's (1998) non-representational theory, with its notion of 'weak ontology', by contrast, along with certain aspects of actor-network theory (Serres and Latour, 1995), are particularly intriguing here. Let me explain.

Gregory (2000) argues, and I think I agree, that both traditional scientific and more existential and phenomenological approaches to ontology are to a substantial degree 'foundationalist'. By this Gregory seems to mean that they are grounded in fairly absolutist notions of 'truth', or at least of what *is*. That which *is* is that which can be shown to exist, and the evidence of existence is usually taken to be some kind of causal efficacy. This is easily seen in the case of traditionally scientific approaches, such as positivism, realism, and idealism, since they define themselves quite straightforwardly in terms of their searches for and isolation of what are taken to be real causal agents. Thus 'place', for example, is conceived as a real, usually material entity possessing causal powers, and its ontological status is confirmed through the detection of its effects. This detection can be either direct (e.g. 'the neighborhood effect' of positivist social science) or indirect (e.g. through theory, contemplation, and reflection, as in transcendental realism). In phenomenological and existential approaches, however, place is seen in both material and more metaphysical terms, and in relation to a broader range of particularly human experiences. Pickles (1985), for example, argues for a 'place-centred ontology of human spatiality', which he describes as the discursive and existential parameters within which knowledge is created. In this sense place becomes a fundamental condition of human existence, albeit one that is potentially flexible and reinventable. While there is much more room, therefore, for place to take on multiple and fluid characteristics in this approach, Pickles' ontology is still characterized by Gregory as foundational in the sense that it is seen as a fundamental frame of reference or causal force in human life, albeit one that is shaped and reshaped in a recursive relationship with human agency.

Non-representational and actor-network theory, by contrast, question this distinction between human agency and place as too steeped in an ontology of representation. Rather than focusing on abstraction and interpretation, Thrift (1996) in particular advocates what he calls a 'weak ontology' that focuses on lived experiences and unmediated *social practices*. Place, then, becomes a conjunction of time- and space-specific material practices, only minimally mediated (if at all) by processes of representation such as abstraction and interpretation. What is intriguing about this formulation is its (inherently spatial) ontology of 'fluid encounters, juxtapositions and divergences' (Gregory, 2000: 564), as well as its critique of elitist representational practices, and its insistence on engagement in the processes being examined by those doing the examining. All of this resonates quite instinctively with my own (and, I am quite sure, other gay men's) identity quests. We are actively engaged in a process of personal reinvention which intrinsically entails examination of ourselves and our surroundings. Hence our ambivalent relationships to place and identity, and our affection for placelessness and movement.

While I am more than a little skeptical about what it might mean for the practices of self-aware beings to be unmediated, I take Thrift's points about how important it is for *practice* to be at the heart of any spatial ontology (or at least of ontologies of place and movement). I also appreciate actor-network theory's broadening of this notion of practice to include the 'agency' generated by networks that include non-human 'mediaries and intermediaries' such as 'nature' and 'environment' (Thrift, 2000). And both critique the inherent elitism of representational acts, while celebrating the political and ethical value of direct engagement. An ontology of place that embraces embodied performativity (Butler, 1990) as every bit as much at the heart of the matter as any rhetorical or other representational articulation, then, seems to me to capture many aspects of how identity quests are actually experienced, as well as of what they mean.

Existing ontologies of placelessness are even less well equipped to deal with the identity quest experience. Geographers have largely conceived of placelessness as place's opposite, which is to say as an absence or a lack rather than as an embodied experience or practice that *is* anything or provides anything positive. But if placelessness is conceived as something active, something practiced, or as an embodied form of human agency, it becomes much more recognizable to gay men and others struggling with issues of identity. Whether because of its perceived homogeneity, its allegedly transitory character, the anonymity it supposedly provides, or its cosmopolitanism, the experience and practice of placelessness can dispense enormous amounts of both pleasure and emotional/ontological security to those of us in such circumstances, particularly if we are marginalized or oppressed. Thrift's 'weak ontology', then, despite its own obvious contradictions (e.g. its anti-representational representations), again offers a way of conceiving of placelessness as *practice* that is very resonant with the identity quest experience. The same can be said of actor-network theory's rhizomatic analogies, in which linkages and traffic through networks constitute fluid and elusive topologies of meaning.

Closely related to placelessness, in my mind, is movement. Yet the two are not identical. If placelessness is a set of practices that dispense particular meanings, movement is a much broader set of practices producing an even wider set of meanings. In geography, the study of movement includes all kinds of spatial activity and interaction at a variety of scales, from the body to the globe. So my comments here focus on just one form of movement that has preoccupied geographers for a long time: diffusion. Most 'scientific' ontologies of diffusion focus, I would argue, on origins, destinations, patterns, pathways, vectors and flows, and see these primarily as static bearers and carriers of objects and information (e.g. Hägerstrand, 1967). The pathways, vectors and flows in particular aren't seen as having quite the same ontological significance as the places and sites that they connect. Moreover they are rarely seen as complexes of social *practice* entailing complicated sets of power and other social relations. They are evaluated still less in terms of what it is that they generate ontologically themselves. Some more phenomenological and postmodern approaches probably do allow the possibility of these pathways, vectors and flows having the ontological status of sites (perhaps Rose's 1993 critique of time geography can be seen this way), and they may even be conceived in terms of social practices (Blaut, 1987; 1992). But these critical perspectives on diffusion have so far done little around the idea that these movements may, through their contingency, fluidity, and incompleteness, constitute practices of reflection and reinvention that are as ontologically significant themselves as the sites they connect, the phenomena they bear, or their physical descriptions and trajectories. In other words, the journeys themselves, as human engagements, are generative of all kinds of important emotional and ontological 'stuff'. And it may be their perceived placelessness itself, as well as their contingency, fluidity, and incompleteness, that is at the heart of this.

Conclusion

Quests for identity such as my own, then, offer several interesting avenues for reconsidering our ontologies of place, placelessness and movement. We can think about these concepts as fluid and 'under construction' projects of becoming, sets of spatial practice that dispense pleasure, security, and empowerment, and not just as origins, destinations, bearers, carriers, lacks, or hegemonic discursive frames of reference. Such conceptualizations have the potential advantages of minimizing the elitism of abstract representation, of honoring the messiness and indeterminacy of human experience, and by focusing on minimally mediated *practices* – especially critical examinations of self and surroundings – of forcing political and ethical engagement. Naturally there are no guarantees that such engagements will be to our liking, but at least they will be made. And for a discipline that has postured in so many ways and for so long as apolitical, that is something.

NOTES

An expanded version of this chapter appeared in *Gender, Place and Culture*, 11: 121–34, under the title 'Ontologies of place, placelessness and movement: The effects of quests for identity on contemporary geographic thought'.

1 I do not wish to enter here into a discussion of how to characterize the current historical moment in terms of 'modernity' or

'postmodernity'. I do, however, wish to acknowledge that one of the many contradictions of the current moment is that this hypermobility itself is unevenly experienced. Capital and commodities circulate much more freely and widely than do most humans, while the humans who experience it are disproportionately drawn from privileged castes, classes, and status groups (including those defined by 'race' and 'gender').

2 The crises to which I refer include everything from the so-called 'crisis of representation' (Barnett, 1997; Duncan and Ley, 1993; Duncan and Sharp, 1993) to struggles over the 'souls' of science and of geography and to contradictions in global capitalism.

3 Compare, for example, Spivak (1990; 1993), Butler (1990), Fuss (1989), Kobayashi and Peake (1994), Gibson-Graham (1996), and Pile (1996).

4 Communities of limited liability are loose affiliations of people that emerge out of relatively narrow sets of common interest and that serve primarily as instruments of individual self-actualization.

References

Barnett, C. (1997) 'Sing along with the common people: politics, postcolonialism and other figures', *Environment and Planning D: Society and Space*, 15: 137–54.

Bech, H. (1997) *When Men Meet: Homosexuality and Modernity.* Chicago, IL: University of Chicago Press.

Blaut, J. (1987) 'Diffusionism: a uniformitarian critique', *Annals of the Association of American Geographers*, 77: 30–47.

Blunt, J. (1992) *The Colonizer's Model of the World: Geographical Diffusionism and Eurocentric History.* New York: Guilford.

Browning, F. (1996) *A Queer Geography: Journeys toward a Sexual Self.* New York: Crown.

Butler, J. (1990) *Gender Trouble.* New York: Routledge.

Duncan, J. and Ley, D. (1993) *Place/Culture/Representation.* New York: Routledge.

Duncan, J. and Sharp, J. (1993) 'Confronting representation(s)', *Environment and Planning D: Society and Space*, 11: 473–86.

Fellows, W. (1996) *Farmboys: The Lives of Gay Men in the Rural Midwest.* Madison, WI: University of Wisconsin Press.

Fuss, D. (1989) *Essentially Speaking: Feminism, Nature and Difference.* New York: Routledge.

Gibson-Graham, J.-K. (1996) *The End of Capitalism (As We Knew It).* Oxford: Blackwell.

Gilroy, P. (1993) *The Black Atlantic: Modernity and Double Consciousness.* Cambridge, MA: Harvard University Press.

Grebinoski, J. (1993) 'Out north: gays and lesbians in the Duluth, Minnesota–Superior, Wisconsin area', presented at the Annual Conference of the Association of American Geographers, Atlanta, April.

Gregory, D. (2000) 'Ontology', in R.J. Johnston, D. Gregory, G. Pratt and M. Watts (eds), *The Dictionary of Human Geography*, 4th edn. Oxford: Blackwell, pp. 561–4.

Hägerstrand, T. (1967) *Innovation Diffusion as a Spatial Process.* Chicago, IL: University of Chicago Press.

Janowitz, M. (1952) *The Community Press in an Urban Setting.* Chicago, IL: University of Chicago Press.

Knopp, L. (1990) 'Some theoretical implications of gay involvement in an urban land market', *Political Geography Quarterly*, 9: 337–52.

Knopp, L. (1995a) 'Sexuality and urban space: a framework for analysis', in D. Bell and G. Valentine (eds), *Mapping Desire*. New York: Routledge, pp. 149–61.

Knopp, L. (1995b) 'If you're going to get all hyped up, you'd better go *somewhere*!', *Gender, Place and Culture*, 2: 85–8.

Knopp, L. (1997) 'Rings, circles and perverted justice: gay judges and moral panic in contemporary Scotland', in M. Keith and S. Pile (eds), *Geographies of Resistance*. New York: Routledge, pp. 168–83.

Knopp, L. (1998) 'Sexuality and urban space: gay male identities, communities and cultures in the U.S., U.K. and Australia', in R. Fincher and J. Jacobs (eds), *Cities of Difference*. New York: Guilford, pp. 149–76.

Knopp, L. (1999) 'Out in academia: the queer politics of one geographer's sexualization', *Journal of Geography in Higher Education*, 23: 116–23.

Knopp, L. (2000) 'A queer journey to queer geography', in P. Moss (ed.), *Placing Autobiography in Geography*. Syracuse, NY: Syracuse University Press, pp. 78–98.

Knopp, L. and Brown, M. (2003) 'Queer diffusions', *Environment and Planning D: Society and Space*, 21: 409–24.

Kobayashi, A. and Peake, L. (1994) 'Unnatural discourse: "race" and gender in geography', *Gender, Place and Culture*, 1: 225–43.

Leap, W. (1995) *Beyond the Lavender Lexicon: Authenticity, Imagination, and Appropriation in Lesbian and Gay Languages*. Langhorne, PA: Gordon and Breach.

Miller, N. (1989) *In Search of Gay America: Women and Men in a Time of Change*. New York: Atlantic Monthly.

Northwest Lesbian and Gay History Museum Project (2002) 'Life stories: from isolation to community', Mosaic 1. Seattle: Northwest Lesbian and Gay History Museum Project.

Pickles, J. (1985) *Phenomenology, Science and Geography: Spatiality and the Human Sciences*. Cambridge: Cambridge University Press.

Pile, S. (1996) *The Body and the City: Psychoanalysis, Space and Subjectivity*. New York: Routledge.

Rose, G. (1993) *Feminism and Geography: The Limits of Geographical Knowledge*. Minneapolis: University of Minnesota Press.

Serres, M. and Latour, B (1995) *Conversations on Science, Culture, Time*. Ann Arbor, MI: University of Michigan Press.

Spivak, G. (1990) *The Post-Colonial Critics: Interviews, Strategies, Dialogues*. London: Routledge.

Spivak, G. (1993) *Outside in the Teaching Machine*. London: Routledge.

Thrift, N. (1996) *Spatial Formations*. London: Sage.

Thrift, N. (1998) 'Steps to an ecology of place', in D. Massey, J. Allen and P. Sarre (eds), *Human Geography Today*. Cambridge: Polity, pp. 295–322.

Thrift, N. (2000) 'Actor-network theory', in R.J. Johnston, D. Gregory, G. Pratt and M. Watts (eds), *The Dictionary of Human Geography*, 4th edn. Oxford: Blackwell, pp. 4–5.

White, E. (1982) *A Boy's Own Story*. New York: Dutton.

21 SPACES AND FLOWS

Janice Monk

Searching for a geographic metaphor that would capture four decades of research, writing, and professional engagement in geography, I settled on 'braided streams'. This fluvial form is characterized by divergent and convergent channels, mostly occurring 'where there are almost no lateral confining banks' (Fairbridge, 1968: 90).[1] Two channels account for the greatest volume of my work – feminist studies and geographic education, primarily as it is related to higher education, though the two often intersect. But others are evident and also overlap with these, including research related to racial/ethnic minorities in white dominant societies and on change in rural communities.

Reflecting movements and encounters in my life course, I have been stimulated to write about Australia, the Caribbean, the European Union, the southwestern United States, and the United States–Mexico border region. My writing in English has been published in Australia, Britain, Canada, the Caribbean, New Zealand, and the United States and appeared in Catalan, Chinese, German, Italian, Japanese, and Spanish.[2] I have conducted field surveys and observations, archival research, oral histories, and textual interpretations. Editorial commitments have also been a major part of my work. Professional connections and related friendships have led to short-term appointments and periods as a visiting scholar or consultant in Australia, Canada, India, Israel, the Netherlands, New Zealand, Spain, and Switzerland. I have held appointments in departments of geography and in an interdisciplinary institute in women's studies and been very active in professional organizations. Across these various sites of endeavor, I see convergences and continuities in my motivations including a consistent concern with social equity, with action as well as research, and with responsiveness to the people and places with which I have been associated. Crossing disciplinary boundaries, working within and to change institutions, and valuing international ties are pervasive in my practice. Others may not see my work as I portray it here and I hope my comments do not appear too self-serving. This account is from the perspective of hindsight and also from one that reflects ideals that I might not have been able to articulate at the time work was undertaken. To a considerable extent, the directions emerged, rather than being planned.

Places, People, and Ways of Knowing

I have often looked to other disciplines, while retaining a deeply rooted geographical commitment to recognizing the importance and specificities of place.[3] My doctoral dissertation on differences among Aboriginal communities in New South Wales exemplifies this position (Monk, 1974). When I approached this topic in the mid 1960s, the boundaries of Australian geography largely excluded Aborigines. Interest was increasingly focused on testing spatial theories developed elsewhere. Neither was there much international interest in geography in minority populations or questions of 'race'. My choice arose from personal experience, not from the literature. Shortly after earning my BA I had worked as a volunteer constructing homes for Aboriginal families in a small town in New South Wales. The project was organized by a church group, with young white men and women (mostly recent graduates and professionals) providing labor, and the state government paying for materials as part of its policy to 'assimilate' Aborigines by moving them into town from reserves on the fringes of white communities. The project raised many questions for me, both ethical and geographical. Almost a decade later, formulating a research topic as a graduate student in the United States during the era of the 1960s Civil Rights Movement, I turned to questions of social and economic relations between Aboriginal and white communities in New South Wales.

Seeking insights, I looked beyond geography to the obvious site of Australian anthropology, but it did not offer work that spoke to my geographical sensibilities. Drawing on traditions of British social anthropology, it presented participant-observational studies of single communities, and interpreted them in relation to the persistence of 'traditional' cultural ways or in terms of the psychology of their being closed, institutional communities. I found more useful the ideas of some anthropologists and sociologists in the United States who brought ecological and material perspectives to the study of cultural change and of race relations. While planning my research I encountered Charles Rowley, a political scientist directing a large-scale project on the history and current conditions of Aborigines for the Social Science Research Council of Australia (Rowley, 1970a; 1970b; 1970c); these studies were associated with subsequent major changes in national policies. He was sympathetic to my interests in material relations, and ultimately incorporated my work in his book *Outcasts in White Australia*.

My dissertation illustrates movement outside the channels that were common in geography at the time, while representing one that has persisted in my work – drawing on personal experience to prompt examination of links between public policies and individuals' experiences of them. Another example is my study of the residential patterns and social networks of Asian professional immigrants to Sydney in the 1970s (Monk, 1983), again at a time when the discipline was not especially addressing issues of race and immigration. The choice undoubtedly reflected the confluence of my experiences of growing up in a society where the 'White Australia' policy was still under debate, where a relatively homogeneous Anglo-Celtic majority population was beginning to change with the substantial influx of immigrants of diverse national origins, and my own situation of living as a foreign student and immigrant junior faculty member in the multi-'racial'/ethnic society of the United States. My subsequent work in the Caribbean was initiated when Charles Alexander and I were assigned to teach a summer field class in Puerto Rico for the University of Illinois. What we saw in the landscape prompted us to ask questions about the impact of changing development policies on rural communities. A physical geographer who had earlier studied cultural historical geography on Margarita Island, Venezuela, he was open to collaboration and to multiple ways of knowing. We integrated survey interview methods, of which I had experience, his expertise in making field observations, his rusty and my beginning Spanish (Monk and Alexander, 1979; 1985). The Puerto Rican research also yielded my first feminist writing, prompted by the strengthening women's movement in society in the 1970s and its impact on academic work across disciplines. We studied the intersections of gender, class, and migration in Puerto Rico and later on Margarita Island (Monk, 1981; Monk and Alexander, 1986).

In the late 1970s, another set of encounters directed my work into new channels, but ones that were compatible with my earlier work on people outside the centers of power. The feminist movement prompted scholars to see more clearly how research and teaching reflected social, cultural, and political values and gender-based inequalities. For feminists in geography, this led to organizing paper sessions, networking and support activities at conferences, developing new courses and teaching materials, and directing new research to women's and gender issues. Bonnie Loyd and Arlene Rengert, learning of funding opportunities under the Women's Educational Equity Act (WEEA) in the United States, approached me about collaborating to write a proposal to support the development and pilot testing of curriculum modules that would introduce feminist content into introductory human geography courses. We submitted a successful request under the auspices of the Association of American Geographers, thus giving professional sanction and recognition to feminist concerns. The result was a booklet of student and instructor materials titled *Women and Spatial Change* (Rengert and Monk, 1982). Bonnie and Arlene also coordinated a special issue of the *Journal of Geography* on 'Women in Geographic Curricula' to which I contributed an analysis of gender biases in the language and roles represented in published simulation games in the discipline (Monk, 1978a). These pieces were intended both to critique existing practices and to advocate a more inclusive human geography. The same goals served as the impetus for a paper with Susan Hanson (Monk and Hanson, 1982) that addressed gender biases in prevailing theories, methods and purposes of geographic research. Since that time, efforts directed towards feminist-inspired curriculum change have been a dominant stream in my work, cutting across disciplines and embracing attention to minority groups and cross-cultural perspectives as well as having a gender focus (e.g. Monk, 1988; 2000; Monk et al., 2000; Lay et al., 2002).

Changing Institutions

In two senses, changing institutions has been a critical part of my professional work. From one perspective, my involvement in educational projects has meant linking research and action to attempt innovation in teaching in higher education. From another, changing the place of my employment from a geography department to a regionally oriented institute in women's studies has markedly influenced my opportunities and obligations.

The educational work began almost by chance when researchers in the Office of Instructional Resources (OIR) at the University of Illinois at Urbana–Champaign approached the Geography Department seeking collaborators for an applied research project that would experiment with alternative approaches to improving assessment of student learning. I was close to finishing my dissertation and the department asked me to take a grant-funded junior faculty appointment, teaching an honors class in physical geography and collaborating in the research. Next, OIR sought further collaboration in efforts to evaluate courses and teaching; this work extended my association. Both assignments required learning new literatures, creating new approaches in the classroom, supervising graduate teaching assistants, and co-authoring publications (e.g. Monk, 1971; Monk and Stallings, 1975; Monk and Alexander, 1973; 1975). They also meant engagement with experimental research designs, quantitative and qualitative methods, illustrating their strengths, limitations, and complementarity. The experience enhanced my awareness of the value of accepting multiple ways of knowing.

Around this time, the Association of American Geographers (AAG), which was engaging in an array of projects to improve college geography, obtained funding from the National Science Foundation to address the preparation of doctoral students for their teaching roles. With my newly acquired expertise and my department's interest in being part of a national project, I took on being the local project director in this multi-university program (Monk, 1978b). It fueled my interest in and commitment to changing practices in higher education through faculty and curriculum development, connected me to a national (and subsequently international) network, and initiated me into participating in large-scale multiperson and multi-institutional projects.

My diverse experiences at this early career stage and my feminist commitments stood me in good stead when I was forced to find a new position at a time when the academic job market in geography was poor. I moved to the University of Arizona to become Associate (subsequently Executive) Director of the Southwest Institute for Research on Women (SIROW). SIROW conducts interinstitutional, interdisciplinary research, educational, and outreach programs focusing on the regional diversity of women or of interest to scholars in the region it serves. My place-sensitive orientations as a geographer, my experiences in multiperson and multifunded projects that crossed disciplinary boundaries, and my engagement with feminist scholarship positioned me well for the new work. But the move also inhibited continuation of personal, field-based research of the type I had carried out in Australia and the Caribbean. I substituted more text-based projects, review essays, and editorial endeavors.

Over two decades at SIROW, I have engaged with colleagues and community organizations concerned with women's health, economic situation, education (especially in science, mathematics, and engineering), and cultural expressions. In some projects my role has mainly been to co-author grant proposals, administer the work, and see that it is disseminated. In others, I have taken a lead. The most sustained effort brought together my geographic interests in the meaning of place with feminist commitments, resulting in the book *The Desert Is No Lady: Southwestern Landscapes in Women's Writing and Art* (Norwood and Monk, 1997) and a film inspired by the book. Vera Norwood, a scholar in American studies at the University of New Mexico, and I put together a team of researchers in literature, anthropology, and art history to explore how Mexican American, American Indian, and Anglo-American women over a century had connected their senses of identity and place and expressed these in their creative work. Our interpretations contrasted with studies of the writing by white men which had dominated scholarship on the Southwest. They were identified with representations of the land as a virgin to be conquered, as a nurturing mother, as a place for development or conversely a wilderness to protect. We saw the women's work as focused on drawing energy from the land and celebrating its wildness and sensuality. We explored how women's representations were inflected by historical contexts, ethnic and cultural differences, and specific geographies. This project took me into new methodological terrain and prompted further writing related to gender and landscape (e.g. Monk, 1992; Norwood and Monk, 1997) as well as collaboration in film making with British colleagues (Williams, 1995). Though my commitments in that endeavor were largely for fundraising and consulting, the project highlighted how greatly representations can be manipulated through editorial processes, and heightened my awareness of the many-faceted aspects of whose voices are represented in research.

Valuing the International

The final stream I would like to discuss involves convergence of the various channels. Starting in the 1980s, I began to look for ways to link what we might now term 'the local' and 'the global' while bringing together feminism, educational efforts, and professional networks. I initiated a series of faculty and professional development programs at SIROW to introduce feminist work into internationally oriented courses across disciplines and international perspectives into women's studies. These programs resulted in a number of consultancies and co-editing of collections to disseminate approaches to linking teaching in women's studies and international studies (Monk et al., 1991; Lay et al., 2002). Since the mid 1990s, this direction has involved collaboration with Mexican colleagues in promoting research, faculty development, and community outreach on the theme of gender and health at the Mexico-US border. Though collaborative work has long been part of my professional life, this cross-border project has been especially informed by feminist thinking that has addressed such questions as who sets the agenda for and benefits from research, and what are the ways in which research and action might be respectfully linked. We have developed approaches to sharing decision-making and resources equitably, and have reflected on the relationships among researchers and those working in community agencies (Monk et al., 2002).

Within geography, I have worked since the 1980s with like-minded colleagues to foster gender scholarship within the International Geographical Union and feminist geography more generally. An important motive has been not only to enhance the visibility of feminist scholarship in geography, but to try to promote perspectives that value visions and voices outside the dominant US and British realms. It led me to co-edit the book series *International Studies of Women and Place* with Janet Momsen, to co-edit two books that include contributors from multiple countries (Katz and Monk, 1993; García-Ramon and Monk, 1996), to write on comparative perspectives in feminist geography (Monk, 1994), to visit and consult in universities in several countries, and to be an active participant in the IGU Commission on Gender and Geography, editing its newsletter since 1988. Whenever I can, I cite work by scholars outside the hegemonic regions and draw attention to issues involved in working across national boundaries in geographic education (Monk, 1997; Shepherd et al., 2000; García-Ramon and Monk, 1997), and promote teaching that attends to human diversity in the US and beyond (Monk, 2000; Monk et al., 2000). These perspectives also informed the directions I took when I had the privilege of serving as President of the AAG in 2001–2, and invited the women presidents of four geographical associations (Australian, Californian, Canadian and Catalan) to speak at the Presidential Plenary at the Annual Meeting on the theme 'Points of View, Sites for Action', also initiating a reception to welcome the many geographers from outside the US who attend AAG meetings.

Looking back, I see my commitments as in some ways reflecting having been on the margins geographically and professionally – an expatriate Australian feminist, a woman who grew up in the 1950s when women did not expect to pursue academic careers – but also as an immigrant who has now lived for almost 20 years near the border of the United States and Mexico, and who is employed in an interdisciplinary feminist institute. These circumstances have contributed to the positionality of my work. The braided streams, while flowing relatively unconstrained by 'lateral banks,' have nonetheless sustained a commitment to a life in geography and a desire to bring my values to its course.

NOTES

1 Other tempting choices which some might apply to me include 'misfit stream' and 'erratics',

but I hope not 'deranged drainage', 'rubble drift', or 'planetary wobbles'.
2 Colleagues have undertaken the translations.
3 This account of my dissertation research is adapted from Janice Monk and Ruth Liepins (2000). I appreciate the permission of my co-author and the journal publisher to draw on it.

References

Fairbridge, R.W. (1968) 'Braided streams', in R. W. Fairbridge (ed.), *The Encyclopaedia of Geomorphology*. New York: Reinhold, pp. 90–3.

García-Ramon, M.D. and Monk, J.J. (eds) (1996) *Women of the European Union: The Politics of Work and Daily Life*. London and New York: Routledge.

García-Ramon, M.D. and Monk, J.J. (1997) 'Infrequent flying: international dialogue in geography in higher education', *Journal of Geography in Higher Education*, 21: 141–5.

Katz, C. and Monk, J.J. (eds) (1993) *Full Circles: Geographies of Women over the Life Course*. London: Routledge.

Lay, M.M., Monk, J.J. and Rosenfelt, D.S. (eds) (2002) *Encompassing Gender: Integrating International Studies and Women's Studies*. New York: Feminist Press.

Monk, J. (1971) 'Preparing tests to measure course objectives', *Journal of Geography*, 70: 157–62.

Monk, J. (1974) 'Australian aboriginal social and economic life: some community differences and their causes', in L.J. Evenden and F.F. Cunningham (eds), *Cultural Discord in the Modern World*. Vancouver: Tantalus, pp. 157–74.

Monk, J. (1978a) 'Women in geographical games', *Journal of Geography*, 77: 190–1.

Monk, J. (1978b) 'Preparation for teaching in a research degree', *Journal of Geography in Higher Education*, 2: 85–92.

Monk, J. (1981) 'Social change and sexual differences in Puerto Rican rural migration', in O. Horst (ed.), *Papers in Latin American Geography in Honor of Lucia Harrison*. Muncie, IN: CLAG, pp. 29–43.

Monk, J. (1983) 'Asian professionals as immigrants: the Indians in Sydney', *Journal of Cultural Geography*, 4: 1–16.

Monk, J. (1988) 'Engendering a new geographic vision', in J. Fien and R. Gerber (eds), *Teaching Geography for a Better World*. Edinburgh: Oliver and Boyd, pp. 91–103.

Monk, J. (1992) 'Gender in the landscape: expressions of power and meaning', in K. Anderson and F. Gale (eds), *Inventing Places: Studies in Cultural Geography*. Melbourne: Longmans/Cheshire, pp. 123–38.

Monk, J. (1994) 'Place matters: comparative international perspectives on feminist geography', *Professional Geographer*, 46: 277–88.

Monk, J. (1997) 'Marginal notes on representations', in H.J. Nast, J.P. Jones III and S.M. Roberts (eds), *Thresholds in Feminist Geography: Difference, Methodology, and Representation*. Lanham, MD: Rowman and Littlefield, pp. 241–53.

Monk, J. (2000) 'Finding a way: a road map for teaching about gender in geography' in *Learning Activities in Geography for Grades 7–11*. Finding a Way: A Project of the National Council for Geographic Education. Indiana, PA: National Council for Geographic Education.

Monk, J.J. and Alexander, C.S. (1973) 'Developing skills in a physical geography laboratory', *Journal of Geography*, 72: 18–24.

Monk, J.J. and Alexander, C.S. (1975) 'Interaction between man and environment: an experimental college course', *Journal of Geography*, 74: 212–22.

Monk, J.J. and Alexander, C.S. (1979) 'Modernization and rural population movements: western Puerto Rico', *Journal of Interamerican Studies and World Affairs*, 21: 523–50.

Monk, J.J. and Alexander, C.S. (1985) 'Land abandonment in western Puerto Rico', *Caribbean Geography*, 2: 1–15.

Monk, J.J. and Alexander, C.S. (1986) 'Free port fallout: gender, employment, and migration, Margarita Island', *Annals of Tourism Research*, 13: 393–414.

Monk, J.J. and Hanson, S. (1982) 'On not excluding half of the human in human geography', *Professional Geographer*, 34: 11–23.

Monk, J. and Liepins, R. (2000) 'Writing on/across the margins', *Australian Geographical Studies*, 38: 344–51.

Monk, J.J. and Stallings, W.M. (1975) 'Classroom tests and achievement in problem solving in physical geography', *Journal of Research in Science Teaching*, 12: 133–8.

Monk, J.J., Betteridge, A. and Newhall, A.W. (eds) (1991) 'Reaching for Global Feminism: Approaches to Curriculum Change in the Southwestern United States', collection, *Women's Studies International Forum*, 14: 239–379.

Monk, J.J., Sanders, R., Smith, P.K., Tuason, J. and Wridt, P. (2000) 'Finding A Way (FAW): a program to enhance gender equity in the K-12 classroom', *Women's Studies Quarterly*, 28: 177–81.

Monk, J.J., Manning, P. and Denman, C. (2002) 'Working together: feminist perspectives on collaborative research and action', *ACME: An International E-Journal for Critical Geographies*, 2(1): 91–106.

Norwood, V. and Monk, J. (1997) *The Desert Is No Lady: Southwestern Landscapes in Women's Writing and Art* (1st edn 1987). New Haven, CT: Yale University Press.

Rengert, A. and Monk, J. (1982) *Women and Spatial Change*. Dubuque, IA: Kendall/Hunt.

Rowley, C.D. (1970a) *The Destruction of Aboriginal Society: Aboriginal Policy and Practice*. Canberra: Australian National University Press.

Rowley, C.D. (1970b) *The Remote Aborigines*. Canberra: Australian National University Press.

Rowley, C.D (1970c) *Outcasts in White Australia*. Canberra: Australian National University Press.

Shepherd, I.D., Monk, J.J. and Fortuijn, J.D. (2000) 'Internationalizing geography in higher education', *Journal of Geography in Higher Education*, 24: 163–77.

Williams, S. (1995) *The Desert Is No Lady*, video recording. London: Arts Council of England.

Part 3
Practices

This part explores the relationships between theory and methodology/methods. Rather than providing a 'how to do guide', these chapters focus on explaining the links between research designs, the development of particular methods and different philosophical approaches to knowledge (for example, what constitutes data or evidence).

In Chapter 22 Stewart Fotheringham makes the case for quantitative methods, arguing that they are important because they provide strong evidence on the nature of spatial processes – much stronger, he claims, than can be produced by other methods. He shows how quantitative methods have their roots in positivistic philosophies, but points out that 'positivism' and 'quantification' are not synonymous even though many people use them interchangeably. For example, he argues that whereas the goal of a positivist would be to uncover the truth about reality in the form of absolute laws, quantitative geographers recognize that it is rare to find such absolutism. Rather, geographers use quantitative methods to build up sufficient evidence on which to make judgements about reality. Fotheringham particularly stresses the role of quantitative methods as a bridge between human and physical geographers, providing them with a language to talk to each other and the basis for joint research. He concludes his chapter by outlining a defence for quantitative methods against some of the common criticisms levelled at them.

In the following chapter Mike Goodchild (Chapter 23) presents a brief history of the introduction of GIS into geographic research. He then outlines two perspectives. The first, widely held among researchers working with GIS, is that the technology is value-neutral, and that its users reflect a wide range of approaches, from the strongly positivist stance of researchers in physical geography and some areas of human geography, to the more human-centric stance of those working in such areas as public participation GIS and critical social theory. The second perspective, which stems from the strong critique of GIS which emerged in the early 1990s, is that GIS is inherently value-laden. Goodchild goes on to trace this tension between these two positions, and to explore efforts at reconciliation and accommodation. Like Fotheringam (Chapter 22), who recognizes the limits of rationality in positivistic approaches and the need to develop quantitative methods to model irrational human behaviour, Goodchild reflects on uncertainty and scientific norms as examples of methodological issues that still surround the use of GIS.

Rather than understanding knowledge as an object to be achieved, tested and verified by an impartial observer, Paul Rodaway's chapter on people-centred methods (Chapter 24) is concerned with how geographers who understand knowledge to be subjective, partial and emergent have developed methodologies to explore the individual relationships that people have with the worlds that they inhabit. Here, he identifies how humanistic geographers such as Douglas Pocock and Graham Rowles have developed, or adapted from other disciplines, methods to look at the meaning of individual experiences and how people interpret their encounters with place. Here, the emphasis is on methods of encounter, engagement and participation – including participant observation, interviewing and interpretive readings of texts.

In Chapter 25 Mike Samers argues that the point of research is not just to understand the world but to change it. In doing so he highlights the fluid and ambiguous nature of the boundary between social theory and activism, and traces some of the opportunities and limitations inherent in trying to move between academic research and the wider world. For Samers, doing research and writing for academic audiences is itself a process of political activism because it is through this process that we learn about and understand the processes of exclusion and marginalization. It is, however, a practice that is undercut by the problem of how to represent 'oppressed' or vulnerable groups (see also Chapter 12). The problems of representation are equally endemic in radical geographers' attempts to make a difference by engaging with audiences outside the academy such as government officials, lawyers and community groups. While Marxist activist geographers are concerned to work with 'vulnerable' people to assist rather than lead their struggles, Samers argues that the boundaries between academic and activist can become very blurred.

The problem of the relationship between researcher and research participants is one shared by feminist geographers. In Chapter 26 Kim England characterizes feminist methodologies as those which are sensitive to power relations between researcher and research subjects. For England, it is irrelevant whether a researcher uses a qualitative or a quantitative method. What defines research as feminist is whether researchers develop a research relationship that is based on empathy and respect and whether they seek to reduce the distance between themselves and those with whom they work. This means consciously seeking to connect or engage with participants rather than attempting to remain detached. Here, England outlines how feminists have theorized this relationship in terms of the concepts of *positionality* (how people view the world from particular embodied locations; and how they in turn are positioned and responded to by others) and *reflexivity* (a process by which a researcher self-consciously reflects on their own role in the research process and their research relationships).

In Chapter 27 the focus shifts from reflections on methods for working with people to textual methods. Here, John Wylie demonstrates the importance of Derrida's notion of deconstruction as well as Foucault's articulation of discourse analysis as forms of geographical practice. Wylie then goes on to look at the more recent influences of Gilles Deleuze and his emphasis on emotion, creativity and transformation. Rather than endless critique, Wylie shows how Deleuze favours experimental modes of writing and expression.

Chapter 28 reflects on how to reconcile philosophic approaches that are critical of the power relations embedded within society and the academy with the need to develop research methods to address pressing social problems in sensitive ways. Here, Paul Robbins reflects on what he terms 'the postcolonial contradiction', in doing research in non-western contexts, namely that any project can inevitably be read in some senses as colonial in opposition to the researcher's effort or desire to do non-colonial research. Robbins does so by drawing on his own research on conservation efforts in West Africa and India, to reflect on his efforts to resolve the colonial legacy with a different kind of environmental research suggested by postcolonialism.

While some of the methods discussed in this part are associated with or have their origins in particular approaches, it is important to recognize that they are not exclusive to them. For example quantitative methods are used by some positivistic geographers and also by some feminist geographers; likewise textual methods may be adopted – albeit in different ways – by those coming from humanistic perspectives or poststructuralist perspectives, and so on. Moreover, as Fotheringham's chapter on the relationship between positivistic philosophies and quantitative methods in particular highlights, there is often a tension between philosophical purity and the need for pragmatism because of the practicalities and everyday constraints (policy context, time, resources, funding contexts, etc.) under which academics (and students) carry out their work.

Finally, it is important to recognize that researchers do not necessarily use one type of method alone but rather can mix methods in their research designs. For example, they might carry out a large-scale questionnaire survey that is analysed statistically, alongside conducting in-depth conversational-style interviews with some key informants, or doing textual analysis of appropriate documents/images.

22 QUANTIFICATION, EVIDENCE AND POSITIVISM

A. Stewart Fotheringham

Setting the Scene

Quantitative geographers do not often concern themselves with philosophy, and although externally we are often labelled (incorrectly in many cases) as positivists, such a label has little or zero impact on the way in which we prosecute research. We do not, for example, concern ourselves with whether our intended research strategy breaches some tenet of positivist philosophy. Indeed, most of us would have scant knowledge of what such tenets are. As Barnes (2001) observes, for many of us, our first experience with positivism occurs when it is directed at us as a form of criticism. We do not continually scrutinize our particular *modus operandi* with some philosophical checklist in hand. Our lack of engagement with the philosophical debates that seem to take up so much of the energy of our colleagues is more than compensated for by our concern with statistical, mathematical and, above all, *geographical* theory. To a large extent, our guiding principle is the question, 'Does what I'm doing provide useful evidence towards the better understanding of spatial processes?' The continuous debates about which particular philosophical approach or 'ism' is best leaves most quantitative geographers shaking their heads and wondering what is happening to the rest of their discipline. The wonderful quotation attributed to the late Richard Feynman that 'philosophy of science is about as useful to scientists as ornithology is to birds' is rather apt here (*inter alia* Kitcher, 1998: 32).

However, in the recent decade, the exponents of various philosophical trends that have been co-opted into human geography appear to be increasingly antagonistic towards the use of quantitative methods for reasons that often seem to be more emotive than substantive. It is especially worrying that some of our colleagues appear unwilling to engage at all with our work, despite its relevance to almost every substantive issue studied by geographers. Instead, they dismiss the whole field because it does not fit in with their particular philosophical credo. It therefore seems that some discussion and defence of the approaches taken by quantitative geographers is appropriate in order to generate a more balanced view of their contributions to the discipline. Consequently, in what follows I attempt to articulate what defines a quantitative geographer, given that it is not that we have all joined some philosophical 'club' with strict membership guidelines about how to and how not to prosecute research. I also try to explain why we have trouble with many of the current 'isms' that abound in human geography. In addition, I also discuss some of the issues surrounding the relative demise of the quantitative approach within geography during the 1980s and 1990s as well as its recent resurgence. All I ask is the impossible: that the reader approach this with an open and unprejudiced mind.

So What Do Quantitative Geographers Do?

As I state with colleagues elsewhere:

> **A major goal of geographical research, whether it be quantitative or qualitative, empirical or theoretical, humanistic or positivist, is to generate knowledge about the processes influencing the spatial patterns, both human and physical, that we observe on the earth's surface. (Fotheringham et al., 2000: 8)**

Towards this goal, the work of quantitative geographers can be grouped into four areas.

1. The reduction of large data sets to a smaller amount of more meaningful information. This is important in analysing the increasingly large spatial data sets obtained from a variety of sources such as satellite imagery, census counts, private companies and local governments. Summary statistics and a wider body of data reduction techniques are often needed to make sense of these very large, multidimensional data sets.
2. The exploration of spatial data sets. Exploratory data analysis consists of a set of techniques to explore data (and also model outputs) in order to suggest hypotheses or to examine the presence of unusual values in the data set. Often, exploratory data analysis involves the visual display of spatial data generally linked to a map.
3. Examining the role of randomness in generating observed spatial patterns of data and testing hypotheses about such patterns. In so doing, we can infer processes in a population from a sample and also provide quantitative information on the likelihood that our inferences are incorrect. For instance, suppose we want to investigate the spatial distribution of some disease in order to examine if there might be an environmental link to that disease. We have to decide first on some method of measuring the spatial clustering of the disease with respect to the at-risk population and then we have to determine the probability that such clusters could have arisen by chance. Third, if the clusters are very unlikely to have arisen by chance, we must look at the relationship between the locations of the clusters and various environmental factors, such as the locations of toxic waste dumps or contaminated water sources. We do not claim such statistical tests would provide us with a definite answer to what caused the disease but we would have a better basis on which to judge the existence of a possible relationship.
4. The mathematical modelling and prediction of spatial processes. The calibration of spatial models provides extremely useful information on the determinants of processes through the estimates of the models' parameters. Spatial models also provide a framework in which predictions can be made of the spatial impacts of various actions: examples include the effects of building a new shopping development on traffic patterns, and the building of a seawall on coastal erosion.

The goal of quantitative geography is therefore a very simple one, but a very important one: *to add to our understanding of spatial processes*. This might be done directly, as in the case of store choice modelling where mathematical models are derived based on theories of how individuals make choices from a set of spatial alternatives. Or, it might be done indirectly, as in the analysis of the incidence of a particular disease, from which a spatial process might be inferred from a description of the spatial pattern of the incidence of the disease.

Quantitative geographers would argue that their approach provides a robust testing ground for ideas about such spatial processes. Particularly in the social sciences, ideas become accepted only very gradually and have to be subject to fairly rigorous critical examination. Quantitative spatial analysis provides the means for strong evidence to be provided either in support of or against these ideas. For these reasons, quantitative geographers have skills which are much in demand in the real world and are sought after to provide inputs into informed decision-making.

Positivism and Quantification

The terms 'positivism' and 'quantification' are not synonymous even though many people use them interchangeably.[1] One can be a quantitative geographer, for example, without necessarily being a positivist. This fundamental misunderstanding has arisen because many geographers I suspect have misconstrued the meaning of one or both of these expressions. It is not difficult to see why this has happened: the term 'positivism', in particular, is laden with ambiguity. As Couclelis and Golledge note:

> **Although there is usually little disagreement as to what argument or piece of work is in the positivist tradition, positivism itself as a philosophy turns out to be extremely difficult to define ... popularized accounts, *ad hoc* reformulations, working philosophies, methodological credos, and several contrasting sets of ontological beliefs, all sought and found a place under the umbrella of 'positivism'. (1983: 332)**

However, it is possible to identify two of the more central tenets of positivism which are (in lay terms) that:

1. The only meaningful items of study are those that can be verified. Strictly speaking this means we should be able to judge absolute truth, which in geographical studies we generally cannot; hence, more loosely, we often relax this condition to mean we can judge a statement to be either true or false.[2]
2. Items of study can be verified only when they can be directly measured and observed.

Taken together, these tenets imply that our ability to generate knowledge is restricted to those things that we can observe in reality. For instance, religious discussions are seen as irrelevant to positivists because the beliefs that people hold can be neither proved nor disproved. Since we cannot measure emotions and thoughts, strict positivists would also exclude these items from investigation. Clearly, empirical testing is an important element of positivism: knowledge cannot be gained on those issues which do not lend themselves to such testing.

When geographers use the word 'positivist', they generally imply a somewhat broader set of beliefs and often ascribe the term to someone who follows scientific principles in his/her research. Gatrell, for instance, defines quantitative geography as that which

> **relies on accurate measurement and recording and searches for statistical regularities and associations. It emphasises, via mapping and spatial analysis, what is observable and measurable. Because it then seeks to establish testable hypotheses, in the same way that a natural scientist would, it has many of the characteristics of a positivist or naturalist approach to investigation. (2002: 26)**

A positivist therefore is perceived as someone whose focus is the search for order and regularities with the ultimate goal of producing universal laws (a weaker version of positivism that is applied to human geography is one that attempts to make generalizations about spatial processes rather than 'laws'). The methods used to achieve this are typically quantitative: data might be analysed, hypotheses might be tested, theoretical relationships might be established, and mathematical models might be formulated and calibrated. Quantitative geographers, being labelled as positivists, are typically perceived as ignoring all the emotions and thought processes that are behind what is sometimes, at least in the case of human geography, highly idiosyncratic behaviour. In essence, positivism, science and quantification are all seen as synonymous, unemotional, mechanistic approaches to the study of geography.

It is therefore tempting to label all quantitative geographers as positivists or naturalists (Graham, 1997) but this disguises some important differences both within quantitative geography and between quantitative geography and positivism. For example, while quantitative geography has some adherents who believe in a search for global 'laws' and global relationships, others recognize that there are possibly no such entities. The latter concentrate on examining variations in relationships over space through what are known as 'local' forms of analysis (Fotheringham, 1997; Fotheringham and Brunsdon, 1999; Fotheringham et al., 2002). Here the emphasis is on identifying areas of exception where something unusual appears to be happening; the approach does not necessarily concern itself with establishing any laws, or even generalities, and hence is contradictory to the spirit of positivism. Local types of investigation are increasingly finding favour among quantitative geographers who recognize that global methods of analysis can mask important local variations and that these variations can shed important light on our understanding of spatial processes.

As well as being less concerned with the search for global laws than some might imagine, quantitative geography is also not as sterile as some would argue in terms of understanding and modelling rather abstract concepts such as human feelings and psychological processes (Graham, 1997). There appears to be a strong undercurrent of thought amongst those who are not fully aware of the nuances of current quantitative geography that it is deficient in its treatment of human influences on spatial behaviour and spatial processes. While there is some validity in this view, quantitative geographers increasingly recognize that spatial patterns resulting from human decisions need to account for aspects of human decision-making processes. This is exemplified by the current interest in spatial information processing strategies and the linking of spatial cognition with spatial choice (see Fotheringham et al., 2000, Chapter 9 for an example) and also by the attempts of researchers such as Openshaw (1997a) to incorporate qualitative issues into modelling through the application of fuzzy logic. Indeed, I suspect it would be a real eye-opener for many critics of quantitative geography to read works such as Openshaw (1997a) and to see the efforts some spatial analysts have made to capture qualitative issues within their research.

Quantitative geographers believe that quantification generally provides strong evidence towards understanding spatial processes – much stronger, for example, than is provided by any other competing methods. However, we recognize that, counter to one of the tenets of positivism, rarely can we prove anything absolutely. The usual goal of quantitative analysis in geography is to accrue sufficient evidence to make the adoption of a particular line of thought compelling. As Bradley and Schaefer note in discussing differences between social and natural scientists:

> the social scientist is more like Sherlock Holmes, carefully gathering data to investigate unique events over which he had no control. Visions of a positive social science and a 'social physics' are unattainable, because so many social phenomena do not satisfy the assumptions of empirical science. This does not mean that scientific techniques, such as careful observation, measurement, and inference ought to be rejected in the social sciences. Rather, the social scientist must be constantly vigilant about whether the situation being studied can be modeled to fit the assumptions of science without grossly misrepresenting it ... Thus, the standard of persuasiveness in the social sciences is different from that of the natural sciences. The standard is the

> compelling explanation that takes all of the data into account and explicitly involves interpretation rather than controlled experiment. The goals of investigation are also different – the creation of such compelling explanations rather than the formation of nomothetic laws. (1998: 71)

Hence, whereas the goal of a positivist would be to uncover the truth about reality in the form of absolute laws, quantitative geographers realize that such absolutism is extremely difficult to find in most instances and they hold to the more acceptable goal of simply accruing sufficient evidence on which to base a judgement about reality that most reasonable people would find acceptable. Judgements or hypotheses about reality that are found unacceptable are discarded so that knowledge accumulates via a process of retention or rejection. Of course, retention does not imply validation but merely that the idea survives that particular test. Ideas in social science tend to become generally accepted only when they survive a large number of such tests.

Related to this issue, and one of the strengths of a quantitative approach, is that of the measurement of error. Suppose we wish to understand the spatial distribution of some phenomenon in terms of a set of explanatory variables that influence this distribution in some way. In most cases there are actually two sets of explanatory variables: those which we know to have an effect on the spatial distribution under investigation and which we can measure; and those whose effects we are unaware of or which we cannot measure. A large advantage of a quantitative approach is that it enables the measurement of the determinants that can be measured (and in many cases these provide very useful and very practical information for real-world decision-making) whilst recognizing that, for various reasons, these measurements might be subject to some uncertainty. We can measure this uncertainty and report it as a guide to the degree of belief one should have in the reported results. If, for instance, the errors in the modelling procedure are large, we would probably conclude that something important has been omitted from the model and that the results are not very reliable. Being able to assess the likelihood that a model of the real world is a reasonable one, and hence the degree of belief to ascribe to the outcomes from such a model, is a big advantage of the quantitative approach. Ironically, it also allows to us to be highly self-critical, a trait that is less obvious in what are termed 'critical' approaches to human geography.

Differences between Quantitative Human Geography and Quantitative Physical Geography

To this point I have ignored the differences between quantitative human geography and quantitative physical geography, and this failing should be redressed. Most, if not all, of the plethora of 'isms' that abound in geography originate from, and dwell entirely within, human geography. Indeed, it could reasonably be argued that the rise of the new 'isms' within human geography, with their non-quantitative or even anti-quantitative stances, has widened the intellectual gap between human and physical geography. This division is now so wide that some physical geographers see little point in remaining in departments that to them appear more rooted in sociology than geography. As Graf notes:

> While their human geographer colleagues have been engaged in an ongoing debate driven first by Marxism, and then more recently by post-structuralism, post-modernism, and a host of other isms, physical geographers are perplexed, and not sure what all the fuss is about ... They do not perceive a need to develop a post-modern climatology, for example,

> **and they suspect ... that some isms are fundamentally anti-scientific. (1998: 2)**

One argument for an increasing emphasis on the quantitative approach within human geography is that it acts as a bridge between human and physical geography rather than as a barrier. The common language and purpose of a quantitative approach allows human and physical geographers not just to talk the same language but also to pursue meaningful joint research.

However, despite the relative similarities in the approach of many quantitative human geographers and their colleagues in physical geography, there are some differences in subject matter that lead to somewhat different analytical approaches becoming prominent in the two areas. At least four such differences are apparent.

For physical geographers their subject matter can sometimes be completely separated from our perceptions and therefore analysed entirely objectively. In some instances this is also true in human geography. For instance, if we are measuring death rates across a region there is no perceptual issue in measuring who is dead and who is not.[3] However, in general, perceptions of the real world are important in understanding much of human geography. Most, if not all, spatial decisions are made on the basis of perceived reality and not reality itself. It might therefore be argued that quantitative approaches that use objective measures of reality to explain human behaviour are inappropriate. Indeed, some would use this argument to defend their own non-quantitative approach to human processes. However, the quantitative approach can be defended in two ways. First, various quantitative models in human geography do take account of people's perceptions of their environment to understand their behaviour in that environment. For instance, quantitative geographers have used information on perceived realities to model spatial choice behaviour such as the choice of shopping destination. People's choices of supermarkets, for example, depend on their perceptions of the various supermarkets they can feasible patronize and on their perceptions of how easy it is to get to each of them. Quantitative geographers have even included within their models people's perceptions of the arrangement of alternatives in space and their mode of making decisions between them. Notions of hierarchical information processing and mental maps are used to produce more accurate models of human spatial decision-making (Fotheringham et al., 2002: Chapter 9). Second, quantitative approaches can be defended by recognizing that in most cases our perceptions of reality closely resemble our objective measurements of reality and that models that use only objective measures of reality are still useful. Arguing that the whole of the quantitative approach should be thrown out because quite often information on perceived reality is not available ignores the fact that models based solely on objective measures of reality are still very useful and far better than any alternative.

The concept of rationality is irrelevant to most physical processes. When we deal with the human world we immediately have to recognize that not everyone will act like automatons and behave in a manner predicted by a mathematical model. Again, some of the adherents of various non-quantitative approaches to human geography seek to use this as an argument against quantification. However, there are two issues that are at odds with such an argument. One is that while we cannot hope to model the actions of each human being, the actions of humans *in aggregate* are often quite predictable. Hence, quantitative models of shopping behaviour by groups of consumers or models predicting population movements between regions are frequently used by private companies and various government agencies. A second is that quantitative models of human behaviour increasingly seek to include seemingly irrational behaviour (see the recent developments in spatial interaction modelling, for example, as described in Fotheringham et al., 2000: Chapter 9). This is a difficult and challenging task but it is one that makes the quantitative approach interesting. It also brings into question whether people ever do act irrationally. Perhaps seemingly irrational behaviour is simply behaviour for which we have not determined the proper set of determinants? For instance, a person in a shopping survey who buys groceries from a store that is 20 miles farther away than another identical one might be deemed to be behaving irrationally. However, the shopping trip might be entirely rational if that person has a relative living close to the selected store and combines a shopping trip with a family visit.

In physical geography there are some fundamental relationships that are the same everywhere. For instance, the rate at which temperature decreases with altitude or the infiltration rate of water through soil are entirely predictable given the right information. This is probably not true in human geography, or at least if there are such fundamentals, we have not yet discovered what they are. Yet again, this seems to have been used as a reason why a quantitative approach to human geography is doomed to failure. However, one only has to look at the huge literature now emerging on local statistics and local modelling techniques to realize that quantitative human geographers have solved this problem by developing techniques that recognize intrinsic local differences in processes (Fotheringham et al., 2002). We now have the tools by which we can measure not only *if* there are local differences but also *what these local differences look like*. The latter provide a very good mechanism for a better understanding of locality as a determinant of human behaviour.

Some results in physical geography can be replicated and in this sense they are truly 'scientific'. Results in human geography are usually not replicable. Due to the nuances of the subject matter in human geography, the calibration of the identical model in two or more different systems generally leads to different results. Fortunately, in some cases these differences are small; in other cases the differences have meaning in terms of the effects of location upon behaviour; in still other cases the results vary because we have a poor model and the variation is therefore a useful diagnostic indicator that we should try to improve the model.

Given the above, it is hardly surprising that quantitative geography has evolved slightly differently in human and physical geography. Typically, human geographers are more concerned with stochastic models because their subject matter is less predictable. Typically, human geographers draw more on concepts from psychology and economics, again because of their subject matter. Some elements of physical geography, such as climatology and meteorology, are more closely allied to physics and others, such as fluvial geomorphology, to engineering.

However, despite these differences there is a great deal of common ground between quantitative human and physical geographers because both groups share an interest in understanding spatial processes and both believe that a better understanding of these processes can be obtained through quantitative analysis. The common framework of quantitative methodology is thus a potentially very powerful mechanism to stop the slow disintegration of geography as a discipline, as the gap between physical geographers and the proponents of various non-quantitative isms in human geography grows ever wider.

What Quantitative Geographers Find Problematic about Some Non-Quantitative Approaches

Until recently, most quantitative geographers tended to view the various non-quantitative approaches within human geography with some bemusement; the currently fashionable 'ism' seemed to change on a 5–10-year cycle and some people seemed to jump on whatever bandwagon rolled by. The only commonality amongst the 'isms' that we could perceive was an anti-quantitative bias. However, the increasingly marginalized treatment given to quantitative methods has produced a reaction not dissimilar to the broader scientific response to the anti-science attacks of the postmodernists. For example, consider the following two quotations:

> **I discern a disturbing implication of relativist accounts of the sciences. A generation ago, the British sociologist Stanislav Andreski depicted the social sciences as sorcery, as gibberish designed to placate special-interest groups (Andreski, 1972). More recently, Alan Bloom compared the humanities to the old Paris flea market: Among the masses of rubbish, one can, by diligent searching, find the occasional under-valued intellectual nugget (Bloom, 1987, p. 371). One's first reaction is to dismiss the authors of the remarks as crabbed reactionaries. But my encounter with cultural studies of science leads me to conclude that such views must be taken seriously. (Sullivan, 1998)**

> **'Whatever the correct explanation for the current malaise, Alan Sokal's hoax has served as a flash point for what has been a gathering storm of protest against the collapse in standards of scholarship and intellectual responsibility that vast sectors of the humanities and social sciences are currently afflicted with ... Anyone still inclined to doubt the seriousness of the problem has only to read Sokal's parody' (Koertge, 1998)**

The Sokal parody referred to can be found in Sokal (1996a; 1996b) and can be downloaded, along with a great deal of other interesting material related to the hoax, at Sokal's website: http://www.physics.nyu.edu/faculty/sokal. For those unaware of it, the paper in question is a complete fabrication published as a hoax in a supposedly highly regarded journal of critical studies. It highlights a major problem that quantitative geographers have with some of the recent 'isms'. If there is no value system whereby research can be assessed, then how does one differentiate 'good' research from 'bad' research?

As an example, try this test. Read the four quotations in Table 22.1. Three of these are either paraphrased or direct quotations from prestigious geographic publications; one of them is complete gibberish. Can you tell which is which? The answer is given in note 4 to this chapter. If you are in *any* doubt about the correct answer, then you probably share at least some of the concerns of quantitative geographers.

The Sokal hoax and the recent attacks on critical studies (see Koertge, 1998 for a sample of these) highlight the current low esteem in which much of the so-called critical theorist school of social science is held by others. It mirrors the view of many quantitative geographers who cannot see what distinguishes good from bad research in much of what now passes for human geography. It is also indicative of the strength of the feelings that have been aroused in some quantitative geographers about what they see as the excessive proliferation of the adherents of anti-science and anti-quantitative views within the discipline. For example, consider this from Openshaw:

> **Maybe human geography is about to experience a new age of extreme technophobic Ludditism advocated by an uncomfortable mix of well-intended scholars and nasty minded voyeurs who are steadfastly intent on navel gazing and nihilistic destruction by seemingly endless and rampant deconstruction of anything for which there is a publication opportunity. (1997b: 8)**

Table 22.1 Sample quotations

To recognize the performativity of discourse is to recognize its power – its ability to produce 'the effects that it names'. But the process of repetition by which discourse produces its effects is characterized by hesitancies and interruptions. Unlike the coherent and rational modernist subject, the poststructuralist economic subject is incompletely 'subjected'. Her identity is always under construction, constituted in part through daily and discontinuous practices that leave openings for (re)invention and 'perversion'.
This reinscription lays bare the constitutive relationship between the conditions that make possible a given phenomenon in the apparent fullness of its identity or meaning, and how these same conditions also mark the impossibility of such phenomena ever being realized in their ideal purity. Deconstruction therefore involves an exposure of conditions of possibility and impossibility. This does not refer to two separate sets of opposed conditions. Rather, possibility and impossibility are doubled up in the same conditions. This doubling of (im)possibility excludes an emphasis solely on the pole of either enabling or disabling conditions.
The dialectic untranslatability between empiricism and 'vanguardist theoreticism' continues to befuddle geographic epistemology. For some, the duality is predetermined as essentially semantic; for others, it represents an oscillation between articulation and disarticulation. However, there is little doubt that the irreducible differences in the hermeneutics of theoretical versus empirical research have created a division that is beyond either ontology or metonymy. The aporetic 'space-between' is an example of a binary conceptualization of methodology that has led to an extreme schism within parts of geography.
The semiotic is where a not-yet subject deals with objects and spaces that are not-yet demarcated. This is a space of 'fluid demarcations' of yet unstable territories where an 'I' that is taking shape is ceaselessly straying, where the not-yet-subject experiences 'above all ambiguity', 'perpetual danger' and is engaged in a 'violent, clumsy breaking away' from the mother. The mapping of the body, then, its initial territorialization, takes place in this primary ('maternal') arena, already social and meaningful (at least partly because the mother is a social subject), but not yet linguistic, that is, prior to the subject's advent into language and the ('paternal') law.

Another problem many quantitative geographers have with some (although not all) of the work done under the banner of various other 'isms' in human geography is that we often cannot see where the 'geography' is in the research. Much of the research published in ostensibly geographical journals looks to us very much like sociology or political science: in many instances, the role of space seems secondary or even non-existent.

If geography is to survive as a discipline, it needs a common theme that separates it from other disciplines. Quantitative geographers make it explicit that we need to investigate *spatial* processes and tend to be quite critical of research conducted under any philosophical banner that purports to be geography but is not concerned with space or spatial issues. We are not convinced that the adherents of some other approaches to the study of geography share our vigilance in this regard.

The final problem quantitative geographers have with the non-quantitative 'isms' concerns the lack of strong and impartial evidence that emanates from these latter domains. Research, for instance, in which only a handful of people might be interviewed, to us does not constitute a reliable sample. Often there is no discussion of how the sample was obtained and what the level of uncertainty is in the inferences drawn from such a small sample (it is undoubtedly extremely high in most cases). In many cases, the writing style resembles more that of a newspaper report in which a few selected quotations are taken (possibly out of context) to support the author's viewpoint. In this way, we see such articles as potentially very biased, inevitably supporting the author's political or cultural stance. Thinking in terms of a court of law, or the quotation from Bradley and Schaefer reported above in which they describe the need to accumulate evidence, to most quantitative geographers a great deal of qualitative 'evidence' just does not stand up to a good cross-examination. In many ways the acceptance of such weak evidence in the various anti-quantitative and anti-science 'isms' is a worrying trend that is mirrored in society by a growing belief in things such as creationism, astrology, angels, alien encounters and faith healing. If there is no logical framework in which to reject false claims, essentially anything becomes acceptable: there is no basis for distinguishing between good research and nonsense – as the Sokal hoax aptly demonstrated.

So Why Has Quantitative Geography Been in Decline until Recently?

It is difficult to say exactly when geographers began to turn to quantitative methods in their search for understanding, but it is generally agreed that it began in earnest at some time in the late 1950s and early 1960s, although much earlier examples of individual pioneering work can be found. Certainly, the decades of the 1960s and 1970s were periods when quantitative methodologies diffused rapidly throughout the discipline. Throughout the 1980s and through much of the 1990s quantitative geography then suffered a reversal of fortune. Elsewhere, my colleagues and I note there are several possible reasons for this (Fotheringham et al., 2000: Chapter 1).

Certainly the growth of many newer paradigms in human geography, such as Marxism, postmodernism, structuralism and humanism (Johnston, 1997; Graham, 1997), attracted adherents united in their anti-quantitative sentiments and their lack of quantitative ability.

There was also the seemingly never-ending desire for some new paradigm or, in less polite terms, 'bandwagon' to act as a cornerstone of geographic research. The methodology of quantitative geography had, for some, run its course by 1980 and it was time to try something new. As de Leeuw observes of the social sciences in general:

> This is one of the peculiar things about the social sciences. They do not seem to accumulate knowledge, there are very few giants, and every once in a while the midgets destroy the heaps. (1994: 13)

Another reason for the relative demise of quantitative geography was that as it developed into a well-established paradigm, it became, inevitably, a focal point for criticism. Unfortunately, much of this criticism originated from individuals who had little or no understanding of quantitative methods. As Gould notes:

> few of those who reacted against the later mathematical methodologies knew what they were really dealing with, if for no other reason than they had little or no mathematics as a linguistic key to gain entry to a different framework, and no thoughtful experience into the actual employment of such techniques to judge in an informed and reasoned way. Furthermore, by associating mathematics with the devil incarnate, they evinced little desire to comprehend. As a result, they constantly appeared to be *against* something, but could seldom articulate their reasons except in distressingly emotional terms. (1984: 26)

A final reason is that quantitative geography is relatively 'difficult', especially for those with limited mathematical or scientific backgrounds. It is perceived by many to be easier to follow other approaches to geographical enquiry and, consequently, they obtain remarkably little exposure to quantitative research even in the substantive areas in which they undertake research. This makes it virtually impossible for many geographers to understand the nature of the debates that have emerged and will continue to emerge within the broad field of spatial analysis. It also makes it tempting to dismiss the whole field of quantitative geography through criticisms that have limited validity rather than trying to understand it. As Robinson states:

> **It can be argued that much of the antipathy towards quantitative methods still rests upon criticisms based on consideration of quantitative work carried out in the 1950s and 1960s rather than upon attempts to examine the more complete range of quantitative work performed during the last two decades. (1998: 9)**

It is true that the early examples of quantitative geography were overly concerned with form rather than with process and with the establishment of nomothetic laws, but the retention of such criticisms as almost an anti-quantitative mantra indicates a woeful lack of understanding of much of the quantitative geography that has been undertaken over the past 20 years. In fact, I suspect many of the criticisms of earlier quantitative work stemmed originally from quantitative geographers themselves who, recognizing some shortcomings within a very youthful part of the discipline, strove to improve it as part of the natural evolution of a vital research area.

The relative difficulty of spatial analysis probably also encouraged some researchers to 'jump ship' from quantitative geography (for some interesting anecdotes along these lines, see Billinge et al., 1984) as they struggled to keep up with the development of an increasingly wide array of techniques and methods. As Hepple (1998) notes:

> **I am inclined to the view that some geographers lost interest in quantitative work when it became too mathematically demanding, and the 'hunter-gatherer' phase of locating the latest option in SPSS or some other package dried up.**

It is difficult to know which of the above reasons explain the actions and attitudes of particular individuals (the last reason is certainly not one to which many will admit), but whatever the cause, it is both a shame and an irony that many geographers choose to remain ignorant about the value of quantitative methods just when there is a rapid and sustained growth in spatial data analysis in other disciplines and in society in general. There is now a strong demand for students who can analyse spatial data and a need for geographers to provide leadership in this area. Unless we increase the understanding and acceptance of quantitative methods within geography, historians of the discipline will probably view this era with some bemusement and a great deal of regret.

Summary

Quantitative methods will always have an important role to play in both human and physical geography, not just for pragmatic reasons but because they provide strong evidence on the nature of spatial processes. Unfortunately, many geography students will never realize the potential of such methods in their own research because they are given outdated and heavily biased views of the use of such techniques, often from individuals who have no direct experience in this area and who seem intent on transmitting their own shortcomings and prejudices to their students. One of the saddest ramifications of the deep philosophical schism within geography currently is that many students are being robbed of a well-rounded geographic education. As Openshaw states:

> **There is a risk of ideological intolerance that has not been so visible previously because the gulf is no longer just philosophical or paradigmatic but is reinforced by serious deficiencies in research training. (1997b: 22)**

Consequently, I pose the following questions that geographers should ask themselves:

- Can I envisage situations where quantitative evidence could be useful to support my own research interests?
- Do I have sufficient knowledge to make a reasoned and impartial judgement about the role of quantitative methods within geography?
- To what extent are my attitudes or those of others towards quantitative methods the result of ignorance and prejudice rather than reason and altruism?

Answering such questions truthfully might eventually lead to a greater awareness and acceptance of what quantitative geography can offer than currently exists.

If geography is to survive as a discipline it needs to demonstrate the following traits:

1. That it has a core subject area, understanding spatial patterns and processes, which defines the subject and differentiates it from all others.
2. That geography is relevant to the world outside academia and that students can gain employment using skills they have learned in geography courses.
3. That human and physical geographers can work together sharing similar goals and methodologies.

Quantitative geography provides all three; some other paradigms currently in vogue in geography appear to provide none, which is why I worry about the future of the discipline.

NOTES

This work has its origins in rather peculiar geographical circumstances. The first draft was written on the island of Rarotonga in the South Pacific. The latitude of the island was matched only by that of the administrator at the University of Newcastle who approved my travel expenses.

1. I will exhibit my philosophical ignorance here (and probably elsewhere) by using the term 'positivism' as shorthand for 'logical positivism', 'logical empiricism', 'scientific empiricism', 'neopositivism', 'logical neopositivism' and probably half a dozen other, related, 'isms'.
2. Following the work of Popper (1959), quantitative geographers tend to follow the principle of critical rationalism: that hypotheses cannot be proven as 'true' but can be demonstrated as 'false'. Strictly speaking this runs counter to the tenet of verifiability and therefore immediately puts most of quantitative geography outside the realms of logical positivism. However, the falsification *modus operandi* is still a very useful process that allows us to be critical of ideas and which is absent in competing philosophies.
3. There is an issue of what the appropriate spatial units are for which death rates are reported, but this is a separate issue.
4. The first quotation is from Gibson-Graham (2000: 104), the second from Barnett (1999: 279) and the fourth from Robinson (2000: 296). The third is a concoction of randomly selected words and phrases from several papers on critical theory and is, as far as I can tell, complete nonsense.

References

Andreski, S. (1972) *Social Sciences as Sorcery*. London: Deutsch.

Barnes, T. (2001) 'Retheorizing economic geography: from the quantitative revolution to the "cultural turn"', *Annals of the Association of American Geographers*, 91: 546–65.

Barnett, C. (1999) 'Deconstructing context: exposing Derrida', *Transactions of the Institute of British Geographers*, 24: 277–93.

Billinge, M., Gregory, D. and Martin, R. (eds) (1984) *Recollections of a Revolution.* New York: St Martin's.

Bloom, A. (1987) *The Closing of the American Mind.* New York: Simon and Schuster.

Bradley, W.J. and Schaefer, K.C. (1998) *The Uses and Misuses of Data and Models: The Mathematization of the Human Sciences.* London: Sage.

Couclelis, H. and Golledge, R. (1983) 'Analytical research, positivism, and behavioral geography', *Annals of the Association of American Geographers,* 73: 331–9.

De Leeuw, J. (1994) 'Statistics and the sciences', unpublished manuscript, UCLA Statistics Program.

Fotheringham, A.S. (1997) Trends in quantitative methods I: stressing the local', *Progress in Human Geography,* 21: 88–96.

Fotheringham, A.S. and Brunsdon, C. (1999) 'Local forms of spatial analysis', *Geographical Analysis,* 31: 340–58.

Fotheringham, A.S., Brunsdon, C.B. and Charlton, M.E. (2000) *Quantitative Geography: Perspectives on Spatial Data Analysis.* London: Sage.

Fotheringham, A.S., Brunsdon, C.B. and Charlton, M.E. (2002) *Geographically Weighted Regression: The Analysis of Spatially Varying Relationships.* Chichester: Wiley.

Gatrell, A.C. (2002) *Geographies of Health.* Oxford: Blackwell.

Gibson-Graham, J.K. (2000) 'Poststructural interventions', in E. Sheppard and T.J. Barnes (eds), *A Companion to Economic Geography.* Oxford: Blackwell, pp. 95–110.

Gould, P.R. (1984) 'Statistics and human geography: historical, philosophical, and algebraic reflections', in G.L. Gaile and C.L. Wilmott (eds), *Spatial Statistics and Models.* Dordrecht: Reidel, pp. 17–32.

Graf, W. (1998) 'Why physical geographers whine so much', *The Association of American Geographers' Newsletter,* 33(8): 2.

Graham, E. (1997) 'Philosophies underlying human geography research', in R. Flowerdew and D. Martin (eds), *Methods in Human Geography: A Guide for Students doing a Research Project.* Harlow: Longman, pp. 6–30.

Hepple, L. (1998) 'Context, social construction and statistics: regression, social science and human geography', *Environment and Planning A,* 30: 225–34.

Johnston, R.J. (1997) 'W(h)ither spatial science and spatial analysis', *Futures,* 29: 323–36.

Kitcher, P. (1998) 'A plea for science studies', in N. Koertge (ed.), *A House Built on Sand: Exposing Postmodernist Myths about Science.* New York: Oxford University Press, pp. 35–56.

Koertge, N. (ed.) (1998) *A House Built on Sand: Exposing Postmodernist Myths about Science.* New York: Oxford University Press.

Openshaw, S. (1997a) 'Building fuzzy spatial interaction models', in *Recent Developments in Spatial Analysis: Spatial Statistics, Behavioural Modelling and Computational Intelligence.* Berlin: Springer, pp. 360–83.

Openshaw, S. (1997b) 'The truth about ground truth', *Transactions in GIS,* 2: 7–24.

Popper, K. (1959) *The Logic of Scientific Discovery.* London: Hutchinson.

Robinson, G.M. (1998) *Methods and Techniques in Human Geography.* Chichester: Wiley.

Robinson, J. (2000) 'Feminism and the spaces of transformation', *Transactions of the Institute of British Geographers,* 25: 285–301.

Sokal, A.D. (1996a) 'Transgressing the boundaries: towards a transformative hermeneutics of quantum gravity', *Social Text,* 46/47: 217–52.
Sokal, A.D. (1996b) 'A physicist experiments with cultural studies', *Lingua Franca*, 6: 62–4.
Sullivan, P.A. (1998) 'An engineer dissects two case studies: Hayles on fluid mechanics and Mackenzie on statistics', in N. Koertge (ed.), *A House Built on Sand: Exposing Postmodernist Myths about Science.* New York: Oxford University Press, pp. 71–98.

23 GEOGRAPHIC INFORMATION SYSTEMS

Michael F. Goodchild

Introduction

Geographic information systems (GIS) are massive software packages providing a range of functions for creating, acquiring, integrating, transforming, visualizing, analyzing, modeling, and archiving information about the surface and near-surface of the earth. They associate locations in space, and often in space–time, with properties such as temperature, population density, land use, or elevation, and are widely used today in support of research in geography, and in any other disciplines concerned with phenomena on or near the earth's surface (for an introduction to GIS see Longley et al., 2001).

The notion that GIS could raise questions of an ethical or a philosophical nature would not have occurred to early developers and users, who tended to see these tools as value-neutral, much as one sees a calculator or kitchen appliance. But in the past 15 years a lively social critique has emerged, and users of GIS have begun to ask new kinds of questions. This chapter is about that critique and its ramifications. The chapter begins with a brief history of GIS. This is followed by a comparison between the value-neutral stance of the early years, and the more human-centric stance that now characterizes the field. Confounding this transformation is another that has occurred in GIS since the popularization of the Internet in the mid 1990s, and which sees GIS as a medium for communication of knowledge, rather than as a personal analytic engine. The chapter reviews the underpinnings of this transformation, and its consequences for the GIS landscape of the new millennium. The final section focuses on the limitations of GIS, and on prospects for further development.

A Short History of GIS

People have been using information technologies to store and handle geographic information for centuries, since well before the advent of digital computers in the twentieth century. Paper is a form of information technology, and paper maps are an ancient way of representing knowledge about the earth's surface. Although a paper map is a fairly cumbersome way of organizing geographic information, it can be reproduced in large quantities very cheaply, and it has significant advantages in providing ready visual access, and in allowing simple measurements of distance (Maling, 1989). The atlas is somewhat more powerful, since its plates can be linked through indexes, but both paper map and atlas technologies have remained essentially limited in their ability to support detailed analysis of mapped information, or accurate measurement.

There are intriguing instances of technical advances on simple paper maps in the historical record: atlases that allowed different plates to be superimposed, in order to compare and combine different themes for the same area, and machines that allowed measurement of area from maps. McHarg (1969) popularized the use of manual map overlay, using simple

BOX 23.1 THE POWER OF GIS

Figure 23.1 shows a simple application of contemporary GIS technology. The area is San Diego, California, showing downtown and Mission Bay, the Mexican border, and part of San Diego County. Several data sets have been accessed and superimposed to create this image. Through an Internet connection, the GIS has accessed the National Interagency Fire Center's GeoMAC Wildfire Information System website and extracted the boundaries of the major fires of October 2003 (the Cedar fire is the large area extending across the top of the image), along with shaded relief and major highways. Census databases have been accessed to obtain the boundaries of census tracts, and these are shown in black. Also available are all of the standard published statistics for these tracts, so the analysis could now proceed to examine relationships between the fire and population density, housing value, and so on. The power of GIS lies in its ability (1) to access and retrieve complex data sets at electronic speed, (2) to superimpose different data sets, allowing comparison and analysis, (3) to apply numerous statistical and other procedures to the data, and (4) to enlist a powerful battery of visualization methods. This illustration was made using ESRI's ArcGIS software (http://www.esri.com) and Geography Network technology (http://www.geographynetwork.com) for discovery and retrieval of data from remote sites.

Figure 23.1 A simple GIS application comparing wildfires with demographic data from the census

transparencies, as a method of combining the various factors involved in complex land use decisions. But the real potential of digital computers to facilitate the analysis of geographic information was not apparent until well into the 1970s. Once the contents of a map are represented in digital form, it becomes a simple matter of programming to obtain accurate measurements of area; to combine and compare different themes; and to conduct detailed statistical analysis.

The actual roots of GIS are complex, and at least four projects, all dating from the 1960s, have at least some claim to the title of originator. Most authorities cite the Canada Geographic Information System (CGIS), designed around 1965 as a means for obtaining summaries and tabulations of areas of land from the Canada Land Inventory, a massive federal-provincial effort to assess the utilization and potential of the Canadian land base. In essence, CGIS was designed to solve two simple, technical problems: to measure accurately the areas of irregular geographic patches of homogeneous utilization or potential, and to overlay and compare different themes. It is remarkable that a detailed cost/benefit analysis clearly demonstrated the net benefit of computerization, despite the very limited capabilities and massive costs of computers at the time. Other contributions to the origins of GIS came from the Chicago Area Transportation Studies, from efforts to computerize the geographic aspects of the 1970 US Census, and from the computerization of McHarg's overlay process by landscape architects (for a comprehensive history of GIS see Foresman, 1998; Maguire et al., 1991). In all of these cases GIS was seen as a cost-effective technical solution to a simple administrative problem.

The period of the 1970s was characterized by rapid invention, and the solution of a wide range of technical issues that stood in the way of successful GIS. The costs of computers remained high, however, and it was not until the early 1980s that GIS started to become popular as a standard computer application in government departments, universities, and private corporations. Two significant innovations led to the first commercial viability of GIS: the development of relational database management systems, which took over the details of data management, allowing GIS designers to concentrate on measurement and analysis; and the near-order-of-magnitude fall in the price of computers with the introduction of the mini-computer. Today, of course, the speed of the average desktop computer is orders of magnitude higher, and the cost is orders of magnitude lower.

Once the foundations for handling a particular type of information have been built, it is possible for programmers to add new functions very rapidly and cheaply, and over the past three decades the power of GIS has grown explosively. Today, a typical GIS is capable of handling all of the major types of geographic information and of performing a vast array of functions, from visualization and transformation to detailed analysis and modeling. This power, and the low cost of entry into GIS, have meant that GIS tools are now widely adopted for purposes ranging from administration to scientific research, and from education to policy formulation. Virtually any field that deals with the surface and near-surface of the earth now accepts GIS as essential to its success, and ranks it with other computer applications such as the statistical packages, email, or word processors as a permanent feature of the information technology landscape.

Underlying this increasing reliance on GIS is a simple proposition: that numerical analysis of geographic data is too tedious, inaccurate, or costly to perform by hand. The computer becomes in effect a calculating device, the personal servant of the researcher or administrator. It would probably have surprised observers in the 1960s that computers designed largely to perform vast numbers of calculations could be useful for the analysis of maps, and it is no accident that the development of GIS coincided with the turn to quantitative analysis in geography. Quantitative geographers were quick to recognize the power of GIS for spatial analysis, just as statisticians had exploited the power of computers to perform statistical analysis (Fotheringham and Rogerson, 1994; Goodchild, 1988; O'Sullivan and Unwin, 2003). The commercial GIS industry is driven by its sales to users, however, and the most lucrative markets for GIS software often turned out to be those that used GIS for inventory and management, rather than for sophisticated analysis and modeling. Thus an important part of the academic agenda in GIS over the past three decades has been to push for more powerful spatial analytic capabilities in commercial software. The academic research community has also developed GIS packages of its own, that are more closely aligned in design with the needs of geographic research.

The Emergence of Critique

To its early users a GIS was simply a tool, designed to perform a series of straightforward mechanical operations. To those who had programmed them, or understood how they were put together, they seemed as value-neutral as the hand calculator, the typewriter, or the toothbrush. But to those outside the field, the fascination with GIS that emerged in the late 1980s seemed much more mysterious and potentially sinister. Users of GIS seemed to be saying that they could see patterns and trends in data, through the medium of GIS, that others could not see. Students in GIS classes seemed to be approaching these systems as black boxes, and not demanding the same kind of detailed understanding that one would expect, for example, with hand calculation or manual measurement.

In the late 1980s a series of papers by Brian Harley examined the notion that maps can reveal the hidden agendas of their makers (many of the papers are available as a posthumously published collection: Harley, 2001). This concept of *deconstructing* the map attracted significant attention, as it addressed an issue that had long simmered under the surface of cartography, and more generally of geography: the role of both disciplines in establishing and modifying power relations. Mapping and surveying were important instruments of the imperial powers of the nineteenth century, and geography has always flourished in wartime, in response to the demands of armies for detailed geographic information and analysis. Harley's theme was taken up by several followers, including Denis Wood, whose book *The Power of Maps* (1992) accompanied an exhibition mounted by the Smithsonian Museum and included a chapter entitled 'Whose agenda is in your glove compartment?'

At first it seemed that GIS would be somehow insulated from these critiques, which were directed more at traditional cartography. But a series of papers in the early 1990s made it clear that these and many other dimensions of critique could be leveled at GIS. Taylor (1990) focused on the perception of GIS as an inventory of facts about the earth's surface, and the distinction between such a concern for facts and the traditional and more fundamental concern of academic geography for more sophisticated forms of knowledge, particularly of how the geographic landscape is impacted by human and physical processes, and contemporary human geography's concern for power relations. To Taylor, GIS seemed mired in the trivial, and unlikely to add to geography's understanding of the world; what were needed instead were geographic knowledge systems, or GKS. Moreover, GIS seemed to be the direct descendant of the concern in the 1960s and 1970s for a scientific approach to geography, under a positivist umbrella, and to be ignoring the extensive critiques of that approach: GIS was 'the positivists' revenge' (but see my response, Goodchild, 1991; and the extended debate between Taylor and Openshaw in Openshaw, 1991; 1992; Taylor and Overton, 1991).

Smith (1992) added more fuel to the critical fire in a paper that addressed the military and intelligence applications of GIS. As noted earlier, geography and GIS have a particular relevance to warfare, and it is possible to trace many of the technical developments in GIS and remote sensing to the military. Remote sensing developed as an intelligence-gathering activity in the Cold War; modern geodesy owes much of its progress to the need for accurate targeting of intercontinental ballistic missiles; much of the technology for the digital representation of terrain was developed in support of cruise missile programs; and so on. Smith noted the distinct absence of open academic literature on military applications of GIS, and a distinct reluctance to acknowledge its military roots, and argued that GIS developers should bear some degree of responsibility for the use of their technology, in an echo of the debates over the Manhattan project among that era's scientific leadership. Do developers of technology bear responsibility for the eventual uses of that technology, or is development inherently value-neutral, independent of eventual use? The debate has become stronger through the 1990s as details have emerged about the close historic links between GIS and the military and intelligence communities. Cloud and Clarke (1999) have studied the Corona spy satellite program of the 1960s, and documented the close relationship between that secret program and civilian mapping, and the extent to which much of the early technical progress on GIS and remote sensing depended on classified research.

The publication in 1995 of John Pickles' book *Ground Truth* was perhaps the most significant event in this early period, as the critique was gathering momentum. Chapters in this edited collection reworked much of the earlier ground, while adding new dimensions, such as the potential role of GIS in surveillance and the invasion of personal privacy (Curry, 1997; 1998). Some of it was clearly misguided, driven by naive assumptions about GIS. For example, much was made of the representation of geographic phenomena as a series of layers, a common icon of GIS but by no means a requirement of GIS representations. Similarly, much was made of the limitations of crisp classification, a practice that was inherited from maps of land use, land cover, or soil type, despite the fact that in principle GIS allows geographers to move beyond such crisp limitations into fuzzy classifications and other approaches that are less distorting of geographic reality (Burrough and Frank, 1996).

In summary, the critiques focused on two themes: the limitations of GIS representations, through an inability to represent aspects of phenomena that are of particular interest to researchers, and the implications of those limitations for power relations; and the misuse of GIS for sinister, malevolent, manipulative, or unacceptable purposes. Both clearly have substance, and in the years since they first appeared many efforts have been made to address them, in various ways. The critiques have also been clarified, and it is probably true to say that naive assumptions about GIS are no longer as common. An extensive discussion of the debate, its highlights, and its outcomes has been published by Pickles (1999), and my interview with Schuurman (1999) also provides an overview. Conferences on the social context and impacts of GIS have been convened, an extensive literature has accumulated, and the topic appears on most published research agendas for GIS (UCGIS, 1996). Today, the old view of GIS as a personal assistant found on the desk of a researcher, and performing tasks in an essentially value-neutral context that the researcher finds too tedious, inaccurate, or time consuming to perform by hand has largely disappeared, to be replaced by a more human-centered view in which users are now much more sensitized to the social context of GIS use, the comparative arbitrariness and non-replicability of many of its tasks, and the unattainability of the pure scientific concept of objectivity.

On the other hand, most GIS users would subscribe to the notion that objectivity is an important goal, and that every effort should be made to adhere as closely as possible to the norms of science. It is only through such norms, they would argue, that the work of one person can be fully understood by another; and accepted by the broader community, or a court of law, as a reasoned and appropriate approach to a problem.

One of the most focused topics to emerge from the debate has been public-participation GIS (PPGIS), the study of the use of GIS in community decision-making. Interest in PPGIS arose from the following argument. In the early days of GIS the necessary hardware and software were available only to government agencies and corporations, because of their high cost. The geographic data needed to populate the GIS were produced by government, through agencies such as the US Geological Survey, again because of the high cost and high levels of expertise needed. The perspectives advanced by GIS were therefore those of the powerful, and GIS became an instrument of that power, and a means of reinforcing it. GIS designs accommodated only a single perspective, which its managers and protagonists held to be the truth. In reality, many geographic variables are vague, and cannot satisfy the scientific norm of replicability; two specialists asked to map independently the soils of the same area will not produce identical maps.

What, then, would evolve if GIS were redesigned from scratch, as a vehicle for the maintenance of multiple viewpoints, rather than the privileging of a single viewpoint? The concept was code-named 'GIS/2', and great interest was expressed in research into its design, while the broader subject of decision-making in diverse communities evolved into the research area of PPGIS. Progress has been made in understanding how geographic information is created within communities; how decisions are made using GIS in complex settings; how multiple perspectives can be identified, stored, and displayed in GIS; and how GIS and other information technologies impact the balance of power within communities (Craig et al., 2002; Jankowski and Nyerges, 2001; Thill, 1999).

The Impact of the Internet

Although the earliest efforts to link computers through wide-area networks occurred in the 1960s, the growth in use of the Internet and the invention of the World Wide Web in the early 1990s came as an almost complete surprise to most observers. In less than a decade, these networking technologies have completely revolutionized the world of computing, creating massive demand where virtually none existed before, and helping to redefine the modern economy. Nowhere has the impact been more dramatic than in GIS, where these new technologies have enabled a vast new industry dedicated to the sharing and dissemination of geographic information (Peng and Tsou, 2003; Plewe, 1997). In the early years of GIS much effort had to be devoted to the task of digitizing, the conversion of paper and photographic records to digital form. Today, few GIS projects require extensive digitizing, since the necessary data are almost always available in some suitable form somewhere on the WWW.

Several new technologies have been developed over the past decade to facilitate data sharing. They include digital libraries, and the special form known as the geolibrary (National Research Council, 1999), defined as a collection of information objects that is searchable by geographic location as the primary key. Since this was essentially impossible in the predigital era, geolibraries are an important innovation of the Internet age. The Alexandria Digital Library at the University of California, Santa Barbara (http://www.alexandria.ucsb.edu) is an instance of a geolibrary, and currently provides access to several million information objects. Standards have been devised for the description of data sets, enabling the creation of vast catalogs of records, and automated procedures for searching across catalogs in distributed archives. The National Geospatial Data Clearinghouse (http://www.fgdc.gov/clearinghouse/clearinghouse.html) is an instance of a geolibrary that uses a standard metadata format to tie together several hundred archives of geographic data, allowing them all to be searched by a single operation. Standards have been devised to support automated requests for geographic information from remote archives (http://www.opengeospatial.org), allowing any GIS user to access remote data as easily as data stored on the user's own hard drive.

Of equal importance to these technical developments have been the institutional changes that have occurred in the past decade. US national policy regarding geographic data production and dissemination is now enshrined in the National Spatial Data Infrastructure (National Research Council, 1993), a collection of arrangements and protocols coordinated by the federal government through the Federal Geographic Data Committee (http://www.fgdc.gov). These include the concept of patchwork: that the production of geographic data should be distributed, and conducted at scales appropriate to local needs, with information technology providing the means to coordinate the patchwork into a consistent whole. Production should be distributed over many levels of the administrative hierarchy, and conducted in partnership with the private sector, through arrangements that largely replace the old, centralized system of production by the federal government.

Sui and I (Goodchild, 2000; Sui and Goodchild, 2001; 2003) have argued that these changes constitute a fundamental paradigm shift in GIS, from the old model of an intelligent assistant serving the needs of a single user seated at a desk, to a new model in which GIS acts as a medium for communicating and sharing knowledge about the planet's surface. The shift of paradigm implies a simultaneous shift of technical focus, from local performance to network bandwidth, and an increasing interest in issues of semantic interoperability in place of earlier concerns with syntactic interoperability: in other words, sharing requires a common understanding of meaning, as well as a set of common standards of format. From this new perspective the earlier concern for spatial analysis is relegated to a subsidiary role, as a means for enhancing the message, and for drawing attention to patterns and anomalies in geographic information that receivers of the information might not otherwise notice.

From the perspective of the social critique, this attention to the communication of geographic information raises a series of interesting questions. Are there types of geographic information that cannot be represented in GIS, or to which GIS is inherently hostile, and that cannot therefore be communicated through the medium of GIS? Is it possible to rank types of geographic information according to their ease of representation and communication? And to what extent does GIS impose itself as a filter, compared to other media for communication, such as speech, or the written word?

Technologies such as the National Geospatial Data Clearinghouse provide a single WWW portal to a distributed collection of archives. This notion that information technology is capable of integrating disparate data sources, and providing a uniform view, has stimulated more far-reaching visions, culminating in the vision of Digital Earth. The concept was originally proposed by Gore (1992), and substantially elaborated during his vice-presidency. The most comprehensive statement of the vision appears in a speech written in 1998 for the opening of the California Science Center, of which the following is a key passage:

Imagine, for example, a young child going to a Digital Earth exhibit at a local museum. After donning a head-mounted

> **display, she sees Earth as it appears from space. Using a data glove, she zooms in, using higher and higher levels of resolution, to see continents, then regions, countries, cities, and finally individual houses, trees, and other natural and man-made objects. Having found an area of the planet she is interested in exploring, she takes the equivalent of a 'magic carpet ride' through a 3–D visualization of the terrain. (http://digitalearth.gsfc.nasa.gov/VP19980131.html)**

Digital Earth poses what one might regard as the grand challenge of GIS: the creation of a single, unified perspective on distributed geographic information, together with the ability to visualize that information in a virtual reality. I have argued (Goodchild, 2001) that Digital Earth is feasible with today's technology, despite the enormous volumes of data potentially involved, and the high rate at which those data would need to be accessed, and today Google Earth (http://www.google.com) exhibits many of the features of the Digital Earth vision. But Digital Earth raises numerous questions of an ethical and philosophical nature, and some of these are reviewed in the next section.

Outstanding Issues

Although the debates of the 1990s led to accommodation and reconciliation on some issues, a number of more fundamental issues continue to haunt the world of GIS, and its relationship with critical social theory. In this section two such issues are reviewed, as highlights of the current state of the GIS critique.

Mirror worlds and uncertainty

To the GIS user seated at a desk, the computer screen provides a window on the geographic world. But the contents of the window are entirely determined by the contents of the database, rather than by reality. In essence, GIS use occurs in a virtual environment that acts as a replacement for physical presence in the real world (Fisher and Unwin, 2002). Since the user has no sensory contact with that real world during GIS use, unless the application concerns the user's immediate environment, it follows that his or her understanding of the world is limited to the contents of the database, together with any information already stored in the user's brain. Moreover, the nature of digital representations, which are constrained to an alphabet of two symbols, necessarily means that the virtual world is far cruder and truncated than the real world that it attempts to emulate. In effect, GIS is a mirror world, a representation that purports to be a faithful copy of certain aspects of the real world.

The contrast between reality and its virtual mirror is nowhere more obvious than in the vision of Digital Earth, as captured in the Gore quotation above. Nothing in the vision suggests that the world the child is exploring is limited in any way; the message is clearly that the child's exploration is the equivalent in every way of real exploration. This cannot be so, but one inevitably wonders if the child will realize this, or will simply accept the magic carpet ride as truth.

Over the past decade the topic of uncertainty has emerged as one of the most significant in the GIScience research agenda. Uncertainty is defined through the relationship between the real and virtual worlds, as the degree to which the virtual world leaves its user uncertain about the real world. Uncertainty arises from numerous sources, including traditional measurement error, the vagueness inherent in many of the definitions that are used in compiling geographic data, the process of generalization which omits excessive detail in many data sources, and the approximations that result from economies in data compilation or representation. Zhang and I (Zhang and Goodchild, 2002) provide a recent review of the state of research in the field.

Despite the inevitable presence of uncertainty in geographic representations, there are few if any accepted methods for making users aware of that uncertainty during GIS use. While most uncertainties are clearly benign, the possibility exists to manipulate the differences between the real and virtual worlds to specific ends. A variety of methods have been proposed for visualizing uncertainty, through the graying or blurring of objects, or through animation (Hearnshaw and Unwin, 1994). But despite the extensive research progress of the past two decades, in practice most GIS applications still present data as if they were the truth, opening the possibility of legal liability, as well as of social critique. The developers of commercial software seem unwilling to implement even the simplest methods for tracking, visualizing, and reporting uncertainty, preferring to wait until such methods are demanded by their customers; and in this situation, the academic community clearly has a responsibility to continue to draw attention to the problems of uncertainty, to educate future GIS users to be aware of them, and to make simple solutions available in the form of software tools and readily implemented methods.

Scientific norms

Mention was made earlier of the difficulty of achieving the elusive scientific goals of objectivity in a world of GIS that must accept degrees of subjectivity in some of its aspects. As Harley and Wood showed, maps are not always the simple results of scientific measurement that many GIS users assume them to be. Cartographers frequently distort contours, or move roads apart, in order to make maps more readable, and to convey general impressions, rather than precise representations. While map readers are used to such practices, and anyway lack the tools to make precise measurements from maps, the precision of GIS often encourages a false sense of objectivity, and a false belief in GIS accuracy. Measurements of area made in a GIS are typically printed to the limits of the computer's precision, and are generally far more detailed than is justified by the true accuracy of the data.

For most types of geographic data there is a clear relationship between the cost of data on the one hand, and their accuracy and level of detail on the other. For example, greater accuracy in the measurement of elevation requires greater expenditure, because fully automated systems such as interferometric radar or LiDAR cannot produce the same levels of accuracy as expensive, ground-based measurement. Greater levels of detail, in the form of denser sampling, similarly add to the cost of data capture and compilation. Clearly such situations require compromises between the scientific desire for the truth and the practical cost of accuracy. But science provides no framework for making such compromises between its pure objectives and the practical realities of decision-making and problem-solving.

Another conflict over scientific norms occurs in the application of GIS to decision-making. Consider an increasingly common situation – the use of a GIS to evaluate alternative options against multiple criteria. Such evaluations often occur in meetings of stakeholders, and several methods have been devised for formalizing the process of determining weights for the various criteria, and obtaining a consensus solution. Saaty's (1980) analytic hierarchy process is one of the best known: comparative ratings of the importance of various factors are elicited from stakeholders, and then analyzed to produce a set of consensus weights using a fairly common form of matrix algebra. In such situations it is clear that the degree of satisfaction of the stakeholders with the process is the primary measure of success. Stakeholders will often express support in principle for the goals of science, and will be satisfied only if they believe the approach to be scientific – but the precise meaning of that term, and the precise process by which the approach was determined to be scientific, is often unclear, as it will be to a participant unfamiliar with basic matrix manipulation.

Finally, consider the traditional scientific standard that results should be reported in sufficient detail to allow replication of the experiment. Much modern science is conducted by teams, each member being a specialist in one field, and no member being a specialist in all fields. Moreover, it is common today for the computer to be in effect a member of the team, programmed by an unknown programmer who is not a member of the team. The levels of documentation commonly available for GIS software often do not provide sufficient detail to allow replication, let alone reprogramming. As a result much modern science, and many GIS applications, fail to satisfy the reporting standard.

It is clear from these examples that modern science, typified by the use of GIS in team research, often deviates from the traditional methodological principles that were established in an earlier era of monastic, single-investigator science underpinned by manual analysis. A new philosophy of science is badly needed that helps today's scientists to operate effectively in a world of massive computation, team research, and complex interactions with the real world of decision-makers.

Conclusions

The critique of GIS that emerged in the early 1990s was in large part well founded, and resulted in a lively and largely productive debate. There is no doubt that GIS is better for the experience, that GIS users are more sensitized to the social implications and social context of what they do, and that much important literature has accumulated. Their knowledge of the real world and of the social critique positions geographers well to lead the application of GIS, and to educate and persuade users from other disciplines to be similarly sensitive to the complex relationships that exist between the technology, the real world, and society. For the most part, however, users of GIS, particularly those outside the scientific research community, will continue to show impatience with issues of scientific philosophy, to echo the comments of Fotheringham (Chapter 22) and Kitchin (Chapter 2) in this book.

Two outstanding issues were identified in the previous section. One, the implications of uncertainty, has been the focus of much research over the past 15 years, but remains a major area of concern. The GIS vendor community has made it clear that it will respond with implemented methods only when the market demands them; one senses a distinct tendency to sweep the uncertainty issue under the carpet. Yet many GIS applications are subject to massive uncertainties, and as a result their outputs are similarly uncertain. A cynic once described soil maps as showing 'boundaries that do not exist surrounding areas that have little in common', and although this comment is decidedly unfair, it contains enough truth that even the staunchest advocates of GIS should be concerned. Uncertainty might be described as the Achilles' heel of GIS; or to use another metaphor, the problem that has the potential to bring down the entire house of cards.

The other issue identified in the previous section concerns the methodology of science, and the danger that when GIS applications move too far away from the context of scientific peer review and rigorous analysis their essential objectivity will be badly compromised. As late as the 1970s it was common for instructors in statistics classes to require that their students first execute every test by hand, before being allowed to use computers; it was felt that hand computation was more conducive to student understanding. Today, of course, this principle and its relatives are long forgotten, in a world brimming with such mental aids as calculators, spelling and grammar checkers, and computerized driving directions. Amid these practical realities, it is surely more important than ever to demand detailed software documentation, and to be aware of what is happening inside the increasingly complex and opaque 'black box'.

Although there are celebrated exceptions, for the most part the technology that is GIS plays a value-neutral role in a science that strives to adhere to principles of objectivity. There is a social context for GIS, and the history of the field depends to a substantial extent on the personalities of its pioneers. But this seems a far cry from the opposite extreme, of treating GIS as a social construction, and giving equal time to all of its possibilities, however far they stray from scientific norms.

References

Burrough, P.A. and Frank, A.U. (eds) (1996) *Geographic Objects with Indeterminate Boundaries.* London: Taylor and Francis.

Cloud, J. and Clarke, K.C. (1999) 'Through a shutter darkly: the tangled relationships between civilian, military and intelligence remote sensing in the early U.S. space program', in J. Reppy (ed.), *Secrecy and Knowledge Production.* Occasional Paper 23. Ithaca, NY: Cornell University Peace Studies Program.

Craig, W.J., Harris, T.M. and Weiner, D. (eds) (2002) *Community Participation and Geographic Information Systems.* New York: Taylor and Francis.

Curry, M.R. (1997) 'Geodemographics and the end of the private realm', *Annals of the Association of American Geographers,* 87: 681–9.

Curry, M.R. (1998) *Digital Places: Living with Geographic Information Technologies.* New York: Routledge.

Fisher, P.F. and Unwin, D.J. (eds) (2002) *Virtual Reality in Geography.* New York: Taylor and Francis.

Foresman, T.W. (ed.) (1998) *The History of GIS: Perspectives from the Pioneers.* Upper Saddle River, NJ: Prentice Hall.

Fotheringham, A.S. and Rogerson, P. (1994) *Spatial Analysis and GIS.* London: Taylor and Francis.

Goodchild, M.F. (1988) 'A spatial analytic perspective on geographical information systems', *International Journal of Geographical Information Systems,* 1: 327–34.

Goodchild, M.F. (1991) 'Just the facts', *Political Geography Quarterly,* 10: 192–3.

Goodchild, M.F. (2000) 'Communicating geographic information in a digital age', *Annals of the Association of American Geographers,* 90: 344–55.

Goodchild, M.F. (2001) 'Metrics of scale in remote sensing and GIS', *International Journal of Applied Earth Observation and Geoinformation,* 3: 114–20.

Gore, A. (1992) *Earth in the Balance: Ecology and the Human Spirit.* Boston, MA: Houghton Mifflin.

Harley, J.B. (2001) *The New Nature of Maps: Essays in the History of Cartography.* Baltimore, MD: Johns Hopkins University Press.

Hearnshaw, H.M. and Unwin, D.J. (eds) (1994) *Visualization in Geographical Information Systems.* New York: Wiley.

Jankowski, P. and Nyerges, T. (2001) *Geographic Information Systems for Group Decision Making: Towards a Participatory Geographic Information Science.* New York: Taylor and Francis.

Longley, P.A., Goodchild, M.F., Maguire, D.J. and Rhind, D.W. (2005) *Geographic Information Systems and Science,* 2nd edn. New York: Wiley.

Maguire, D.J., Goodchild, M.F. and Rhind, D.W. (1991) *Geographical Information Systems: Principles and Applications*. Harlow: Longman.

Maling, D.H. (1989) *Measurement from Maps: Principles and Methods of Cartometry*. New York: Pergamon.

McHarg, I.H. (1969) *Design with Nature*. Garden City, NY: Natural History Press.

National Research Council (1993) *Toward a Coordinated Spatial Data Infrastructure for the Nation*. Washington, DC: National Academy Press.

National Research Council (1999) *Distributed Geolibraries: Spatial Information Resources*. Washington, DC: National Academy Press.

Openshaw, S. (1991) 'A view on the GIS crisis in geography: or, using GIS to put Humpty Dumpty back together again', *Environment and Planning A*, 23: 621–8.

Openshaw, S. (1992) 'Further thoughts on geography and GIS: a reply', *Environment and Planning A*, 24: 463–6.

O'Sullivan, D. and Unwin, D.J. (2003) *Geographic Information Analysis*. Hoboken, NJ: Wiley.

Peng, Z.R. and Tsou, M.H. (2003) *Internet GIS: Distributed Geographic Information Services for the Internet and Wireless Network*. New York: Wiley.

Pickles, J. (ed.) (1995) *Ground Truth: The Social Implications of Geographic Information Systems*. New York: Guilford.

Pickles, J. (1999) 'Arguments, debates and dialogues: the GIS – social theory debate and the concern for alternatives', in P.A. Longley, M.F. Goodchild, D.J. Maguire and D.W. Rhind (eds), *Geographical Information Systems: Principles, Techniques, Management and Applications*. New York: Wiley, pp. 49–60.

Plewe, B. (1997) *GIS Online: Information Retrieval, Mapping, and the Internet*. Santa Fe, NM: OnWord.

Saaty, T.L. (1980) *The Analytic Hierarchy Process: Planning, Priority Setting, Resource Allocation*. New York: McGraw Hill.

Schuurman, N. (1999) 'Speaking with the enemy: an interview with Michael Goodchild, January 6, 1998', *Environment and Planning D: Society and Space*, 17: 3–15.

Smith, N. (1992) 'Real wars, theory wars', *Progress in Human Geography*, 16: 257–71.

Sui, D.Z. and Goodchild, M.F. (2001) 'Guest editorial: GIS as media?', *International Journal of Geographical Information Science*, 15: 387–9.

Sui, D.Z. and Goodchild, M.F. (2003) 'A tetradic analysis of GIS and society using McLuhan's law of the media', *Canadian Geographer*, 47: 5–17.

Taylor, P.J. (1990) 'GKS', *Political Geography Quarterly*, 9: 211–12.

Taylor, P.J. and Overton, M. (1991) 'Further thoughts on geography and GIS', *Environment and Planning A*, 23: 1087–90.

Thill, J.C. (1999) *Spatial Multicriteria Decision Making and Analysis: A Geographic Information Sciences Approach*. Brookfield, VT: Ashgate.

UCGIS: University Consortium for Geographic Information Science (1996) 'Research priorities for geographic information science', *Cartography and Geographic Information Systems*, 23: 115–27.

Wood, A.D.B. (1992) *The Power of Maps*. New York: Guilford.

Zhang, J.X., and Goodchild, M.F. (2002) *Uncertainty in Geographical Information*. New York: Taylor and Francis.

24 HUMANISM AND PEOPLE-CENTRED METHODS

Paul Rodaway

Introduction

Humanism has variously impacted upon geographic practice, most notably in the second half of the twentieth century as in large part a reaction against positivist social science. Manifest as 'humanistic geography', it gave geographers the opportunity to reassert the importance of human experience, that is a concern with the individual and the unique, the subjective experience of people and place, a geography of feeling and emotion, involvement and participation. Humanistic geographers developed a distinct research strategy and a series of people-centred methods (or adaptations of existing methodologies) to explicate a detailed and reflective understanding of the relationship between people and place, a geography of the world as home (Tuan, 1974; 1977; Relph, 1976; 1985; Seamon and Mugerauer, 1985). Reflecting back over more than a decade of humanistic geography, Pocock was to note the key characteristic: 'a sensitivity towards, respect for, empathy – or communion – with the people and places being studied' (1988a: 3).

Humanistic geographies were underpinned to varying degrees by the philosophies of idealism, existentialism and in particular phenomenology (especially the work of Heidegger, Husserl, and Merleau-Ponty). In contrast to positivist geographers (with their itemization of facts, search for general laws, and causal explanations), humanistic geographers were more concerned with the subjective experience, the particular and the unique. The focus was an interpretation and reflection on the meaning of what it is to be human and living in a world (Tuan, 1979; Relph, 1985). Humanistic approaches sought to explicate the meaning of individual (and social) experience, and the sense of place (and a world) as it were from within or with the flow of the living wholeness of being, that is our being-in-the-world or dwelling (see Heidegger, 1983). The researcher is always and already embedded in the world he/she studies and this study impacts on one's own sense of self, as well as one's understanding of the world (see Pocock, 1988a: 6). There can be no 'facts' unaffected by the personal values of the researcher (Olsson, 1980).

It has been observed that 'methodologically, however, humanistic ideas of phenomenology and existentialism did not translate easily into practice' (Hubbard et al., 2002: 41). Nevertheless, a number of leading geographers did successfully develop practical research methodologies (e.g. Rowles, 1976; Seamon, 1979; Pocock, 1992). Some humanistic geographers adopted self-reflective strategies which were in essence 'researcher centred', or relied exclusively upon the interpretation by the researcher of cultural texts, images and practices. This focus on a close reading and critical reflection aimed to discern the essential and unique character of particular places and communities (e.g. Tuan, 1974; 1979; 1993; Pocock, 1981; Seamon, 1985). The interpretation is ultimately a personal one, and tells us as much about the researcher as what is researched perhaps (Monaghan, 2001).[1]

People-centred methodologies take a more empirical approach and have been developed by humanistic geographers to explicate and explore the geographical experience of individuals and communities, in particular places and times. Here methods of direct encounter, engagement and participation were developed (e.g. Rowles, 1976; Seamon, 1979; Pocock, 1996; Meth, 2003). The present essay concentrates upon this humanism as a practical approach (Harper, 1987).

Characteristics of a Humanistic Strategy

Despite reference to humanistic philosophy, geographers tended not to adopt specifically philosophical techniques but to more pragmatically adapt existing research methodologies in geography, social science and the humanities. The key characteristic is therefore one of approach, the choice of subject matter to study, the definition of evidence and the basis of truth claims.

In particular, three key and interlinked ideas underpin this research strategy:

- A desire to avoid imposing preconceived ideas (concepts, theories) and to get 'back to the things themselves' (see Husserl, 1983). The phenomenological philosopher Heidegger describes this approach or attitude as 'to let that which shows itself be seen from itself in the very way in which it shows itself from itself' (1983: 58). Fundamental to this attitude is humility, respect and a kind of empathy with the world (see Pocock, 1988a; Relph, 1985; Tuan, 1971).
- An essentially anthropocentric or people-centred approach in the sense that all meaningful knowledge is understood as that which begins and ends with human 'intentionality', that is subjectivity. In other words, knowledge of the world derives from human consciousness and our relationship to other things (objects, people, places) that make up our everyday individual and social environment (or 'lifeworld'). Our knowledge of the world arises though our conscious relationship to that world – feelings and emotions, memories and expectation. It is not possible to sustain an objective and detached view of the world. What we study affects us and we affect it.
- Understanding is essentially holistic, but partial and implicative, as research seeks to appreciate the complexity of our 'being-in-the-world' or 'dwelling' (see Heidegger, 1983). Humanistic geographers are therefore interested in complex and dynamic wholes – experiences, places, lifeworlds.

As Pocock summarizes this strategy:

> **the humanist rejects the dualism of an outer, objective world and an inner, subjective world or representation. The world is the 'lived world' and *is* what it seems – which is not to admit solipsism, which is where the mind creates its own world. There are multiple-emergent worlds or realities which can only be studied holistically. Again, the humanist rejects that the knower and the known constitute a discrete dualism; rather, they are inseparable, interacting and influencing each other. Consequently, any enquiry is value-bound. (1988a: 2)**

In many ways, humanism interpreted in this way generates an approach or attitude rather than a set of specific and unique people-centred methods. The emphasis is on holism, participation, empathy, explication, induction, authentication and trust.

Humanistic geographers have therefore adapted and developed a number of existing methodologies – notably participant observation, in-depth interviewing, and group-based approaches – supplemented with the interpretive reading of texts, images and cultural practices and critical reflection on their own involvement in the research process itself as a 'participant' in the world being studied.

Empathy and Experiential Knowing

Over a number of years, Douglas Pocock has explored place evocation through a series of empathetic and experiential research strategies. Examples include producing a tape recording of the *Sound Portrait of a Cathedral City* (Pocock, 1987; commentary, 1988b), engaging with the tourist experience of 'Catherine Cookson Country' (Pocock, 1992) and developing practical fieldwork exercises to excite students' experiential engagement and understanding of the environment (Pocock, 1983). Underlying his long-term commitment to a practical and people-centred approach has been a belief that 'an epistemology of the heart concerns knowledge acquired by union or communion. It is founded on an intimate engagement, an intimate sensing' (Pocock, 1996: 379).

This personal approach is perhaps most strongly illustrated in his extended study of place evocation in the Galilee Chapel in Durham Cathedral (Pocock, 1996). His approach is grounded in phenomenology, but is practical, eclectic and self-reflective. Central to this project was over two decades of being 'engaged on a personal odyssey, compiling a personal diary' where 'discourse and text have thus been a personal affair, representing a struggle to give outward expression to a myriad of inner feelings and promptings' (1996: 384). Through engagement with the literature about the history and meanings associated with the Chapel, repeated personal visits and contemplation, and observations of and conversations with other visitors, Pocock has sought to engage with the 'unique character of this place'.

This empathy and experiential knowing requires respect, patience and critical reflection, drawing on personal experience, literature and observation. It is a seeking for authentic knowledge. Pocock's approach is holistic, experiential, intimate (sensually and emotionally) and (self-) reflective (see also Pocock, 1983). It does not seek to impose or test prior theories. It does not seek an objective detachment, but rather seeks to 'participate in' or engage with phenomena, to find place evocation through facilitating or allowing it to be revealed through the researcher's own personal engagement. Tellingly in the important afterword to his paper, Pocock writes:

> **The Galilee is my world ... My sketching represents a personal experience of coming to know through understanding gained reciprocally: the world gave itself to me in so far as I opened myself to it. The result of such sharing is a social construct. I am part of a common humanity, sharing a particular culture and language: the Galilee is in turn a world suspended in the webs of significance spun over eight centuries. (1996: 384)**

Although the researcher in this process becomes a kind of channel through which the phenomenon – here the place character of the Galilee Chapel – reveals itself to us, the research draws upon considerably more evidence than that of personal experience of the space. Pocock has drawn upon texts about the history and traditions of the Galilee Chapel, and observation of other actors or visitors to the space and their reactions. However, in this approach the researcher is very much subjectively involved and 'it was I who was changed and made richer' (1996: 384). The Chapel is not so much an object for study, and certainly not set at an objective distance, but more a subject with which the research becomes engaged, or involved intimately, 'a subject to which I was happily subject'. The relationship is therefore intimate, participatory and reciprocal, involving respect and patience to allow 'the Chapel to disclose itself on its terms, in its time' (1996: 384).

The researcher seeks not to verify an *a priori* theory, but to authenticate a particular knowledge of the place. Therefore, Pocock describes his text as variously a sketch, a translation or an interpretation, and argues that its success is judged by 'the extent to which it conveys, convinces or authenticates a sense of being there. A comparison of word and world is the challenge, then, to both author and reader' (1996: 384). He refers to his authorship as 'perhaps firstly talking to himself, with a text revealing as much about himself as the world described' (1996: 385–6). In common with other humanistic geographers (notably Tuan, 1999; Rowles, 1976), he sees the approach as fundamentally as much a journey of self-discovery as discovery about the world. In a sense the process of writing up the research is contradictory and potentially destructive as it inevitably gives a degree of fixity, or privilege, to a particular (partial or moment of) insight, and objectifies. In other words, for the phenomenologically inspired engagement with place, dynamic, authentic knowledge and understanding, emerge through intimacy and ongoing reflection. Writing, whilst offering one route to sharing such knowing and understanding, is less authentic than the actual engagement.

Pocock argues that the author shares their authority not only with the subject (as noted above) but also 'with the reader who engages with and activates the text' (1996: 386; also 1988a). Truth is therefore always presented as emergent, implicated or partial, and any written record (research paper, for instance) cannot be an authoritative statement in itself, but must be a sketch, and a kind of tool to engage the reader in the journey of reflection and authentication. This work illustrates knowledge as process rather than product, and the fundamental humility and respect required of the researcher/author and reader/interpreter. Methodologically central is an intimate engagement experientially with the space, through repeated visits to the Galilee Chapel, observation, quiet contemplation and reflection. This is not a quick process, but one involving repeated visits, time spent submerged in the 'feel' of the place, observing and reflecting on its 'lifeworld'. It is supplemented by reading of texts about the history and meanings of the place, not to identify 'theories' to test, not to crowd preconceptions into the view, but to follow up possible lines of reflection and understanding. Judgement is suspended to allow the phenomenon to reveal itself, to 'speak of itself to us'. In place of specific techniques of observation and interpretive method, we have principles of intimate sensing – patience, respect, observing, listening, reflection, and authentication.

Interpersonal Knowing and Participatory Approaches

Pocock's work illustrates an essentially personal experiential engagement and self-reflective strategy. Several geographers have sought to adopt a more social approach and engage with individuals and communities. These participatory approaches have adapted in-depth interviews, group discussion, reflective diaries, and participant observation techniques as primary sources of evidence (see Rowles, 1976; 1978; 1988; Seamon, 1979; Rodaway, 1988; Harper, 1987). Here an authentic understanding is sought through a concept of shared knowledge, or 'interpersonal knowing'.

Graham Rowles (1978: 175–6) notes a dilemma: subjective knowing is ultimately inaccessible as it lies at the level of individual feelings, and objective knowledge is often overly abstracted, generalized, and so at a remove from actual experience. Drawing on Maslow, he argues that there is a third mode of knowing: interpersonal knowing. Examples might be a friend knowing a friend, two persons loving each other, a parent knowing a child or a child knowing a parent (and it is not always reciprocal). The key characteristic here is that knowing is emergent and partial, subsisting in a relationship between knower and world (objects or beings). It is lived through day-to-day activities, interpersonal relationships, conversations, gestures and actions. The knower is involved with what they know. They participate in, are influenced by and influence what is known. There is not a distance or a detachment, but an intimacy and an involvement. It is a sensual and emotional relationship. What is known may be as much a feeling as something which is or can be expressed. Essential to the notion of interpersonal knowing is the idea of empathy, intuition and feeling. In Rowles' approach the traditional hierarchy between researcher (knower) and subject (to be known) is broken down and the research process becomes one of partnership. Trust and friendship become important components of the researcher's relationship to the participants. Rowles describes his relationship to one participant, Stan, as becoming 'close friends', involving doing the shopping together and having a drink in the bar, as much as more formal conversations in Stan's own home to further 'the research'. This intimacy is reflected in the impact which Stan's death had upon Rowles himself, a conflict between 'my human sensibility and my scholarly purpose' (1976: 19).

In *Prisoners of Space: Exploring the Geographical Experience of Older People* (1976), Graham Rowles used an explicitly humanistic and people-centred methodology. In particular, his work challenged the notion of the researcher as the knowing subject and the elderly as an object to be studied and written about. In contrast, he worked alongside/ with a group of individual elderly people, using in-depth interview techniques, engaging in conversations, sharing reflections, shadowing everyday lives, and being an observant participant. He sought to bracket out the preconceptions of the research literature on the geographical (and social) experience of the elderly, and to seek to understand their experience as they perceived and lived it themselves. His interest was not only in an authentic understanding through a form of interpersonal knowing, but also in a holistic approach through a study of individuals in their own homes and neighbourhoods – that is 'lifeworlds'. Key to the research process was friendship and trust. In writing up his research he chose to present a series of vignettes on the distinctive characters, lifeworlds and experiences of each of the five participants.

Rowles (1976) described his research approach as inductive, the piecing together of the evidence of conversations and observations of everyday lives and participants' reflections upon them. In seeking to authenticate his understanding of their lives, Rowles did not seek to match up his insights to a prior academic literature (or theories). Instead he took two basic strategies. He looked for consistencies and shared themes across the participants' experiences and reflections, *and* sought to share his interpretations with the participants to refine, develop or negotiate an authentic understanding. In other words, the research process was not only inductive but also negotiated and reflective, involving internal feedback and refinement, to identify salient themes. He sought to derive his insights solely or primarily from the 'text' of his encounters, the experiences and the reflections of the elderly participants, each person and their individual 'lifeworld'. To reduce the distortion of researcher interpretation and translation he used participants' own words or phrases, and presented 'results' back to participants for feedback and refinement. Rowles describes his role as translator – a term also used by Pocock. He describes the process as an open relationship, with no attempt to minimize 'contamination', and with emphasis placed on sharing ideas. The research interaction became 'a mutually creative process ... my job therefore would be one of translating this text and distilling the essential geographical themes within a coherent conceptual framework' (1976: 39).

In seeking to take his research beyond the insights gained into individual lifeworlds, he developed a perceptual model. This consisted of the identification of four modalities of experience – a meshing of space and time, physical and cognitive, contemporary neighbourhood, and vicarious participation in displaced environments. These experiential modalities were overlapping and configured dynamically and uniquely in the lifeworlds of each participant. However, in identifying the modalities of action, orientation, feeling and fantasy, Rowles presented these not as some kind of abstract theory or model, but as a dynamic framework for reflection on his individual character vignettes and a point of potential lines of enquiry for future reflective engagements with people and place. Summarizing his approach, Rowles writes: 'a quest for intimacy is the essence of the research strategy. I developed strong interpersonal relationships with five older people. My conclusions result from almost two years' contact with these individuals' (1976: xviii). Whilst the most intense interaction with his participants was within broadly a six-month period, the extended nature of the engagement is typical of this kind of participatory, explicative and authenticating approach to research.

Reflections on a Humanistic Strategy

Pocock writes of the humanistic approach as relying 'heavily on the intuition and initiative of the researcher. Success – or failure – is therefore reliant on personal ability and personality to an extent unthinkable in conventional studies. Humanistic work is individually, and not socially, reflective. It is also person-centred, elevating individual or human agency over societal structure' (1988a: 5). Whilst not relying on structural explanation, through interpersonal knowing and group reflective approaches (e.g. Seamon, 1979; Rodaway, 1987; 1988), it could be counterargued that the person-centred approach can also be socially reflective. However, fundamentally, the people-centred approach of humanistic strategies means that knowledge, or knowing, is primarily subjective, and subsists in the relationship between researcher and what is researched. Although humanistic geographers make claims to 'authentic' knowledge through critical self-reflection, empathy, interpersonal knowing and negotiation with subjects, ultimately much relies upon the quality of the human relationships sustained in the research and the integrity and honesty of the researcher.

For humanistic geographers, knowledge is not produced, nor is the object achieved and understanding tested against repeatable phenomena. Rather, knowledge is a process; it is knowing (or coming to know) through engagement. It is always and already subjective, partial and emergent. The researcher must engage in continued critical reflection on the research process and their role within it, refining the specific methods employed to the context of the study.

Writing up research findings is also problematic since it is a process which objectifies, gives a certain fixity to what has come to be 'known', and is separated from or disconnects from the living flow of the phenomenon that was engaged with during the research. Several strategies have been developed to militate against this. Some researchers seek to feed back their 'findings' to subjects, to involve them in the refinement of interpretations and the 'authentication' of the research 'report' (Rowles, 1976; Rodaway, 1987). Other researchers have sought to 'authenticate' research findings through extended self-reflection and a call to the reader to continue the process of reflection and authentication against their own experience (Pocock, 1988a; 1996).

People-centred approaches as developed by humanistic geographers have been criticized for their explicit subjectivity. However, as a number of humanistic geographers have noted, what is important is that 'appropriate tests of proof' (Pocock, 1988a: 6) are applied to evaluating research 'findings'. Since knowledge is defined as a continuous and incomplete process of knowing, and subsists in a web of relationships between the researcher and the phenomena studied (including research subjects and the reader), any evaluation of the 'findings' must be both contextual to this and partial ('interim' might be a useful word). The positivist science abstraction of verification against a preconceived set of concepts, theories and previous research findings (conducted at a different time and place by a different agent, etc.) is inadequate to the evaluation of the evidence of empathetic, experiential and interpersonal knowing. Evaluation needs to take into account the character and performance of the researcher, the nature of the relationships established between the researcher and his research phenomena (environment, subjects, etc.), and the particular design and operation of the methodology employed to generate the 'findings' (as well as how the interpretations were compiled and presented).

From phenomenology, humanistic geographers have also made much use of the term 'authenticity'. Here the notion of 'proof' or truth is asserted in terms of the very subjective human relationships of a lifeworld, the practical engagement in interpersonal knowing, and the coherence of 'making sense' in the practice of everyday lives. Authenticity is not easily defined, but rather corroborated in lived experience, in a genuine sense of empathy or tuning in to the phenomenon as it really is, in itself. Pocock describes it as achieved 'intersubjective corroboration' and the 'ultimate test is not only whether findings are plausible (the common sense factor), but informative, confirmatory (the aha! factor), and whether they *do* something to the reader' (1988a: 7). In other words, do the findings or research report move the reader to a kind of emotional engagement with, or recognition of, the people (characters), spaces (places) and experiences revealed, which gives that deeper sense of 'reality', akin to aesthetic enlightenment? More technically, the 'test' of authenticity is not the verification of an abstract fact or causal relationship, but a kind of confirmation or assertion, and ultimately a sharing of an insight into the wholeness, the character or essence of a place, its people, its lifeworlds. The ultimate test of authenticity lies not in abstract research findings (and in particular theories), but in the response of those studied (whether it rings true to their lives and situation) and of the readers (especially other researchers who have also engaged in such detailed and in-depth empathetic, experiential and interpersonal research).

The focus on the particular, the unique and the emergent inevitably has consequences for any generalizations and claims to wider understanding of the human condition and geography. Yet, humanists do seek to identify wider insights which might be 'authenticated' against other situations. For example, Seamon (1979) identified a continuum of environmental engagement from immersive to detached; Rowles (1976) summarized his insights as four modalities of perception; Relph (1976) developed a typology of insideness/outsideness of place experience. These generalizations are, however, not attempts to identify laws or causal relationships, nor are they theories to be tested; rather they act as summaries to inform future critical reflection. This might on the surface run the risk of setting up frameworks of presuppositions for future research, but each of these researchers was keen to emphasize the particular context of their insights and the provisional nature of their 'findings'. From a humanistic perspective, all claims to truth are subjective, and relative to particular conditions of their realization. Positivists might infer 'truth' from statistical inference, notably the repeatability of phenomena. Humanists are more likely to relay 'logical inference' from the case study. Pocock suggests: 'it is not the case of the particular being representative of the generalisation, but rather, of the generalisation not applying to the particular' (1988a: 6).

Beyond Humanistic Geographies

The legacy of humanistic geographies continues in contemporary interests in people-centred methodologies – the use of self-reflection and biographical strategies (e.g. Kruse, 2003), participant observation approaches (e.g. Mabon, 1998), in-depth interviews and focus groups (various in *Area*, 1996).

Feminist geographers have drawn on a diversity of methods, both quantitative and qualitative, but also have made much use of self-reflection and empathetic strategies (e.g. Women and Geography Study Group, 1997). 'Activist' geographers have made radical commitment to people-centred methodologies, seeking direct engagement with people and places, identifying a political role for the researcher as agent of change (e.g. Ticknell, 1995; Routledge, 1997). However, this work is rooted in different and diverse philosophies and research traditions, e.g. situated knowledge (Rose, 1997), grounded theory and actor-network theory.

Whilst these 'new' geographies have abandoned reference to phenomenology and other humanistic philosophies, in many ways the humanistic geographers opened the door to the value of a focus on subjective experience, feelings and meanings. A range of new directions were legitimized in part by this earlier engagement with people-centred methods:

- critical reflection on the role of the researcher as involved agent
- deliberate eclecticism and adaptation of methodologies to meet the specific needs of given research contexts
- the potential of research as a process of empowerment to subjects
- engagement with knowledge as process (implicative, contextual, situated, political).

Referring back to Rowles' work, Widdowfield (2000) has more recently reminded geographers of the importance of the reflexivity and emotion in relationship between the researcher and what they study. She argues that:

> **despite the trend in recent years towards more qualitative and reflexive research, discussion and critiques of the research process to date have rarely involved an explicit examination of how the researcher's emotions may impact upon that process. It is almost as if we have become so concerned about how far or indeed whether we can speak for and articulate the experience and emotions of 'others' that we have forgotten or dismissed the ability to speak for ourselves. (2000: 205)**

People-centred methodologies involve coming to terms with a subjectivity in how research is conducted, an explicitness about people's involvement, and an emphasis on how research is communicated and shared.

NOTE

1 Less common in humanistic geography have been self-reflective essays of a more biographical kind where geographers reflect upon their own experience (e.g. Hart, 1979; Tuan, 1999).

References

Area (1996) 'Focus Groups', Special issue, 28(2): 113–14. Introduction by J. Goss, with following articles by Goss, Zeigler, Burgess, Jackson, Longhurst.

Harper, S. (1987) 'A humanistic approach to the study of rural population', *Journal of Rural Studies*, 5:161–84.

Hart, R. (1979) *Children's Experience of Place.* New York: Irvington.

Heidegger, M (1983) *Being and Time.* Oxford: Blackwell.

Hubbard, P., Kitchin, R., Bartley, B. and Fuller, D. (2002) *Thinking Geographically: Space, Theory and Contemporary Human Geography.* London: Continuum.

Husserl, E. (1983) *Ideas Pertaining to Pure Phenomenology*, trans. F. Kersten. The Hague: Martinus Nijhoff.

Kruse, R.J. II (2003) 'Imagining Strawberry Fields as a place of pilgrimage', *Area*, 35: 154–62.

Mabon, B. (1998) 'Clubbing', in T. Skelton and G. Valentine (eds), *Cool Geographies: Youth Cultures.* London: Routledge, pp. 266–86.

Meth, P. (2003) 'Entries and omissions: using solicited diaries in geographical research', *Area*, 35: 195–205.

Monaghan, P. (2001) 'Lost in place', *The Chronicle*, 16 March. www.chronicle.com/free/v47/i27/27a01401.htm.

Olsson, G. (1980) *Bird in Egg: Eggs in Bird.* London: Pion.

Pocock, D. (ed.) (1981) *Humanistic Geography and Literature.* London: Croom Helm.

Pocock, D. (1983) 'Geographical fieldwork: an experiential perspective', *Geography*, 68: 319–25.

Pocock, D. (1987) *A Sound Portrait of a Cathedral City*, tape recording. Presented to the Institute of British Geographers Annual Conference, 1987, Department of Geography, University of Durham.

Pocock, D. (1988a) 'Aspects of humanistic perspectives in geography', in D. Pocock (ed.), *Humanistic Approaches to Geography.* Occasional Paper (New Series) 22, pp. 1–11. Department of Geography, University of Durham.

Pocock, D. (1988b) 'The music of geography', in D. Pocock (ed.), *Humanistic Approaches to Geography.* Occasional Paper (New Series) 22, pp.62–71. Department of Geography, University of Durham.

Pocock, D. (1992) 'Catherine Cookson Country: tourist expectation and experience', *Geography*, 77: 236–43.

Pocock, D. (1996) 'Place evocation: the Galilee Chapel in Durham Cathedral', *Transactions of the Institute of British Geographers*, 21: 379–86.

Relph, E. (1976) *Place and Placelessness.* London: Pion.

Relph, E. (1985) 'Geographical experience and being-in-the-world: the phenomenological origins of geography', in D. Seamon and R. Mugerauer (eds), *Dwelling, Place and Environment.* Dordrecht: Nijhoff, pp. 15–32.

Rodaway, P. (1987) 'Experience and the everyday environment: a group reflective strategy'. Unpublished PhD Thesis, University of Durham.

Rodaway, P. (1988) 'Opening environmental experience', in D. Pocock (ed.), *Humanistic Approaches to Geography.* Occasional Paper (New Series) 22, pp. 50–61. Department of Geography, University of Durham.

Rose, G. (1997) 'Situating knowledges: positionality, reflexivity and other tactics', *Progress in Human Geography*, 21: 305–20.

Routledge, P. (1997) 'The imagery of resistance', *Transactions of the Institute of British Geographers*, 22: 359–76.

Rowles, G. (1976) *Prisoners of Space: Exploring the Geographical Experience of Older People*. Boulder, CO: Westview.

Rowles, G. (1978) 'Reflections on experiential fieldwork', in D. Ley and D. Samuels (eds), *Humanistic Geography: Problems and Prospects*. London: Croom Helm, Chapter 11.

Rowles, G. (1988) 'What's rural about rural ageing? An Appalachian perspective', *Journal of Rural Studies*, 4: 115–24.

Seamon, D. (1979) *A Geography of the Lifeworld*. London: Croom Helm.

Seamon, D. (1985) 'Reconciling old and new worlds: the dwelling-journey relationship as portrayed in Vilhelm Moberg's "Emigrant" novels', in D. Seamon and R. Mugerauer (eds), *Dwelling, Place and Environment*. Dordrecht: Nijhoff, pp. 227–46.

Seamon, D. and Mugerauer, R. (eds) (1985) *Dwelling, Place and Environment*. Dordrecht: Nijhoff.

Ticknell, A. (1995) 'Reflections on activism and the academy', *Environment and Planning D: Society and Space*, 13: 235–8.

Tuan, Yi-Fu (1971) 'Geography, phenomenology and the study of human nature', *Canadian Geographer*, 15.

Tuan, Yi-Fu (1974) *Tophophilia*. Englewood Cliffs, NJ: Prentice Hall.

Tuan, Yi Fu (1977) *Space and Place: The Perspective of Experience*. London: Arnold.

Tuan, Yi Fu (1979) *Landscapes of Fear*. Minneapolis, MN: University of Minneapolis Press.

Tuan, Yi-Fu (1993) *Passing Strange and Wonderful: Aesthetics, Nature and Culture*. Washington, DC: Island/Shearwater.

Tuan, Yi Fu (1999) *Who Am I? An Autobiography of Emotion, Mind and Spirit*. Madison, WI: University of Wisconsin Press.

Widdowfield, R (2000) 'The place of emotion in academic research', *Area*, 32(2): 199–208.

Women and Geography Study Group of the Institute of British Geographers (1997) *Feminist Geographies: Explorations in Diversity and Difference*. Harlow: Longman.

25 CHANGING THE WORLD: GEOGRAPHY, POLITICAL ACTIVISM, AND MARXISM

Michael Samers

The philosophers have only *interpreted* the world, in various ways, the point is to *change* it. (Marx, *Theses on Feuerbach*, number XI)

Introduction

This chapter explores the relationship between geography, activism, and Marxism. While few geographers would consider themselves Marxists today, more than a handful probably see themselves as activists, or at least inspired by the activism of others. My purpose here, then, is to show how Marxism can, does, and should inform practical interventions in the world. However, I do not offer a 'blueprint' for potential activists, nor do I outline the contours of Marxism (this is accomplished in Chapter 5 of this volume). And while there is a cavernous literature on activism in general, I focus on only a limited number of contributions by geographers who are or have been working from a Marxist tradition. Thus, in the first section of this chapter, I review some general criticisms of Marxism and then provide a defence. This is followed in the subsequent section by a discussion of what David Harvey calls 'dialectical, spatiotemporal utopianism' in order to provide a foundation for a renewed Marxist geographical activism. This in turn lays the groundwork for honing a definition of Marxist geographical activism in the following section, and I discuss some of the promises and pitfalls of the difficult movement between academic work and activism.

Defending Marxism

Marxism and after-Marxism in geography

My purpose in this section is to focus on *some* objections by the 'after-Marxist' or 'postmodern' left to the 'old' or 'modern' left – that is, those who subscribe to one degree or another to certain principles of Marxism, and who tend to focus on class inequalities and issues of political economy (see e.g. Castree, 1999a; Chouinard, 1994; Corbridge, 1993; Fraser, 1995; Laclau and Mouffe, 1987; Sayer, 1995). After reviewing these objections, I then provide a certain defence of Marxism.

First, those associated with the after-Marxist/postmodern left are critical of the old/modern left for what they perceive to be the obliteration of 'difference' (this word usually refers to cultural differences between (groups of) people) through class-alone discourses. They object not to the importance of class *per se*, but to the notion that class typically implies the white male, heterosexual, able-bodied working classes (often working in factories), to the exclusion of other kinds of oppressions and 'identity politics' such as those based around 'race', ethnicity, gender, sexuality, age, (dis)ability, and so forth, and occurring in different sites (e.g. the 'home' rather than simply the factory). In fact, in terms of the focus on factories, the after-Marxist/postmodern left was equally critical of the undue focus on production rather than consumption (the latter would tie in particularly well with questions of identity). To paraphrase the political theorist Nancy Fraser (1995), social theory in general became preoccupied with questions of recognition (that is recognizing particular 'cultural' identities or interests) to the detriment of redistribution (inequality and the redistribution of wealth, or in Marxist terms, surplus value). Second, in terms of postmodern environmental thought, many associated Marxism with a (western) philosophy that – instead of a more 'sustainable' dialogue with nature – entailed the domination over the non-human world and an obsession with 'productivism'. The latter refers to a political philosophy of workers that favours a strategy of increasing production to provide work for themselves, rather than a more revolutionary stance. Third, they argued that theories such as Marxism involved a 'god trick' (a term used by the feminist historian of science Donna Haraway) by which the world could be understood through a dubiously detached and disembodied 'scientific objectivity'. In other words, many geographers wanted to dismiss metatheory (that is 'grand' or 'totalizing' theory that attempted to explain everything through a single theoretical lens). Rather, knowledge should be 'situated' (so-called 'situated knowledges') in which the researcher's position should be explicit and 'locatable'. As Merrifield puts it, 'Under such circumstances, knowledge is always embedded in a particular time and space; it doesn't see everything from nowhere but *sees something from somewhere*' (1995: 51). Fourth, Marxism became associated negatively with 'actually existing socialism' – meaning the sort of political economic systems that were to be found in China, Eastern Europe, and the former Soviet Union prior to 1989, and their record of non-democratic, authoritarian, and ultimately repressive regimes symbolized by the horrors of Stalinism. And thus, as Smith writes, 'many no longer saw revolutionary transformation as achievable, realistic, or even necessarily desirable' (2000: 1019).

A response to critics and a short defence of Marxism

Most non-Marxist human geographers (and certainly others) were quite right to highlight the limits of Marxism, but in doing so they managed to jettison what remains useful in Marxism for building an activist geography. In the discussion below, then, I provide a defence against some of the criticisms of Marxism discussed above. To reiterate, the first criticism of Marxist analysis centred on 'class-alone' explanations, and at the same time, human geographers seemed to write less and less about class. As a consequence, the economic geographer Ray Hudson argued that: 'It is of the utmost importance to stress that we live in a world in which capitalist social relations *are* dominant, the rationale for production *is* profit, class and class inequalities *do* remain, and that wealth distribution *does* matter' (2001: 2). Hudson leaves no doubt as to the state of the world, but his reference to 'class' and 'class inequalities' deserves at least some attention. Indeed, we need to focus on class as a concept, if only because the notion of a (potentially) 'classless' society is so pervasive, and because of the objections to class-alone discourses, as discussed above.

The question of class and class struggle

Marx never offered any systematic theory of class conflict, nor did he have a fixed notion of class; nor, at least as Smith (2000) claims, did he privilege class in his analysis, despite this image of his work as 'class centred'. It would be left to Marxists, rather than Marx himself, to expound on this matter. In this sense, the Hungarian Marxist György Lukács distinguished between 'class-in-itself' and 'class-for-itself'. Class-in-itself implies that people constitute a 'class' (such as the 'working class') by their relationship to the 'means of production', their control of the labour process, and their position in terms of the extraction of surplus value (see Chapter 5 in this volume). These are otherwise known as their 'objective class conditions', regardless of whether they demonstrate politically as a 'class'. The concept of class-in-itself is by no means accepted uncritically. First, as I have noted above, Marxist class analysis has been the subject of sustained attack by feminist geographers for its neglect of gender and sexuality. Indeed, women who perform unpaid domestic labour ('housework') do not constitute a class in traditional Marxist terms, and therefore have no place in Marxist analysis.

There are certainly other weaknesses or 'silences' in Marxist class analysis, and in this regard the geographer Richard Walker and the sociologist-geographer Andrew Sayer argued in their book *The New Social Economy* (Sayer and Walker, 1992) that the study of class has neglected the 'division of labour'. (The division of labour can be defined simply as work specialization, either *between* families, firms, groups of people, and so on – in other words, the 'social division of labour' – or *within* specific firms – commonly called the 'technical division of labour'). Sayer and Walker maintain that the division of labour, in all its forms, has acted upon class formation, and class in turn has shaped the division of labour throughout the history of capitalism. Thus, 'One cannot settle on a tidy definition of class that stops history in its tracks' (1992: 22). This has undeniable implications for the concept of 'class-for-itself', since the ever-changing division of labour has divided workers by age, gender, ethnicity, and so forth, which means they may have very different political interests – a point which Marxists had either neglected or denied. Today, most Marxists would presumably agree that the global working classes are diverse and fragmented, and that gender, ethnicity, and so forth are not simply to be added to 'class'. Rather, they acknowledge that class is constituted (i.e. constructed) through these different dimensions in the first instance (Blunt and Wills, 2000; Sayer and Walker, 1992). That is, despite the temptation to separate class and these other dimensions analytically, we can argue that one is disadvantaged partly *because* one's body is marked through racial and sexual coding, by national citizenship, regional accent, manner of dress and so forth.

With this in mind, let us turn toward the second of Lukács' concepts: 'class-for-itself'. By this, Lukács meant that the 'working classes' recognize themselves as such and organize politically on these grounds. In this sense, many may perceive the concept of 'class' to be obsolete because of the decline of traditional industrial (class) struggles. And there is more than sufficient evidence to argue that the kind of production politics (that is the mobilization of workers at particular sites of production) that punctuated industrial plants across the advanced economies prior to the late 1980s has waned. But it has hardly disappeared (academics in my own university are planning a strike as I write this) and geographers such as Jane Wills and Andrew Herod have shown just how resilient such struggles can be (Wills and Waterman, 2001; Herod, 2001a). And there are other reasons to doubt the end of class struggle. To begin with, much (industrial) strike activity against multinational corporations has simply been *displaced* to the 'global south' (the so-called 'Third World') while the command, control, advertising, marketing, and much of the distribution of multinational products has remained in the advanced economies. The result is that to observers in the advanced economies, industrial struggles under capitalism (the 'bad old days of the 1970s' according to the UK Prime Minister Tony Blair) *seem* to have disappeared. Second, class struggles can assume many forms, other than those we associate with traditional production politics. They can involve multi-issue, anti-globalization/anti-capitalist protests (such as those that transpired at Seattle and Genoa, or the traditional May Day protests in London). Similarly, traditional class struggle is increasingly indistinguishable from so-called 'social movements'. Social movements refer to grassroots activity around a particular set of political interests such as environmental degradation, affordable housing, gay/lesbian rights, the protest of domestic migrant workers, and so on. Nonetheless, the rise of such social movements is also affecting the strategies of what is considered to be the traditional class struggle – that is transnational workerist activism. Indeed, the academic activist Kim Moody (1997) has called this 'social movement unionism', by which he means such activism is moving towards a 'looser, more inclusive, more grassroots way of working' (Castree et al., 2003: 224–5).

The problem of metatheory and the 'god trick'

Another objection to Marxist theory by the postmodern and feminist left focused on Marxism's ideological veil of objectivity – what Haraway calls the 'god trick' – and such a critique is actually vital for a reinvigorated Marxist activism. Let me explain. To recall, Haraway's notion of 'situated knowledges' insists on the partial nature of understanding the world ('seeing something from somewhere', rather than 'everything from nowhere') that is always open, and makes no claim to absolute privileged knowledge. As Merrifield puts it, 'There are always different and contrasting ways of knowing the world, equally partial and equally contestable' (1995: 51). And yet, this can lead us down a path of hopeless relativism that Haraway herself acknowledges. Indeed, for Haraway relativism and objectivism/absolutism (the 'god trick') are mirror images. Merrifield argues then:

> **To this extent a committed and situated knowledge offers a corrective to the god-tricks of positivism and some post-modernism: situatedness implies that an understanding of reality is accountable and responsible for an *enabling* political practice. Ultimately then, the realm of politics conditions what may count as *true* knowledge. (1995: 51)**

Thus, Merrifield shows how (in his discussion of a radical expedition in Detroit in the late 1960s, which I consider later in the chapter) situated knowledges can be deployed in an activist geography that is able to 'circumvent the current paralysis within "strong" post-modern critical theory' (1995: 52) – strong in the sense that some versions of postmodernism relativize all knowledge.

On the death of 'actually existing socialism' and the rise of 'critical geographies'

The last of the criticisms of Marxism I discuss consisted of a damning assessment of the regimes of China, the former Soviet Union and the Soviet bloc countries. I do not have the space here to elaborate on the nature of these regimes, but criticisms of authoritarianism and repression are more than justified. However, their failure as socialist projects has contributed (along with potent doses of what has come to be called 'neoliberal' ideology) to a lack of political imagination about alternatives to capitalist social relations, or at least to 'neoliberalism'. As a consequence, interest in Marxism has declined as an analytical lens and socialism as a practical project. Indeed, the former Prime Minister of Britain, Margaret Thatcher, proclaimed 'the end of society' and 'there is no alternative' (to capitalism). Thus, in reference to the Seattle protests against the

WTO and global neoliberalism, Merrifield laments: 'protesters are denounced as idiotic, juvenile, naïve: listen up, wise up, grow up. There is no alternative' (2002: 133). As a consequence, the Marxist geographer David Harvey (2000) proclaims, utopianism has developed a bad name.

The response to such conservative cynicism is not an easy one, since many academics on the 'left' can no longer envision an alternative to capitalist social relations, and Marxist (or radical) geography seems to have been superseded by a wider hue of what has come to be called critical geographies. So much so, that we might say that 'all geographers are critical geographers now' (The term 'critical geography' even appeared in the subtitle of one of Harvey's most recent books *Spaces of Capital*). Curiously, the distinction between Marxist and critical geography is one that has somehow escaped substantial debate in geography (but see Castree, 2000a; 2000b), and it begs the question whether one ought to distinguish between the two at all. If we do not, some burning questions remain, such as: do struggles that are not defined by class really challenge capitalism *as a system*? In other words, are 'new' critical geographies simply liberal and reformist in contrast to previous and (apparently) more revolutionary Marxist geographies? Kathy Gibson and Julie Graham (Gibson-Graham, 1996; Community Economies Collective, 2001) have even questioned whether capitalism is a system at all, and stress the need to see the world as a composition of a variety of 'economies'. In such a way, they seek to emphasize and celebrate non-capitalist spaces (communal enterprises, local currency initiatives, and so forth) in order to construct a non-capitalist imagination that moves from the local to the global, and does not rely on an overwhelmingly difficult blueprint for socialism. In a way, Kathy Gibson, Julie Graham, and the Community Economies Collective have reawakened an anti-capitalist sensibility without having us suffocate under the heavy weight of a largely unimaginable socialism.

Nonetheless, what began with the radical journal *Antipode* in 1969 continues with a chorus of geographers who may not wear the 'Marxist badge' so prominently, but who are nonetheless committed to a rainbow of oppositional social change. For example, Don Mitchell (currently a Professor at Syracuse University in the United States and in fact more unwilling to jettison the Marxist badge) established in 1999 the People's Geography Project' – a network of geographers and scholars engaged in projects that are about contesting existing power relations (see http://www.peoplesgeography.org). At the same time, oppositional geographers held the first 'International Conference of Critical Geography' in Vancouver in 1997 as a response to more status quo academic institutions, and out of this, some of its principal organizers launched a free-access online journal *ACME: An International E-Journal for Critical Geographies* in 2002 (see http://www.acme-journal.org). Apparently, the flame of anti-capitalism has yet to be extinguished.

Dialectical, spatiotemporal utopianism

Beginning with Harvey's 'historical geographical manifesto' in 1984 (from which the People's Geography Project draws its inspiration), we find the root of what Harvey (2000) has recently called 'a dialectical and spatiotemporal utopianism'. In other words, he seeks a utopian project – not one that is based on 'pious universalisms', not one divorced from the materiality of the social world, but one that is forged with a vision of space *and* time, and from the material circumstances in which we find ourselves. Spatiotemporal dialectical utopianism calls for the integration of the 'particular' (particular spatial and temporally defined interests) and the 'universal' (perceived or envisioned commonalities that are produced through space and time).

A short example will suffice to illustrate what Harvey means. While Harvey occupied the prestigious Halford Mackinder Chair at Oxford University, he, along with a local activist (Teresa Hayter), published a book entitled *The Factory and the City* (1994). In this work, Harvey recounts the problems and disagreements of their involvement in a campaign to save the Rover plant in Cowley, near Oxford. In particular, Hayter questioned Harvey's allegiances. Did he really support the Cowley workers, or was he just a 'free-floating intellectual' musing about the international but abstract cause of socialism? Harvey responded that he supported the Cowley workers (the particular if you will) but was also concerned about overcapacity in the European automobile industry and what eventual implications this had for all automobile workers (the universal).

Yet for Harvey, his version of utopia must also involve certain 'closures'. That is, against the after-Marxist/postmodern left, he argues that universal absolutes are both unavoidable and necessary. But Harvey demands openness as well for his utopia – the kind of 'heterotopia' imagined by the social and political theorist Michel Foucault, and so praised by the postmodernists. Certainly, Harvey's utopia raises questions about 'commonalities' or 'universals' and how these are supposed to be constructed through time and space, and it is to some of these thorny issues that we now turn.

Building universals

Just as white-male-led trade unionism failed to engage (and sometimes continues to actively exclude) working-class people of colour (e.g. immigrants) in their struggle for economic justice, so too have *some* critical and oppositional movements based on allegedly non-class-based issues (e.g. affluent feminists) failed to engage with other struggles by working-class feminists and therefore did not galvanize a broader coalition of people. Similarly, Smith (2000) recounts the limitations of the anti-AIDS struggles in New York City in the form of the ACT UP organization. He argues that the otherwise remarkably imaginative and successful campaign of ACT UP (led primarily by wealthier gay men) eventually collapsed because of its failure to engage with a wider set of struggles, namely those of working-class activists, intravenous drug users, people of colour, and lesbians. That is, the levelling off of HIV infection among white gay men meant that this particular group had less reason to carry on the struggle, and therefore missed an opportunity for a much wider oppositional movement.

As Harvey and Smith argue, the point is that the interests of certain groups (the 'particular') must be aligned with a search for commonalities (the universal). In fact, locally based struggles, situated as they are within a wider political economy, can actually undermine more global forms of activism. The search for commonalities is premised upon the idea that 'universals are socially constructed not given' (Harvey, 2000: 247), and we 'know a great deal about what divides people but nowhere near enough about what we have in common' (2000: 245). It is only in seeing commonalities – or what Castree et al. call, in the context of wage workers, 'contingent universals ... That is, commonalities that appear to be "inherent" to all wage workers but which are, in fact, strategically useful concoctions' (2003: 242) – that collective struggle can be carried out. The importance of vision for Marxist or socialist struggle is one that is emphasized by a number of Marxist geographers (see especially DeFilippis, 2001; Merrifield, 2002). Nonetheless, not all geographers working from a Marxist or anti-capitalist tradition are so comfortable with universals. As I suggested above, the Community Economies Collective (2001), for example, is more concerned with moving from local oppositional movements to a wider scale of anti-capitalist struggle without the construction of universals.

Coping with scale

The Marxist cultural geographer Don Mitchell (2000) writes that any strong notion of worker in/justice must be 'fully situated within a politics of scale'. Thus, from Marx's famous dictum at the close of the *Communist Manifesto* – 'WORKING MEN OF ALL COUNTRIES UNITE!' – it has been tacitly assumed that trade unions for example (as representatives of the working classes) have to organize internationally or globally in order to combat global capitalism. Perhaps not coincidentally, in the 1980s – a time when Marxism became increasingly discredited – the aphorism 'Think globally, act locally' emerged as popular on the bumper stickers of American cars). In other words, one had to have a cosmopolitan, multicultural, global understanding of injustice, but any political action could not realistically take place (or so the bumper stickers proclaimed) at the global level. In short, one had to act locally for any meaningful change to be fruitful. Who was right, Marx or that ubiquitous slogan?

Certainly, not all geographers agree with what might be considered 'Marx's implicit politics of scale'. I will discuss two such objections here. First, Herod (2001b) provides an analysis of two disputes between workers and management, the first at Ravenswood, West Virginia and the other at a General Motors plant in Flint, Michigan. While Herod concedes that in the case of Ravenswood a more translocal and global strategy worked, in the case of the latter a local strategy proved successful owing to the particular importance of Flint's factory to GM's national production system. As he writes, the success of the Flint strike 'is testimony to the power that locally focused industrial actions on the part of workers may sometimes have in an increasingly integrated global economy' (2001b: 114). Nonetheless, his paper concludes that neither strategy among workers, 'going global' or 'going local', is necessarily more successful. Similarly, James DeFilippis (2001) stresses the importance of the local and 'community' in acting up against global capitalism. Here, I can do no better than repeat a passage he cites from Naomi Klein:

> **There are clearly moments to demonstrate, but perhaps more importantly, there are moments to build the connections that make demonstration something more than theatre. There are clearly times when radicalism means standing up to the police, but there are many more times when it means talking to your neighbor. (cited in De Filippis, 2001: 5)**

For DeFilippis, his objection to *only* focusing on global-level interventions rested on how certain peoples and places were left out of the protesting equation. As he notices, 'One of the key critiques leveled against the anti-globalization protesters is that they are overwhelmingly white and middle class' (2001: 5). Global protest had not included the poor of the global south, nor disenfranchised African Americans and Latinos, for example. However, like Herod, he is certainly not dismissing the spectacular forms of anti-capitalist/anti-globalization protests that transpired in Seattle in November 1999. That is, he is not rejecting the protests against the capitalist system as a whole (sometimes associated with the global scale). Rather, he echoes Harvey who claims: 'the choice of spatial scale [concerning politics or protest] is not "either/or" but "both/and", even though the latter entails confronting serious contradictions' (2001: 391).

Towards a Marxist Geographical Activism

Defining a Marxist geographical activism

There is no unambiguous or uncontested definition of activism, let alone Marxist geographical activism. Let us, however, define the latter tentatively as active engagement with social movements against capitalist oppression through *collective* struggle. Collective implies a tendency to universalism without neglecting particularism. It means that Marxist activists need not be restricted to oppression defined in strictly 'class' terms, and that such activism against class oppression should not exclude other protest movements. Indeed, as Smith suggested earlier, '"Back to class" in any narrow sense is its own self-defeating cul-de-sac' (2000: 1028).

At the very least, this distinguishes Marxist geographical activism from both participatory research and status quo policy-making. Let me elaborate on this. First, Marxist geographical activism should not be confused with all 'participatory action (PA) research'. For Rachel Pain, the common element of PA research is that 'research is undertaken collaboratively with and for the individuals, groups or communities who are its subject. Participatory approaches present a promising means of making real in practice the stated goals and ideals of critical geography' (2003: 653). If we ignore her failure to define what is meant by 'critical geography' – a crucial omission – Pain notes that PA research has grown in part because of the need by professional geographers to look towards non-academic bodies for funding, or because the research should have 'user relevance' (2003: 651). On the one hand, this careerist rationale for PA research probably does not rank high among the reasons for which the politically committed Marxist geographer becomes involved (after all, 'user relevance' could also mean developing GIS applications for military oppression). On the other hand, PA research can certainly rest on more Marxist motivations for action and does not necessarily exclude a Marxist perspective.

Second, activism *can* indeed involve policy-making. As Merrifield writes, through activism, 'academic geographers can articulate the "collective will" of a people (Gramsci's phrase) by gaining access and speaking to power elites, or by giving evidence at public inquiries and the like' (1995: 62). And as David Harvey (2001) has maintained, the issue is not so much whether Marxists should be involved in policy-making but *what kind* of public policy is necessary. Thus, activist research should not be deemed 'good' if it is grassroots and bottom up, and 'bad' if it is top down and state policy oriented (Pain, 2003). The question of the relationship between activism and the state has long historical antecedents. During the nineteenth century, an agonizing divide existed between Marxist anarchists (like Kropotkin) and those Marxists who believed the control of the state remained key to securing a better future for the world's proletariat.

This debate more or less repeated itself in the 1990s, when Nick Blomley (1994) published an editorial calling for a renewal of the activist tradition. In an editorial reply, Adam Tickell (1994) argued for an activist engagement *with the state* (absent in large measure from Blomley's editorial). Such engagement remained indispensable, Tickell claimed, because it meant a way of building a reformist 'meso-level social order' in response to the ravages of neoliberal capitalism. Moreover, he asserted, neoliberalism was being meted out at the national and international scales, and therefore activist geographers had to intervene at these scales, rather than at the level of 'local communities' that Blomley seemed to privilege. Blomley replied in the same issue and, while tending to agree with Tickell, also cautioned him that the boundary between the state and community was not always clear and that activism through the state, rather than through Blomley's own experience of activism – that is through local community organizing – 'can all too easily be blunted, and even redefined and positioned for conservative ends' (1994: 240).

Third, for academics, activism is generally associated with doing something 'out there', but universities can be oppressive places as well, and therefore (Marxist geographical) activism should also be turned 'inwards' towards universities themselves. Castree (1999b) calls this 'domesticating critical geography'. Such activism could begin, for example, by student and academic involvement in fighting for higher wages for those cleaners and other workers who maintain university campuses or low-paid teaching assistants and/or those on temporary contracts.

Fourth, Maxey (1999) points out that the media have popularized a view of activism that has 'emphasized dramatic, physical, "macho" forms of activism with short-term public impacts'. Using feminist theory, he argues instead that activism should seek to 'inspire, encourage, and engage as many people as possible' (1999: 200). Fifth, Maxey insists on the value of 'reflexivity' for activism, and seeks to emphasize the false divide between theory and activism (see also the special issue of *Area*, 1999 and the papers from the 'Beyond the Academy: Critical Geographies in Action' conference[1]). It is to this question of theory (and writing) and activism to which we shall now turn.

Theory, writing and activism

When geographers theorize or write from a Marxist perspective, they are Marxist activists. In other words, theory and writing are forms of activism. Indeed, as Frank Lentricchia wrote, 'struggles for hegemony are sometimes fought out in (certainly relayed through) colleges and universities; fought undramatically, yard for yard, and sometimes over minor texts of Balzac: no epic heroes, no epic acts' (cited in Barnes, 2002: 9).

But there are two issues here. One is the relationship between theory and activism and the other is the question of writing. First, the border between academic theory and practice is fluid and ambiguous, as many activist geographers insist. For example, what happens, asks the activist James DeFilippis, if the 'community' is wrong?[2] Indeed, can a 'community' be wrong? In general, the geographer Paul Routledge (1996) refers to the challenge of moving between the academy and the 'outside' as 'critical engagement'.

An early example in geography of the difficulty of 'critical engagement' is witnessed in William Bunge's famous (and at that time surely infamous) Detroit Geographical Expedition. During the 1960s, Bunge remained a theoretical geographer mostly concerned with abstract mathematical modelling. However, expelled from Detroit's Wayne State University in 1967, Bunge created the Detroit Geographical Expedition and Institute (later the Society for Human Exploration) because he became enraged at the striking social differences between the white suburbs of Detroit and the 'black' 'inner city'. Young black kids were literally starving and being run over by cars in front of their homes. Infant mortality had reached alarming rates. The Society for Human Exploration sought to engage local people as both students and professors in an 'expedition' to inner city Detroit. (Bunge used the term 'expedition' apparently as a way of subverting more traditional geographical expeditions and any given starting point for the expedition in Detroit would become the 'base camp'.) By 1970, the education wing of the Society offered 11 alternative courses throughout southern Michigan educational institutions to black students – tuition often being paid from voluntary contributions by professors. In 1972, after moving to Toronto, Bunge established CAGE (Canadian–American Geographical Expedition) involving both Toronto and Vancouver. However, Bunge points out that the Vancouver expedition failed because it 'lacked a true community base and was never self-critical about its democratic failings' (cited in Merrifield, 1995: 63). Furthermore, by 1973 the Detroit institute eventually fell apart as activist students graduated and left the university and Bunge himself departed for Toronto, in part because local universities no longer tolerated the 'alternativeness' and volunteer nature of the project. (For an extended discussion of Bunge and the Detroit Geographical Expedition, see Bunge, 1977; Horvath, 1971; Merrifield, 1995; Peet, 1977.) And yet Peet recounts the significance of Bunge's efforts:

> Expeditions provided an alternative source of information and planning skills to help low income communities bargain for power over their own affairs. Likewise advocate geographers and planners offered their professional experience to disenfranchised groups to help them deal with powerful institutions and eventually to shift power to the presently powerless. (1997: 15)

Similarly, Merrifield claims:

> expeditions become more than an attempt to learn *about* the impoverished: they become an effort to learn *with* them the oppressive reality that confronts ordinary people in their daily lives. (1995: 63)

In this context, the task of writing is not immune to difficulty. But there can be two interrelated 'moments' of writing. The first consists of writing for an academic or student audience usually 'post-expedition' (although as I suggested all oppositional writing is a form of activism and need not be generated from the so-called 'field'). When writing for such an audience, the problem of how to represent the 'oppressed' is a chief issue. On the one hand, Richard Peet asserts: 'We have to get over the blockage to action formed by a reluctance to speak for others' (2000: 953). This may be necessary to galvanize the more reflective of us into action, but the problem of representation is unlikely to go away so easily. On the other hand, Merrifield (1995) seeks to employ a 'pedagogy of the oppressed' (after Paolo Freire), in which one writes *with* the 'oppressed'. At any rate, writing only for an academic or student audience has its limits because of a limited readership. This is not to denigrate the seriousness, purpose, and effect of academic writing (including this chapter!) but, as Bunge put it, we need less 'citing' and more 'siting' (Merrifield, 1995).

The second – which would appeal more to Bunge – is writing for an audience beyond the academy (public or governmental officials, lawyers, etc.) in the form of reports, petitions, and so forth. In this case too, the problems of representation do not disappear. Even writing in a collective with other non-academic community activists raises problems, since they may not represent the opinions of the whole 'community'. Furthermore, while academic papers often demand theoretical nuance, and an exclusive scholarly lexicon, such complexity and 'jargon' would be usually out of place in, for example, an oppositional report addressed to public officials.

In sum, Marxist activist geography is not simply concerned with an *ethnography* of the poor (that is through a close dialogue with the oppressed), nor is it just participatory action research; rather it is about *working with* people in a non-patronizing way to *assist* – rather than *necessarily* direct – their organization and protest against (collectively determined) oppression. This may involve writing and careful planning with an inspirational vision (as DeFilippis reminds us), non-violent work such as cyber-protests, petition gathering, marching, and boycotting, or (despite the contentions of Maxey) spontaneous and violent protest as we saw at Seattle, Genoa, or Quebec City (Merrifield, 2002). In the last case, the distinction between scholar and activist becomes very blurry indeed.

Conclusions

While few geographers would label themselves 'Marxists' today, this does not diminish the value of the Marxist legacy. What seems to have changed among Marxist geographers is that 'class-*alone*' arguments – either theoretically or as a basis for action – have been relinquished in favour of nurturing an imagination for a broader coalition of contestation against capitalist social relations. (I use this term, instead of the more systemic word 'capitalism', after Gibson-Graham.) This does not mean that 'class' as a social relation (rather than a category of people) has become any less important for Marxists. Indeed, geographers working in the Marxist vein may be more concerned with class exploitation and questions of economic justice than with what appears to be non-economic (cultural) forms of justice.

Nonetheless, Marxist geographers do not see local, particularist or cultural struggles as necessarily inadequate to the task of fighting capitalist oppression (although local struggles can potentially undermine more global forms of activism, as Harvey argues). Rather, they see a contradiction between the particular and the universal and the need to reconcile these, *somehow*. Thus, the Marxist geographical activist treads a difficult path between a hopefulness about an emancipatory future – a collectivist vision if you will – and a realization that the world is a complex place.

Yet complexity, and the difficulty of constructing commonalities or universals beyond those strictly defined by class, should not seduce us into a hopeless relativism or a cynical scepticism about the necessity of activism, and the possibility of socialism or some other emancipatory form of society (see e.g. Corbridge, 1993). Again, referring to Bunge's Detroit Geographical Expedition, Merrifield points out that 'through expeditions it is incumbent upon the geographer to become a person of action, a radical problem-raiser, a responsible critical analyst participating *with* the oppressed. That said, the ambit of the geographer's responsibility is always ambiguous' (1995: 63).

No doubt, amongst this ambiguity, there is potentially a substantial price to pay for most professional geographers. One's activism may be deemed too 'radical'. Furthermore, to be actively involved in such anti-capitalist movements takes *time*, and this may clash with professional expectations like publishing in recognized academic journals (Castree, 2000a; Routledge, 1996). Nonetheless, Marxist *activist* geographers tend not to be primarily concerned about career advancement, and at any rate, active engagement *may* be as fruitful and rewarding for one's professional life as it can be damaging. In any case, somewhere in the space of a geographer's professional life (teaching, publishing, earning grant money) there is limited room for activism. Professional obligation need not mean all or nothing.

In fact, inaction means capitalist social relations will perpetuate inequality and the bulk of the globe will suffer. There is too much injustice for us to be relativist about this. How can we be, in a world in which some live comfortably – even lavishly – while others die of toil, starvation and disease? This is precisely why the Marxist critique of capitalist social relations is so vital for understanding and acting in the world, and for sharing a geographical project that is less oppressive than the one in which we currently live.

NOTES

1 http://online.northumbria.ac.uk/faculties/ss/gem/conferences/timetable.html
2 Personal communication, February 2004.

References

Area (1999) 'Research, Action and "Critical Geographies"', theme issue, 31: 195–246.
Barnes, T.J. (2002) 'Critical notes on economic geography from an aging radical: or, radical notes on economic geography from a critical age', *ACME: An International E-Journal for Critical Geographies*, 1: 8–14.

Blomley, N. (1994) 'Activism and the academy', *Environment and Planning D: Society and Space*, 12: 383–5.

Blunt, A. and Wills, J. (2000) *Dissident Geographies*. Harlow: Prentice Hall.

Bunge, W. (1977) 'The first years of the Detroit Geographical Expedition: a personal report', in R. Peet (ed.), *Radical Geography: Alternative Viewpoints on Contemporary Social Issues*. Chicago, IL: Maaroufa.

Castree, N. (1999a) 'Envisioning capitalism: geography and the renewal of Marxist political economy', *Transactions of the Institute of British Geographers*, 24: 137–58.

Castree, N. (1999b) '"Out there"? "In here"? Domesticating critical geography', *Area*, 31: 81–6.

Castree, N. (2000a) 'Professionalisation, activism, and the university: whither "critical geography"', *Environment and Planning A*, 32: 955–70.

Castree, N. (2000b) 'What kind of critical geography for what kind of politics?', *Environment and Planning A*, 32: 2091–270.

Castree, N., Coe, N., Ward, K. and Samers, M. (2003) *Spaces of Work: Global Capitalism and Geographies of Labour*. London: Sage.

Chouinard, V. (1994) 'Reinventing radical geography: is all that's Left Right?', *Environment and Planning D: Society and Space*, 12: 2–6.

Community Economies Collective (2001) 'Imagining and enacting noncapitalist futures', *Socialist Review*, 28: 93–135.

Corbridge, S. (1993) 'Marxisms, modernities, and moralities: development praxis and the claims of distant strangers', *Environment and Planning D: Society and Space*, 11: 449–72.

DeFilippis, J. (2001) http://comm-org.utoledo.edu/papers2001/defilippis.htm, accessed 26 December 2003.

Fraser, N. (1995) 'From redistribution to recognition? Dilemmas of justice in a "post-socialist" age', *New Left Review*, 212: 68–93.

Gibson-Graham, J.K. (1996) *The End of Capitalism (As We Knew It)*. Oxford: Blackwell.

Harvey, D. (2000) *Spaces of Hope*. Edinburgh: University of Edinburgh Press.

Harvey, D. (2001) *Spaces of Capital: Towards a Critical Geography*. Edinburgh: University of Edinburgh Press.

Harvey, D. and Hayter, T. (1994) *The Factory and the City: The Story of the Cowley Automobile Workers in Oxford*. London: Mansell.

Herod, A. (2001a) *Labor Geographies: Workers and the Landscapes of Capitalism*. New York: Guilford.

Herod, A. (2001b) 'Labor internationalisms and the contradictions of globalization: or, why the local is sometimes still important in the global economy', in P. Waterman and J. Wills (eds), *Place, Space and the New Labour Internationalisms*. Oxford: Blackwell.

Horvath, R. (1971) '"The Detroit Geographical Expedition and Institute" experience', *Antipode*, 3: 73–85.

Hudson, R. (2001) *Producing Places*. New York: Guilford Press.

Laclau, E. and Mouffe, C. (1987) 'Post-Marxism without apologies', *New Left Review*, 166: 79–106.

Maxey, I. (1999) 'Beyond boundaries? Activism, academia, reflexivity and research', *Area*, 31(3): 199–208.

Merrifield, A. (1993) 'The Canary Wharf debacle: From TINA there is no alternative to THEMBA there must be an alternative', *Environment and Planning A*, 25: 1247–65.
Merrifield, A. (1995) 'Situated knowledge through exploration: reflections on Bunge's geographical expeditions', *Antipode*, 27: 49–70.
Merrifield, A. (2002) 'Guest editorial: Seattle, Quebec, Genoa: *Après le Déluge* ... Henri Lefebvre', *Environment and Planning D: Society and Space*, 20: 127–34.
Mitchell, D. (2000) *Cultural Geography: A Critical Introduction*. Oxford: Blackwell.
Moody, K. (1997) *Workers in a Lean World*. London: Verso.
Pain, R. (2003) 'Social geography: on action-oriented research', *Progress in Human Geography*, 27: 649–57.
Peet, R. (1977) 'The development of radical geography in the United States', in R. Peet (ed.), *Radical Geography: Alternative Viewpoints on Contemporary Social Issues*. Chicago, IL: Maaroufa.
Peet, R. (2000) 'Celebrating thirty years of radical geography', *Environment and Planning A*.
Routledge, P. (1996) 'The third space as critical engagement', *Antipode*, 28: 398–419.
Sayer, A. (1995) *Radical Political Economy*. Oxford: Blackwell.
Sayer, A. and Walker, R. (1992) *The New Social Economy: Reworking the Social Division of Labour*. Oxford: Blackwell.
Smith, N. (2000) 'What happened to class?', *Environment and Planning A*, 32: 1011–32.
Tickell, A. (1994) 'Reflections on "Activism and the academy"', *Environment and Planning D: Society and Space*, 13: 235–7.
Wills, J. and Waterman, P. (2001) *Place, Space, and the New Labour Internationalisms*. Oxford: Blackwell.

26 PRODUCING FEMINIST GEOGRAPHIES: THEORY, METHODOLOGIES AND RESEARCH STRATEGIES

Kim England

For some time now feminist geographers have been engaged in a lively debate about research methods and methodology. Debates are an important part of knowledge production, and feminist geographers have not (and do not) all agree on the best way to produce feminist understandings of the world. In part this is because there is not a singular feminist geography but rather several strands existing simultaneously. So there are multiple, even competing visions of feminist geographies, with disagreements, negotiations and compromises around different approaches to practicing and producing feminist geographies. That said, there are commonalities among the strands of feminist geography. At the heart of feminist geographies are analyses of the complexities of power, privilege, oppression and representation, with gender foregrounded as the primary social relation (although gender is increasingly understood as constructed across a multiplicity of social relations of difference). Feminist geographers expose the (often 'naturalized') power relations in past and contemporary constructions of gender. And feminist geographers share the political and intellectual goal of socially and politically changing the world they seek to understand.

Feminist research challenges and redefines disciplinary assumptions and methods, and develops new understandings of what counts as knowledge. In this chapter I discuss one of the most important aspects of 'the feminist challenge': our debates about methods (techniques used to collect and analyze 'data') and methodologies (the epistemological or theoretical stance taken towards a particular research problem).[1] The task of the first feminist geographers was to recover women in human geography and to address geographers' persistent erasure of gender differences. Thus early feminist scholarship closely focused on challenging male dominance, making women's lives visible and counting and 'mapping' gender inequalities. Debates about methods and methodologies were about the usefulness for feminists of existing (gender-blind, sexist, malestream) methods of inquiry, especially quantitative methods, standardized surveys and 'traditional' interviews conducted 'objectively'. Debates focused on 'Is there a feminist method?' and 'Which method is most feminist?' 'Feminist' here is *adjectival* in the sense of whether certain research methods are 'feminist' in they way that some are 'quantitative' or 'qualitative'. Qualitative methods, especially interactive interviews, were generally considered best suited to the goals and politics of feminist analysis (Reinharz, 1979; Oakley, 1981; Stanley and Wise, 1993). In their recollections about these early feminist debates, Liz Stanley and Sue Wise (1993) assert that they were not really about method as such, but about sexist method*ologies* and competing epistemologies. In fact they and others argue that there is nothing inherently feminist in either quantitative or qualitative *methods*, but that what is 'feminist' is the *epistemological stance* taken towards methods and the uses to which researchers put them. No single method provides privileged access to the 'truth', and as it becomes less imperative (and less expedient) to associate certain methods with particular epistemologies, there has been a move towards the choice of appropriate method depending on the research question being asked.

The argument I make in this chapter is that feminists' contributions to research practices in human geography are generally more about epistemology (ways of knowing the world), methodology, and politics than about inventing new research methods. In the first section I discuss the various epistemological claims feminist scholars make about research methods and methodologies. In the second section I turn to the methods feminist geographers rely on to produce and represent feminist understandings of the world.

Methods and Methodology in Feminist Geographies

Since its inception in the 1970s, feminist geography has deconstructed the 'taken-for-granted' and offered profound and influential critiques of conventional concepts and categories in human geography (see Chapter 4). Across the academy, feminist scholars challenged conventional wisdom that 'good research' requires impartiality and 'scientific' objectivity. Since then feminist scholars have continued to challenge conventional wisdom and to develop feminist approaches to knowledge production. Feminists have produced a sizeable literature about feminist methods and methodologies, and in the last several years geographers have published many book chapters and journal articles on this topic (e.g. McDowell, 1992; special issues of *The Canadian Geographer*, 1993; *The Professional Geographer*, 1994; 1995; Gibson-Graham, 1994; Jones et al., 1997; Moss, 2001; 2002; ACME, 2003). In this section I describe the major elements of the discussion regarding the epistemological claims and politics of practicing feminist research that are entwined in the ongoing process of feminist knowledges creation and feminists' commitment to progressive research practices. (Much of my description of this discussion is about face-to-face research encounters, but similar arguments are made about other methods: see Gillian Rose, 2001 on visual cultures and methodologies; and Mona Domosh, 1997 on feminist historical geography.)

Critique of positivism and situated knowledges

The 'western industrial scientific approach values the orderly, rational, quantifiable, predictable, abstract and theoretical: feminism spat in its eye' (Stanley and Wise, 1993: 66). Early on, feminists raised suspicions that 'good research' could be produced only by unbiased 'experts' seeking universal truths by using value-free data where 'the facts speak for themselves'. Research informed by the 'western industrial scientific approach' is anchored by a positivist epistemology of objectivity. Positivism, what it values (rationality, etc.), and the search for universal truths and all-encompassing knowledge constitute an approach Donna Haraway describes as an all-seeing 'god trick … seeing everything from nowhere' (1991: 189). No research inquiry, whether positivist or indeed humanist or feminist, exists outside the realms of ideology and politics; research is never value-free (even 'hard science' research). Instead, feminists understand research to be produced in a world already interpreted by people, including ourselves, who live their lives in it. By becoming 'researchers', whether physicists or feminists, we cannot put aside common-sense understandings of the world. Instead, feminists argue, 'good research' must be sensitive to how values, power and politics frame what we take to be 'facts', how we develop a particular research approach and what research questions we ask and what we see when conducting research.

Since the early 1990s, feminist critiques of objectivity have been enriched by feminist scholars of science and technology studies (e.g. Fox-Keller, 1985; Harding, 1986; Haraway, 1991). Evelyn Fox-Keller argues that traditional western thought rests on an ontology (a theory of what is, and the relations between what is) of self/other opposition, and a binary opposition between (male) objectivity and (female) subjectivity. Her alternative is a *feminist relational ontology* of self–other mutuality and continual process (rather than stasis). Haraway argues for an *embodied feminist objectivity* where both researchers and participants are appreciated for their *situated knowledges* and *partial perspectives*. Situated knowledges means that there is no one truth waiting to be discovered; and those knowledges are situational, marked by the contexts in which they are produced, by their specificity, limited location, and partiality.

Researcher–researched relationship and power relations

Sensitivity to power relations lies at the very heart of feminists' discussions about method/methodology. Traditional objectivist social science methods (be they quantitative or qualitative) position researchers as detached omniscient experts in control of the research process, the (passive) objects of their research, and themselves (remaining unbiased by being detached, uninvolved and distant). Feminist relational ontology and embodied feminist objectivity challenge this strict dichotomy between object and subject. In feminist research, especially in face-to-face fieldwork, the researched are not passive, they are knowledgeable agents accepted as 'experts' of their own experience. Instead of attempting to minimize interaction (in order to minimize observer bias), feminists deliberately and consciously seek interaction. Feminist researchers try to reduce the distance between ourselves and the researched by building on our commonalities, working collaboratively and sharing knowledge. By seeking research relationships based on empathy, mutuality and respect, feminists focus on the informant's own understanding of their circumstances and the social structures in which they are implicated (rather than imposing our explanations). In practice this usually means being flexible in question asking, and shifting the direction of the interview according to what the interviewee wants to, or is able to talk about. As a research strategy this may provide deeper understandings of the subtle nuances of meaning that structure and shape the everyday lives of informants, and politically it grounds feminist knowledge and politics in women's everyday experiences.

More recent poststructural feminist theorizing sees researchers and the researched as caught up in complex webs of power and privilege. Much feminist research is about marginalized groups, and there is a great deal of social power associated with being a scholar. Thus research strategies based on an embodied feminist objectivity have the potential to minimize the hierarchical relationship between researcher and interviewee, and to avoid exploiting less powerful people as mere sources of data. At the same time, the research encounter is now understood as being structured by both the researcher and the researched, both of whom construct their worlds. Poststructural understandings of the researcher–researched relation see it as one where both discursively produce 'the field' and create a co-produced project. This idea is also useful when considering the power relations and research relation in feminist geographers' interviews with elites (see for example McDowell, 1998; England, 2002). In this case, we are in positions of less power relative to the researched who are accustomed to having a great deal of control and authority over others, but nevertheless the researcher and researched are still engaged in co-produced research.

Positionality and reflexivity

Among of the most influential elements in feminists' theorizing about the research process are the concepts of positionality and reflexivity (England, 1994; Rose, 1997; Falconer Al-Hindi and Kawabata, 2002). These concepts raise important questions about the politics and ethics of research, and have been highly influential concepts both within and beyond feminist geography. Positionality is about how people view the world from different embodied locations. The situatedness of knowledge means whether we are researchers or participants, we are differently situated by our social, intellectual and spatial locations, by our intellectual history and our lived experience, all of which shape our understandings of the world and the knowledge we produce. Positionality also refers to how we are positioned (by ourselves, by others, by particular discourses) in relation to multiple, relational social processes of difference (gender, class, 'race'/ethnicity, age, sexuality and so on), which also means we are differently positioned in hierarchies of power and privilege. Our positionality shapes our research, and may inhibit or enable certain research insights (see Moss, 2001 where geographers discuss their autobiographies in relation to their research). Positionality has been further extended to include considering others' reactions to us as researchers. As researchers we are a visible, indeed embodied and integral part of the research process (rather than external, detached observers). So both our embodied presence as researchers and the participants' response to us mediate the information collected in the research encounter.

In a research context, reflexivity means the self-conscious, analytical scrutiny of one's self as a researcher. Within feminist methodologies, reflexivity extends to a consideration of power and its consequences within the research relationship. Gillian Rose (1997) raises concerns about a possible emerging feminist orthodoxy about reflexivity. She argues that being reflexive cannot make everything completely transparent and we cannot fully locate ourselves in our research, because we never fully understand (or are fully aware of) our position in webs of power. Her concerns remind us to constantly interrogate our assumptions and remember that knowledge is always partial, including that about ourselves. Nevertheless, reflexivity gets us to think about the consequences of our interactions with the researched. For instance, is what we might find out actually worth the intrusion into other people's lives? Are we engaging in appropriation or even theft of other people's knowledges? However, while reflexivity can make us more aware of power relations, and asymmetrical or exploitative research relations, it does not remove them, so we alone have to accept responsibility for our research.

Politics and accountability

Feminist geographers argue that we must be accountable for our research, for our intrusions into people's lives, and for our representations of those lives in our final papers. We still need to acknowledge our *own* positionality and our locations in systems of privilege and oppression, and be sure that we write this into our papers. As Lawrence Berg and Juliana Mansvelt argue, 'The process of writing constructs what we know about our research but it also speaks powerfully about who we are and where we speak from' (2000: 173). We need to be accountable for the consequences of our interactions with those we research (and many university ethics review boards require it). This is acutely important where our research might expose previously invisible practices to those who could use that information in oppressive ways, even when our goal was to make systems of oppression more transparent to the oppressed (Katz, 1994; Kobayashi, 1994). For example, some researchers studying lesbian communities do not reveal the location of their fieldwork (see for example Valentine, 1993), out of respect both for the participants' desire not to be 'outed', and for the participants' concerns about reprisals, including physical attack.

Feminist geographers share the political and intellectual goal of socially and politically changing the world our research seeks to understand. The increased popularity of reflexivity and positionality in research has raised serious political and ethical dilemmas, a crisis even, about working with groups we do not belong to. This raises difficult questions about the politics of location (both socially and geographically, including white women from the global north researching women from/in the global south). This has been an especially contentious and even painful debate for feminists, both inside and outside the academy. Some academic feminists have abandoned research projects involving groups to which they have no social claim, leaving them to those with 'insider status'. This discussion escalated just as (or because?) feminist geographers are becoming increasingly committed to taking account of the diverse positionings of women (and men and children) across a wide range of social and spatial settings.

This impasse is especially troubling for those feminists wishing to address multiple and cross-cutting positions of privilege and oppression, and who are committed to effect change. However, Audrey Kobayashi argues that 'commonality is always partial … [and so] field research and theoretical analyses have more to gain from building commonality than from essentializing difference' (1994: 76). This is not to suggest that problematizing essentialism (reifying categories and naturalizing difference) means ignoring difference or dismissing the experiences of marginalized groups; quite the opposite. Rather it means building on the notion that everyone is entangled in multiple webs of privilege and oppression, so that there are really few pure oppressors and pure oppressed. Materially engaged transformative politics can emerge from accepting that privilege results from historical and contemporary conditions of oppression, and people are variously located in the resulting webs of power. This means for instance that whether I acknowledge it or not, as a white woman I participate in and benefit from white privilege. For those of us with more social privilege (including being scholars), rather than agonizing over our culpability, it may be more productive to address our complicity, to make our lives as sites of resistance and to work hard to unlearn our privilege (Peake and Kobayashi, 2002). Feminists argue that we are committed to the political and intellectual goals not only of exposing power and privilege, but also of transforming them. An important part of that is to understand how the world works, and to theorize how power operates and expose it, because this means we are better able to gauge the possibilities for transformation, and provide situated knowledges that can most effectively produce change.

Producing Feminist Understandings

In this section I discuss how methods are employed by feminist geographers to produce and represent feminist understandings of the world. Generally, methods are described as either qualitative or quantitative, so I begin with broad definitions of each. Then I will provide some examples of how methods are used in feminist geographers' research (see Box 26.1). Finally I address the so-called 'quantitative–qualitative debate'.

Quantitative, qualitative and mixed methods

Quantitative research focuses on questions like 'how many?' and 'how often?' and seeks to *measure* general patterns among *representative samples* of the population. *Statistical techniques* are used to analyze data, for example, descriptive statistics, spatial statistics and geographic information systems (GIS). The data are often secondary data (usually collected in an 'official' capacity, like the census) and are based on standardized measures (again like those in the census). Primary data may also be used; the researcher collects their own data usually based on large samples using highly *structured questionnaires* containing easily quantifiable categories. For examples, see Box 26.1

Qualitative research focuses on the question of 'why?' and seeks to decipher experiences within broader webs of meanings and within sets of social structures and processes. Techniques are interpretive- and meanings-centered and include *oral methods* (e.g. semi-structured interviews, focus groups, and oral histories), *participant observation* and *textual analysis* (of, for example, diaries, historical documents, maps, landscapes, films, photographs, and print media). Samples are usually small and are often *purposefully selected* (to relate to the research topic), and if oral methods are used it is not uncommon for researchers to ask informants to help find other participants (known as snowballing). For examples see Box 26.1

In some instances, feminist understandings of the world are best produced with a politically informed combination of research methods, variously described as *mixed methods*, *multimethods* or *triangulation*. In human geography we commonly think of mixed methods as mixing qualitative and quantitative methods as complementary strategies. For example in their extensive study of gender, work and space in Worcester, Massachusetts, Susan Hanson and Geraldine Pratt (1995) sought to understand more fully the complex links between domestic responsibilities, occupational segregation, job search and residential choice. Their research involved statistical analyses and mapping of census data; and the quantitative and qualitative analyses of semi-structured questionnaires gathered in interviews with 700 working-age women and men from across Worcester, and 150 employers and 200 employees in four different Worcester communities.

'Mixed methods' also refers to mixing methods or a variety of 'data' within a broadly qualitative or quantitative research project. In the examples in Box 26.1, Richa Nagar's Dar es Salaam project includes oral histories, interviews and participant observation; and Sara McLafferty's breast cancer project involved statistical techniques and GIS. Mixed methods can also involve a research design with different investigators coming at the research question from different fields of research or epistemological positions. For example, Susan Hanson and Geraldine Pratt describe how their collaboration was based not only on their 'shared interest in feminism and urban social geography but in our differences: one of us having roots in transportation and quantitative geography; the other in housing and cultural geography' (1995: xiv). Mixed methods can allow all of these sorts of differences to be held in productive tension, and may keep our research sensitive to a range of questions and debates.

The quantitative–qualitative divide?

The sorts of epistemological claims I described in the previous section mean that feminists do tend to use qualitative rather than quantitative methods. But Liz Stanley and Sue Wise (1993: 188) point out that even in the early 1980s, few feminist scholars called for an outright rejection of quantitative methods,

BOX 26.1 SOME EXAMPLES OF QUALITATIVE AND QUANTITATIVE RESEARCH IN FEMINIST GEOGRAPHY

An example of qualitative feminist research is Richa Nagar's work on the gendered and classed communal and racial politics in South Asian communities in postcolonial Dar es Salaam. Richa's fieldwork in Tanzania included analyzing *documents* from Hindi and caste-based organizations and the Tanzanian government; gathering 36 *life histories* and 98 shorter *interviews* with Hindi and Ithna Asheri (Muslim) women and men of South Asian descent; and conducting *participant observations* of communal places, homes and neighborhoods. In her paper 'I'd rather be rude than ruled' (Nagar, 2000), she tells the stories of four economically privileged women and focuses on their spatial tactics and subversive acts against the dominant gendered practices and codes of conduct in communal public places. Another example of qualitative feminist research is Gillian Rose's research about interpreting meanings in landscapes and visual representations. Gillian's recent work investigates visual culture, especially contemporary and historical photographs (see her 2001 book on reading visual culture). In a recent paper (Rose, 2003) about family *photographs* she explores the idea that the meanings of photographs are established through their uses, in this instance being a 'proper mother' and the production of domestic space that extends beyond their house to include, for example, relatives elsewhere (in other places and other times). She conducted *semi-structured interviews* with 14 white, middle-class mothers with young children. The women showed Gillian family photos, and she took note of where and how the women stored and displayed the photos.

Both Richa and Gillian are posing 'why?' questions, and seek to understand meanings within broader social processes and structures. Richa looks at the creation and modification of social identity in a context of rapid political and economic change, while Gillian explores the multiple meanings of mothering, 'the' family and domestic space. In each project the samples are small (four women in Richa's case, 14 in Gillian's) and the research strategies were based on the participants' own understanding of their circumstances, which Richa and Gillian interpret in relation to broader social structures and processes. Also, as is common in qualitative research, they write about the research using extensive quotes from the participants, and detailed textual descriptions of the cultural codes and webs of significance evident in and beyond the research setting.

In feminist geography, quantitative methods are frequently employed in what can broadly be described as accessibility studies (such as access to child care, jobs and social services). For example, in a series of papers, Orna Blumen uses *census data* for Israeli cities to *measure quantifiable* aspects of gendered intra-urban labour markets (e.g. commuting distances). In a paper with Iris Zamir (Blumen and Zamir, 2001), Orna used *census categories* of occupations to look at social differentiation in paid employment and residential spaces. They analyzed the data using a weighted index of dissimilarity (a commonly used *measure* of occupational segregation that captures the segregation of one group relative to another) and smallest space analysis (to produces a *graphic presentation* – a sort of map – of relative occupational segregation).

(Continued)

A second example of employing quantitative methods in feminist geography is Sara McLafferty's research on geographic inequalities in health and social wellbeing in US cities. In a piece with Linda Timander (McLafferty and Timander, 1998), Sara explored the elevated incidence of breast cancer in West Islip, New York, using *individual-address-level data* from a *survey* of 816 women (39 with a history of breast cancer, 777 without). The survey data were collected by a group of women in West Islip. Sara and Linda employed *statistical techniques* (chi-squared and logistic regression analysis) to analyze the relationship between breast cancer and 'known risk factors' (such as a family history of breast cancer). Then for those women where 'known risk factors' did not explain the incidence of breast cancer, Linda and Sara used *GIS* to analyze *spatial clustering* to see whether local environmental exposure was a factor.

In these two examples, the authors' ask 'how many?' Orna counts how many people are in particular occupations in the employment and home spaces of Tel Aviv; Sara asks how many women have breast cancer in specific locations on Long Island. They also show how quantitative techniques can be applied to primary (an individual-level, large survey) and secondary (standardized census categories) data. Each study involves measuring some quantifiable occurrence (occupational segregation and the incidence of breast cancer) and employing spatial statistics and mapping.

and they urged feminists to use any and every means possible to produce critically aware feminist understandings of the world. Since the mid 1990s there has been a spirited discussion among feminist geographers about employing quantitative methods, but adapting them as appropriate (*The Professional Geographer*, 1995; *Gender, Place and Culture*, 2002). Vicky Lawson argues that 'feminist scholars can and should employ quantitative techniques within the context of relational ontologies *to answer particular kinds of questions*' (1995: 453, emphasis in the original). Some feminist geographers argue that certain long-standing feminist critiques of quantitative methods need reconsidering. For example, one criticism is that quantitative research can only analyze a particular cross-section in time (e.g. the census), whereas qualitative research captures changing historical and social contexts. Damaris Rose (2001) suggests that recent innovation in quantitative techniques blurs this distinction. For example, event history analysis involves longitudinal studies and documents the historical sequencing of events to predict statistical probabilities of a particular event generating a particular action. Others claim that critically aware quantitative methodologies are possible. For instance, Sara McLafferty (2002) describes how she was approached in West Islip, New York by women for help in analyzing their breast cancer survey and conducting further statistical analysis (see Box 26.1). Thus, Sara argues, GIS has potential as a tool for feminist activism and women's empowerment. And Mei-Po Kwan (2002a; 2002b) makes a case for feminist GIS (especially 3D visualization methods), arguing that converting quantitative data into visual representations 'allows, to a certain extent, a more interpretative mode of analysis than what conventional quantitative methods would permit' (2002b: 271).

My descriptions of qualitative and quantitative methods at the beginning of this section were represented as dichotomies, which is often the way they are represented in method/ methodology debates. Potent dichotomies structure our concepts of research (object/ subject, researcher/researched) and for feminist geographers an enduring dualism is the quantitative–qualitative divide. But disagreements over methods are often really disagreements about epistemology and methodologies, and the use to which the methods are put. Quantitative and qualitative methods do have different strengths and weaknesses, but rather than a clear epistemological break between quantitative and qualitative methods, there is a fundamental link between the two, because, for instance, one often involves an element of the other. For example, interview data can be coded using both qualitative and quantitative techniques and the same data set can be analyzed using qualitative and quantitative analyses (e.g. Susan Hanson and Geraldine Pratt's project described above). Rather than assuming that qualitative and quantitative research methods are mutually exclusive, it is more productive to think of methods forming a continuum from which we pick those best suited to the purpose of our inquiry. So although qualitative methods continue to be favoured by feminist scholars, since the early 1990s feminist geographers have employed an increasing range of methods, including quantitative techniques, and 'newer' qualitative techniques such as textual and visual analysis.

Conclusion

Today, feminist geography strides confidently across human geography's terrain. Because of feminist theorizing, it is now common (and increasingly expected) for all human geographers to locate themselves socially, politically and intellectually within their research. Human geographers are now likely to consider themselves producing partial, embodied, situated knowledges rather than fixed, universal truths. Feminist geographies have transformed human geographies. Feminist reconceptualizations have also transformed our understandings of ways of knowing and seeing the world. So feminist geography has not only extended human geography's research agendas, but redefined what human geographers do and how they do it. Looking to the future, feminist geographers will continue to produce new understandings and to engage politically in the progressive use of research. But we do need to be more open to 'negative' findings and to evidence that runs counter to our point of view. Like Susan Hanson, I hope 'to see us devise methods and methodologies that maximize the chance that we will see things we were not expecting to see, that leave us open to surprise, that do not foreclose the unexpected' (1997: 125). By thinking critically about epistemologies, methodologies, and methods, feminist geographers have already created richer, more complex human geographies; and feminist meditations on the research process have transformed the way human geography is practiced, produced and taught. By asking incisive questions and by seeking to develop the very best approaches to knowledge production in the future, the explanatory power of feminist geography will become even stronger and more compelling.

NOTE

1 I choose to use 'we' and 'us' throughout this chapter, but not because I speak for all feminist geographers (or you the reader!). I am also

mindful of concerns raised by, for instance, women from the global south/Third World who argue they are excluded from the 'we' of many feminisms (for example see Mohanty et al., 1991). I avoid the third person because it distances me from what I am writing, and I am certainly not (and do not wish to be) disconnected from feminist knowledge creation (Berg and Mansvelt, 2000).

References

ACME: An International E-Journal for Critical Geographies (2003) 'Practices in Feminist Research', collection, 2(1): 57–111.

Berg, L. and Mansvelt, J. (2000) 'Writing in, speaking out: communicating qualitative research findings', in I. Hay (ed.), *Qualitative Research Methods in Human Geography*. Oxford: Oxford University Press.

Blumen, O. and Zamir, I. (2001) 'Two social environments in a working day: occupation and spatial segregation in metropolitan Tel Aviv', *Environment and Planning A*, 33: 1765–84.

Canadian Geographer (1993) 'Feminism as method', collection, 37: 48–61.

Domosh, M. (1997) '"With stout boots and a stout heart": historical methodology and feminist geography', in J.P. Jones III, H. Nast and S.M. Roberts (eds), *Thresholds in Feminist Geography: Difference, Methodology, Representation*. Lanham, MD: Rowman and Littlefield.

England, K. (1994) 'Getting personal: reflexivity, positionality and feminist research', *The Professional Geographer*, 46: 80–9.

England, K. (2002) 'Interviewing elites: cautionary tales about researching women managers in Canada's banking industry', in P. Moss (ed.), *Placing Autobiography in Geography*. Syracuse, NY: Syracuse University Press.

Falconer Al-Hindi, K. and Kawabata, H. (2002) 'Toward a more fully reflexive feminist geography', in P. Moss (ed.), *Placing Autobiography in Geography*. Syracuse NY: Syracuse University Press.

Fox-Keller, E. (1985) *Reflections on Gender and Science*. New Haven, CT: Yale University Press.

Gender, Place and Culture (2002) 'Feminist geography and GIS', collection, 9(3): 261–303.

Gibson-Graham, J.K. (1994) '"Stuffed if I know": reflections on post-modern feminist social research', *Gender, Place and Culture*, 1: 205–24.

Hanson, S. (1997) 'As the world turns: new horizons in feminist geographic methodologies', in J.P. Jones III, H. Nast and S.M. Roberts (eds), *Thresholds in Feminist Geography: Difference, Methodology, Representation*. Lanham, MD: Rowman and Littlefield.

Hanson, S. and Pratt, G. (1995) *Gender, Work and Space*. London: Routledge.

Haraway, D.J. (1991) *Simians, Cyborgs, and Women: The Reinvention of Nature*. New York: Routledge, Chapman and Hall.

Harding, S. (1986) *The Science Question in Feminism*. Bloomington, IN: Indiana University Press.

Jones, J.P., Nast, H.J. and Roberts, S.M. (eds) (1997) *Thresholds in Feminist Geography: Difference, Methodology, Representation.* Lanham, MD: Rowman and Littlefield.

Katz, C. (1994) 'Playing the field: questions of fieldwork in geography', *Professional Geographer*, 46: 67–72.

Kobayashi, A. (1994) 'Coloring the field: gender, "race", and the politics of fieldwork', *Professional Geographer*, 46: 73–80.

Kwan, M.-P. (2002a) 'Feminist visualization: re-envisioning GIS as a method in feminist geographic research', *Annals of the Association of American Geographers*, 92: 645–61.

Kwan, M.-P. (2002b) 'Is GIS for women? Reflections on the critical discourse in the 1990s', *Gender, Place and Culture*, 9: 271–9.

Lawson, V.A. (1995) 'The politics of difference: examining the quantitative/qualitative dualism in post-structural feminist research', *Professional Geographer*, 47: 449–57.

McDowell, L. (1992) 'Doing gender: feminism, feminists and research methods in human geography', *Transactions of the Institute of British Geographers*, 17: 399–416.

McDowell, L. (1998) 'Elites in the City of London: some methodological considerations', *Environment and Planning A*, 30: 2133–44.

McLafferty, S. (2002) 'Mapping women's worlds: knowledge, power and the bounds of GIS', *Gender, Place and Culture*, 9: 263–9.

McLafferty, S. and Timander, L. (1998) 'Breast cancer in West Islip, NY: a spatial clustering analysis with covariates', *Social Science and Medicine*, 46: 1623–35.

Mohanty, C., Russo, A. and Torres, L. (eds) (1991) *Third World Feminism and the Politics of Feminism.* Bloomington, IN: Indiana University Press.

Moss, P. (ed.) (2001) *Placing Autobiography in Geography.* Syracuse, NY: Syracuse University Press.

Moss, P. (ed.) (2002) *Feminist Geography in Practice: Research and Methods.* Oxford: Blackwell.

Nagar, R. (2000) 'I'd rather be rude than ruled: gender, place and communal politics in Dar es Salaam', *Women's Studies International Forum,* 5: 571–85.

Oakley, A. (1981) 'Interviewing women: a contradiction in terms?', in H. Roberts (ed.), *Doing Feminist Research.* London: Routledge and Kegan Paul, pp. 30–61.

Peake, L. and Kobayashi, A. (2002) 'Policies and practices for an antiracist geography at the millennium', *Professional Geography*, 54: 50–61.

Professional Geographer (1994) 'Women in the Field: Critical Feminist Methodologies and Theoretical Perspectives', collection, 46: 426–66.

Professional Geographer (1995) 'Should Women Count? The Role of Quantitative Methodology in Feminist Geographic Research', collection, 47: 427–58.

Reinharz, S. (1979) *On Becoming a Social Scientist.* San Francisco, CA: Jossey-Bass.

Rose, D. (2001) *Revisiting Feminist Research Methodologies.* Ottawa: Status of Women Canada, Research Division.

Rose, G. (1997) 'Situating knowledge, positionality, reflexivities and other tactics', *Progress in Human Geography*, 21: 305–20.

Rose, G. (2001) *Visual Methodologies.* London: Sage.

Rose, G. (2003) 'Family photographs and domestic spacings: a case study', *Transactions of the Institute of British Geographers*, 28: 5–18.

Stanley, L. and Wise, S. (1993) *Breaking Out Again: Feminist Ontology and Epistemology* (1st edn 1983). London: Routledge.

Valentine, G. (1993) '(Hetero)sexing space: lesbians' perceptions and experiences of everyday spaces', *Environment and Planning D: Society and Space*, 11: 395–413.

27 POSTSTRUCTURALIST THEORIES, CRITICAL METHODS AND EXPERIMENTATION

John W. Wylie

Introduction

Poststructuralist methods are above all *critical methods*. That is, they enable a perspective from which we can make critical assessments of, for example, existing social institutions, cultural beliefs and political arrangements.

But straightaway we need to be careful in making such a claim. First, poststructuralist thought, writing and practice are characterized by a profound suspicion of bald statements and simple explanations for things. Poststructuralism is profoundly suspicious of anything that tries to pass itself off as a simple statement of fact, of anything that claims to be true by virtue of being 'obvious', 'natural', or based upon 'common sense'. As a philosophy and a set of methods of doing research, poststructuralism (see Chapter 10) exposes all such claims as contingent, provisional, subject to scrutiny and debate.

Second, we need to be careful because, while poststructuralism may be reasonably called a 'critical philosophy', it is very different from the types of critical or radical geographies, sociologies, histories and so on that are ultimately rooted within Marxist philosophies as elaborated in Chapter 5. Poststructuralism in fact offers a *more* radical and critical approach than these, because it is not based upon one particular diagnosis of how the world is organized (such as Marxism), and because it does not offer a systematic alternative (as Marxism is held to do). The writer Michel Foucault captured this aspect of a poststructural approach well when he stated that, 'nothing is fundamental. That is what is interesting in the analysis of society' (in Rabinow, 1984: 247). Thus, because it is based, as this quote suggests, upon principles of plurality and complexity, poststructuralism in a sense demands that we be endlessly critical. And as one consequence of the radical way in which it urges us to incessantly question our most rooted assumptions about who we are and how the world is, poststructuralism necessarily pushes us towards inventive and experimental ways of researching and writing.

One particular thing that poststructuralist writers have tended to be critical of is the way in which academic or scholarly knowledge tends to be produced, organized and communicated within both specific institutions such as universities, and education systems more generally. What tends to be produced, and what students tend to expect because they have been socialized into such systems, is *structured* knowledge: ideas boiled down to their 'essence', bullet points, lists of 'key ideas', clear statements of what the issues are, fairly definite conclusions. The notion that the entire purpose of academic study is to make an opaque reality clearer, a complex world more graspable, is very deeply entrenched within western culture. Poststructuralism, however, is very suspicious of this notion, and especially of the systems it entails. For example, textbooks designed for consumption by an undergraduate audience tend to be organized by a system in which a body of knowledge (e.g. human geography) is divided and classified into discrete approaches, topics and methods and then packaged into manageable chunks (chapters) in which key points are identified and summarized. From a poststructural perspective, this type of reductive, systematic procedure is precisely the sort of thing that stunts our ability to think critically and radically. This, of course, means that my task in this chapter is in a sense an ironic, contradictory and dubious one. However, as with the other chapters of this volume, the aim here is not to provide a 'how-to' guide, or a recipe that if followed correctly will magically produce a 'poststructural' piece of research or writing. The aim instead is to spotlight two of the main theoretical avenues down which poststructuralism travels – deconstruction and discourse analysis – and, in doing so, to identify some of the substantial topics that poststructuralist geographers have unearthed and written about.

Deconstruction, or, What Happens

Deconstruction. This term, forever associated with the work of the French-Algerian philosopher Jacques Derrida, was originally confined to use in technical, specialist and scholarly circles. More recently it has to an extent entered common usage. Today, for instance, it can sometimes be found in book, film or music reviews in broadsheet newspapers. In such contexts, 'deconstruction' often seems to simply mean 'analyse' or 'scrutinize', without any specific or technical connotations; for example, 'if we deconstruct X's album, we can see their influences are ...'. Alternatively, the term 'deconstruction' is sometimes used to designate the style or characteristic feel of a book, film, or album. The term is particularly applied if a book, film, album, etc. is perceived to be genre blurring, hybrid in form, or, in the most general sense, ambiguous, complex or difficult to understand. For example, 'this film deconstructs our conventional notions of the cowboy', or, 'the book's plot is a deliberate deconstruction of standard narrative devices'.

These quasi-popular usages only skim the surface of the potential meanings and ramifications of deconstruction. But they do serve to highlight that deconstruction may usefully be thought of (and deployed within research projects) in two complementary ways. First, in the sense conveyed by 'analyse' or 'scrutinize', deconstruction is a particular *method* which may be used in studying any topic. More specifically, deconstruction is a way of reading and writing about things; a way of reading and writing which is based upon an original understanding of how language, and the meanings and messages conveyed by language, works. Second, thought of as a characteristic of things in themselves (of objects, artworks, transport systems, whatever), deconstruction is not merely a research method; it is an actual process which actually occurs in the world, and may thus be witnessed and documented. Deconstruction, in other words, is *what happens* (Royle, 2003). One interpretation of the works of Jacques Derrida (1976; 1978) is that deconstruction is already going on 'out there'. It is a process which is (though this word is rather inappropriate) 'inherent' in language, in the way humans communicate with each other, think about themselves and others – in the general ways cultural and material worlds operate.

Some care is needed in making these claims, though. Deconstruction does not and cannot aver to be the methodological key to unlocking the one and single truth of how things 'really are'. This is because its aim is, precisely, to oppose and undermine claims to truth, certainty and authority. To explain this we may, following Derrida, adopt the language of ghosts. While most methods rest upon an *ontology* – a set of beliefs and assumptions about what is real, what can be taken as self-evidently true – deconstruction might be described as a 'hauntology' (Derrida, 1994; Royle, 2003). Deconstruction is an uncanny, ghostly process. Like a ghost, it is incessant, forever in a process of appearing, but, 'never present as such' (Derrida, 1994; xviii). It is always both there and not there. In this way, in a sense, the *raison d'être* of deconstruction is, parasitically, to haunt, inhabit and contest claims to truth. Deconstruction destabilizes notions of truth, clarity and certainty through a spectral logic: it differentiates, disturbs, unsettles.

Given the importance of relations between theories and methods, the remainder of this section focuses upon deconstruction as a method of reading, analysing and writing. Having outlined the rudiments of the deconstructionist approach – a quixotic venture given the space available – I will briefly discuss some of the main areas in which human geographers have applied deconstructive analyses.

Deconstruction sets out from the principle that language is a system of differences. Take, for example – as does Eagleton (1983), whom I follow here – the English word 'cat'. A first, small step onto the deconstructive track is to recognize that 'cat' is 'cat' because it is not 'mat', 'car', 'cut' and so on. Equally, 'mat' is 'mat' because it is not 'man', 'car' is 'car' because it is not 'bar', 'cut' is 'cut' because it is not 'but', and so on, and so on – *endlessly*. Rather than words possessing meaning because they correspond to actual things or mental images (e.g. a small, furry, four-legged animal), words acquire meaning via being caught up within an infinite series of differential relations. This, crucially, is not merely a peculiarity of language of interest only to academic specialists. It is a process at work within actuality. Its most glittering consequence is that the meaning of something is constituted by what that thing is *not*. Meaning is not *inside* a word – or an object, a thing, a process – inherent to it, uniquely owned by it. The meaning of something is constituted instead by what it is *not*. To put this another way, the *presence* of a thing, its existence, identity, validity, etc., is constituted by what is *absent* from it, or what is excluded from it.

Consider the couplet male/female. 'Common sense' might seem to say that the terms 'male' and 'female' refer to real and unchanging qualities. There are, we tend to assume, uniquely 'masculine' and 'feminine' characteristics, attributes and tendencies in people, animals and things. However, the application of a deconstructive reading reveals that male/female might equally be written not-female/female. In other words, the meaning of the term 'male' is defined and produced through the meaning of the term 'female'. The definition of a series of 'feminine' qualities constitutes by opposition those that are held to be 'masculine'. To put this the other way round, the definition, in a particular place or time, of what it is to be a 'man', is achieved via the exclusion from this definition of all those things that are held to constitute what it is to be a 'woman'. But, in the deconstructive process that is always already occurring within any such attempts at definition, the trace of the 'feminine' is incessantly returning to haunt the definition of the 'masculine' from within. Therefore the terms 'male' and 'female' are never stable in their meaning. They never acquire presence, meaning or validity in themselves. The terms 'masculine' and 'feminine' can only ever describe a territory of meaning that is shifting, incessantly undercutting itself.

Our culture is littered with binaries like male/female – mind/body, me/you, us/them, natural/artificial, west/east and so on. Their impact upon the way we think, how we relate to each other, and how society and politics in general operate, is incalculable. What deconstructive analysis reveals is not that such binaries are unreal, but that they are never pure or coherent: the two sides of the coin are not produced in isolation from each other but are rather always inextricably intertwined. Derrida calls such binaries *violent hierarchies*, because as a rule one of the two terms is understood to be superior to the other, to be the original, or the norm. An example is heterosexuality/homosexuality. In this case, in many societies past and present, 'heterosexual' is thought of as primary and normal; this positions 'homosexuality' as a secondary and deviant condition. Again, however, once we recognize deconstruction, we can see that, far from being a deviation or aberration from a norm, what it is to be homosexual is in fact at the very heart of the meaning of what it is to be heterosexual.

Hopefully the *critical* potential of deconstruction can be sensed in what I have written thus far. At issue is the fact that the sorts of binary distinctions under discussion are linked to violence both real and symbolic; they may be witnessed at work within social, political and economic inequalities and injustices at all scales; they are hardwired into beliefs and assumptions at both the popular and the intellectual levels. Many current examples come to mind: the so-called 'clash of civilizations' between the west and Islam, 'genuine' and 'bogus' asylum seekers, 'organic' versus genetically modified foodstuffs, globalization and anti-globalization. These are fairly polarized and unsubtle examples. But deconstruction is a critical method because it can be applied to all attempts to presence, centre, purify, divide, classify and exclude. It exposes the fragility, arrogance and indeed the impossibility of such attempts. And, although there is no explicit political agenda here, deconstruction leads us towards a recognition of the differential impacts and outcomes occasioned by our dreams of presence and absence, identity and difference.

Finally, it is worth stressing again that the term 'deconstruction' refers to both a way of analysing, a 'method', *and* an actual process, what is actually happening in academic texts, in paintings, music, conversations, government policies, scientific reports, etc. Deconstruction is most definitely *not* about inferring in such things meanings that are 'not really there', or about reading a situation wilfully against the grain. This misconception arises because deconstruction often sidesteps or ignores what the 'obvious' meaning of things seems to be, what 'common sense' would seem to tell us a situation is saying to us. In fact, deconstruction is about reading texts, events, situations, processes and so on with very close attention, in an effort to be as faithful as possible to them. Its aim is maximum fidelity. At the same time it would be futile to pretend that deconstructive writing – with Derrida's own as a paragon – is anything other than complex and challenging, and does often seem to morph and twist texts in startling, even implausible ways. This seeming paradox, of simultaneously representing something faithfully *and* altering it beyond recognition, is captured well by Royle (2003: 21) when he states that deconstruction both 'describes and transforms'. As he goes on to say, 'in a sense [Derrida] does little more than describe what happens when reading, say, a passage of Shakespeare or a dialogue of Plato' (2003: 26). But the point is that, given that deconstruction *is* what happens, such a description must necessarily be also and always already a transformation. Deconstruction, to be faithful to the world, must bear witness to the hauntedly deconstructive nature – the slipping, spectral, supplemental, instable actuality – of the world itself.

Examples of the adoption and application of such an analysis are legion within human geography, as they are across the social sciences and humanities. Over time, the principles of Derrida's philosophy have become almost mainstream, and the language of deconstruction is now commonly, even casually deployed within human geographical research. It is as well to note here that deconstruction is usually associated with a series of phrases and themes that have quickly become *passé*; the 'cultural turn', postmodernism, the politics of identity, the 'celebration of difference' and so on. However, while most of these have come and gone, deconstruction has persisted, as the depth and rigour of Derrida's analyses have become more obvious to Anglophone academics. Indeed it may even (and appropriately) be the case that deconstruction *has yet to happen* within human geography, and that the complexities and possibilities of deconstruction are only now beginning to be explored (e.g. see Rose, 2004).

Alongside the slogans mentioned above, deconstruction entered human geography in the late 1980s and early 1990s. To begin with, its implications for geographical writing and practice were the focus of interest, with especial attention being paid to issues of epistemology and language in geography (Doel, 1992; Olsson, 1991; Barnes, 1994). More widely, deconstructive analyses have been enrolled alongside feminist and radical analyses of the constructedness of gendered and sexualized identities (for early examples see Domosh, 1991; McDowell, 1991). This is a wide-ranging and ongoing project; its review is beyond the purview of this chapter (see Chapter 10), but it is worth noting that much recent research on gender, identity and performance, for example (see Bell and Valentine, 1995; Gregson and Rose, 2000), has been inspired by the work of Judith Butler (1990; 1993) which is itself in large part indebted to Derrida's work. However, as a broadly poststructural or, more accurately, constructivist agenda – one focused upon the critical analyses of geographical knowledges and identities – quite quickly spread through the 1990s, then so the strategies of deconstruction have become widely used across different branches of geography.

It is important to note that this has involved not only the application of a 'new' method to traditional topics, but also, in most instances, the development of new areas of study. It is further important to be aware that in many cases the insights of a deconstructive analysis implicitly rather than explicitly inform study and interpretation; such has been the extent to which deconstruction has suffused critical analysis. Within political geography, for example, deconstruction has in part inspired and impelled the development of a new *critical geopolitics*. This has taken as its substantial focus the construction of geopolitical imaginations at a variety of levels, from the evolution of governmental foreign policies and stratagems (e.g. O'Tuathail, 2000; Dalby, 1991) to the representation of global political and military issues within news reportage (O'Tuathail, 1996) and print media (Sharp, 1993; Dodds, 1996). Deconstructive strategies are perhaps particularly apt with regard to critical analysis of the fields of geopolitical reasoning and propaganda, given that these often tend to paint a distinctly divisive and polarized picture of global interests and relations.

In the widespread turn towards discussion of the construction and deployment of geographical knowledges in different times and spaces, historical geography has been revolutionized by the advent of poststructural approaches in general. The critical manoeuvres and avenues opened up in part by a deconstructive approach to issues of knowledge production have led to a sustained interrogation of geography's chequered past, as an intellectual tradition deeply imbricated, via practices of mapping, surveying, exploring and so on, within colonial and imperial histories. For example, in what is a now classic discussion on 'deconstructing the map', Harley (1989) demonstrates how a Derridean analysis of cartography works to unravel claims to objectivity, transparency and innocence in map-making by making visible the rhetorical and metaphorical elements which in actuality constitute any map's meaning. Pratt (1992), Barnett (1998) and Ryan (1996), amongst others, further supply deconstructive readings of the accounts of nineteenth-century European explorers in South America, Africa and Australia, demonstrating that the 'silencing' of indigenous voices in such texts was an essential element in the process of establishing the authority and centrality of European geographical knowledges, embodied in the figure of the explorer. Equally, Blunt (1994) and Clayton (2000) deploy a mixture of archival scholarship and deconstructive analysis to, respectively, the nineteenth-century African travels of Mary Kingsley and the eighteenth-century Pacific navigations of Cook and Vancouver, in each case demonstrating the ambivalences, ambiguities and contradictions at the heart of imperial travels which sought, in contrast, purity, identity and certainty. In the wake of a generation of postcolonial theorists inspired by Derridean thought (Spivak, 1988; Bhabha, 1994), and in an area of overlap between such historical geographies of empire and exploration and contemporary geographies of the developing world, 'postcolonial geographies' (e.g. Sidaway, 2000; Crush, 1995; Blunt and McEwan, 2002) have emerged as a set of critical analyses of the languages, texts and silences of colonial relations past and present.

As these brief highlights hopefully reveal, deconstructive strategies have become part of the lingua franca of human geography, and I could go on at length to discuss the use of deconstruction within social geographies, cultural geographies and so on. But in conclusion I want to reiterate that the implications of deconstruction for the practice of a subject such as geography are radical. Deconstruction is not *just* a 'useful' method we can deploy; it hauntingly demands questioning of normalized assumptions and procedures, and perhaps above all entails a rethink of how academics such as geographers *write*. The work of Marcus Doel (1992; 1994; 1999) is exemplary in its insistence upon exploring the potential of deconstruction in this regard – but few others have, so far, taken up this challenge.

Discourse Analysis

Like deconstruction, the term 'discourse' requires careful definition. This section focuses upon discourse analysis as a poststructural method most commonly associated with the work of the French historian and philosopher Michel Foucault. As with Derrida, his near-contemporary and one-time student, Foucault's writings, on issues as varied as epistemology, madness, punishment, power and the histories of science and sexuality, have been enormously influential across the social sciences and humanities. They have inspired not only new approaches but new fields of study. Indeed, Derrida's and Foucault's works are commonly yoked together, at least rhetorically, in the plethora of books and papers which set out to 'deconstruct discourses' of race, gender, sexuality, the state, nature, landscape, and so on *ad infinitum*.

Foucault's understanding of the term 'discourse', central to his various inquiries, is much more complex and multifaceted than the dictionary definition: '*noun*. 1. Conversation. 2. (foll. by *on*) speak or write about at length' (*Collins Pocket English*, 1996). For Foucault, a discourse, while retaining connotations of dialogue and speech, refers more broadly to the totality of utterances, actions and events which constitute a given field or topic. To clarify this further, we can consider two definitions of discourse by geographers.

Gregory states that discourse 'refers to all the ways in which we communicate with one another, to that vast network of signs, symbols and practices through which we make our world(s) meaningful' (1994: 11). This definition alerts us to two things. First, a discourse is not just a set of written texts. A discourse encompasses texts, speeches, dialogues, ways of thinking *and* actions; bodily practices, habits, gestures, etc. Second, a discourse is *not* a series of things that are said and done regarding a pre-existing thing – gender, say. A discourse of gender is not 'about' gender: instead it *creates* gender, makes it really, actually exist as a consequential and meaningful set of beliefs, attitudes and everyday practices and performances. Furthermore, as Barnes and Duncan note, 'discourses are both enabling and constraining ... they set the bounds on what questions are considered relevant or even intelligible' (1992: 8). In other words, a discourse defines both what *can* and what *cannot* be said or done, what appears to be true, legitimate or meaningful and what is dismissed as false, deviant or nonsensical. Take gender again: as a discourse, this both enables some 'male' and 'female' ways of thinking and behaving and constrains or proscribes others. In this way, the political and ethical import of discourse begins to clarify: discourses establish some behaviours and identities as normal and natural and establish others as unusual, marginal or unnatural.

Discourses must thus be considered in terms of power. To grasp anything about Foucault's thought and method it is necessary to engage with his innovative understanding of how power operates. Common sense would suggest that power operates in a top-down manner. In other words, some people, organizations or states 'have' power: they possess it, and use it to influence and dominate others, who are, comparatively, powerless. In this view, power is (1) concentrated in the hands of a minority, (2) exercised over life rather than being part of life, and (3) negative rather than creative in its effects – power is exercised to constrain, limit, forbid, detain, etc. Foucault, however, disagrees with this view, arguing that 'we must cease once and for all to describe the effects of power in negative terms ... In fact, power produces; it produces reality' (1977: 155). Power, in other words, is what *creates* new identities, new social, economic and political systems, not what prevents change or 'progress'. Moreover this creative power operates in a diverse and dispersed manner; it does not emanate from a single source. As Foucault notes, 'power is everywhere, not because it embraces everything, but because it comes from everywhere' (1981: 93). The sites where power is produced, and where its effects are felt, worked through and transformed, are multiple and heterogeneous.

Two crucial points regarding the use of discourse analysis as a critical method devolve from this discussion of power/discourse. The first is that Foucault is most definitely *not* seeking to argue away the palpable existence of inequalities, injustices and repressions in the societies we live in. He is rather attempting to develop, via his understanding of how power operates, a more distinctive and subtle way of analysing how inequalities acquire such concrete existence. Again the gender example is instructive. Foucault is arguing that certain ways of being male and female in our society are viewed as normal and appropriate *not* because some mysterious central 'power' is forcing us to conform to them, but because we ourselves exercise power over ourselves, in the sense that we continually self-regulate and monitor ourselves and others. Gender norms are sustained via a multitude of small, local, specific practices. Discourse *is* everyday practice – not an invisible web of 'powerful' ideas imposing themselves from above.

The second crucial point – a radical consequence of the first – is that we are ourselves the effects of discourse. Various discourses (of, say, manliness, Irishness, fatherhood) are not external layers of meaning enclosing an inner, unique self. Rather, for Foucault, the very idea that every person is unique, that we are individuals with inner thoughts, feeling and attitudes, is in fact a relatively recent invention. He argues that modern societies are characterized by processes in which humans are individualized, both as objects to be studied by academic disciplines such as human geography, and as selves (subjects) to be experienced and nurtured. The various knowledges and categories through which we know ourselves and others are, in other words, culturally and historically specific. Foucault's most notable achievement, perhaps, is to have shown that categories often assumed to be universal, natural or objectively definable – categories such as sane/mad, healthy/ill, normal/deviant – are in fact the products of particular discursive practices; are in fact, as the saying goes, 'socially constructed'. This does not mean that they are somehow false or unreal. They have *become* real and meaningful within particular cultural and historical contexts.

Hopefully this discussion points clearly towards the critical potential of Foucault's concepts of discourse and power. Hopefully also, the partial similarities between a Foucauldian and a Derridean, deconstructive analysis are evident. Discourse analysis is a critical method which seeks to describe how certain identities and narratives are produced, privileged, sometimes naturalized, and asserted over identities and narratives which are comparatively marginalized, excluded or silenced. Discourse analysis seeks to describe in detail, with close attention to particular events, episodes and practices, how certain behaviours, attitudes and beliefs come to be sedimented and reproduced through continual repetition. As with the language of deconstruction, Foucault's approach and method have permeated much geographical research over the past 15 years, and again I want to briefly highlight some of the main directions that have been pursued.

Foucault's own interests in practices of observing, punishing and confining have to an extent inspired new research agendas in human geography. For example, inspired by classic texts such as *Madness and Civilization* (Foucault, 1967) and *Discipline and Punish* (Foucault, 1977), new geographies of asylums, prisons and other institutions of care and confinement have arisen (e.g. Driver, 1985; 1993; Philo, 1995; Philo and Parr, 1995). Equally, Foucault's famous analyses of the visual gaze which isolates, objectifies and classifies that which it gazes upon has proved especially fertile within human geography. It has inspired, for instance, work on surveillance and CCTV within urban areas (Davis, 1990). And in relation to cultural geographies of art, nature and the visual, Foucault's concept of the gaze has further influenced writings on landscape, voyeurism and the objectification of women as 'nature' (e.g. Rose, 1993; Plumwood, 1993; Nash, 1996; Pollock, 1988).

The impact of the 'discursive turn' has also been notably evident within medical geographies, where it has occasioned an at least partial shift from an almost exclusive focus upon the enumeration and mapping of disease and illness within populations to an agenda which explores the manifold ways in which issues of health and illness discursively construct bodies and identities (e.g. Kearns, 1993; Dorn and Laws, 1994; Butler and Parr, 1999). More specifically here we can point to work on the geographies of disability (e.g. Chouinard, 1997), which, without neglecting issues of access, mobility and so on, explores nuances of definition and self-definition through which 'disabilities' are classified and performed. This further feeds into geographies of mental illness, exploring the institutional, therapeutic and policy spaces through which 'mentally ill' subjects are produced and enacted (e.g. Parr, 1999).

As with Derrida, Foucault's discursive methodology, with its clear emphasis on knowledge, power and the relations between theories and practices, and in particular its historical and archival bent, has been especially influential for cultural and historical geographies. An outstanding example here has been Said's (1978) *Orientalism*, which (though Said was not a geographer) heralded and informed a range of studies of 'geographical imaginations', in particular those associated with colonialism and imperialism, and with the representation of non-European 'others' by European 'selves' (Said's particular focus was upon the historical place of the Near and Middle East in the western imagination). Numerous geographers have further pursued Said's adoption of Foucault as a critical historian of western thought and practice, in particular scrutinizing the discursive practices of exploration (Driver, 2001), travel (Duncan and Gregory, 1999) and geography itself as an intellectual tradition (Gregory, 1994).

I could go on to chronicle at length the role of discourse analysis in forming and informing a variety of studies by cultural geographers, for example the study of the consumption of 'exotic' foodstuffs (May, 1996), or the representation of masculinity within contemporary 'lad's mags' (Jackson et al., 1999). In concluding this section, however, I would suggest that while the language of 'discourse', like that of deconstruction, has become commonplace, the concept has perhaps been too often taken to mean something akin to 'structure'. It is worth reiterating that a discourse is not a structure of 'powerful' or influential ideas, towering over and dominating individuals; nor is it akin to a chessboard of predetermined identities and positions which individuals are forced to choose amongst. A discourse consists precisely in, and is malleable through, the manifold of everyday, repeated, sometimes 'habitual' practices and performances – textual, imagined and literally enacted – that weave together the fabric of life. A good example of this approach within geography, one especially indebted to Foucault's conception, is David Matless' *Landscape and Englishness* (1998), which details how in mid-twentieth-century England elite notions of citizenship, regional planning policies and emerging ecological perspectives combined with the evolution of suburban lifestyles and the increasing popularity of countryside leisure pursuits to produce complex discursive articulations of the relationships between self, society and landscape.

Conclusion: Experimentation

I have concluded both main sections of this chapter by offering some perhaps quite mild criticisms of the manner in which the poststructural approaches of Derrida and Foucault have been taken up and applied within human geographies. In both cases I have the feeling that geographers (along with perhaps most Anglophone social scientists) have in a sense too quickly incorporated poststructural insights. In consequence, poststructuralism has been rather grafted onto previous, more conventional, 'structuralist' and empiricist understandings of topics such as power, language, identity and meaning. Strictly speaking it entails a rejection of these and a movement towards something quite distinctively new. The radical questioning and destabilizing implied by poststructural approaches has had relatively little impact, for example, upon how academic papers and books are written; for the most part these continue to view themselves as incremental additions to already existing, established bodies of knowledge, and as commentaries upon an external, objective world (see Law and Benchop, 1997).

Business as usual. In this brief chapter I have chosen to focus upon deconstruction and discourse analysis, and thus Derrida and Foucault, because these are the writers who have clearly had the biggest impact upon human geography in the past 15 odd years. This has meant excluding some significant others in the poststructural pantheon, for example Julia Kristeva, Jean-François Lyotard and Jean Baudrillard. It has also meant excluding a third writer whose work is often bracketed with that of Foucault and Derrida, and who may well in time prove as decisively influential: Gilles Deleuze.

I want to conclude this chapter with some comments on Deleuze and emergent trends within human geography. At the start I noted that the radical and unsettling implications of poststructural philosophies necessarily entailed questions regarding academic thought and practice, and at the least implied a need for experimental and creative approaches to academic writing, and modes of representation generally. This is a lodestone of Deleuzian thought. Deleuze's philosophy emphasizes creativity, vitality and transformation, not just as values to be cherished, but as immanent, ontological features of life in general. While he is characteristically poststructural in his rejection of 'grand' theories and appeals to a single explanatory key to unlocking the secrets of life (e.g. appeals to God, 'social structures', free markets, the self, or the psyche), his response, unlike that of Derrida or Foucault, is not to undertake painstaking, forensic criticism aimed at revealing the true plurality and complexity of things. Deleuze rather seems to advocate resolutely experimental modes of writing and performance. He is especially critical of the way in which much academic writing sees its function as *representing* the world. For Deleuze (1994; 1990), representation is essentially a negative procedure, in so far as (1) attempting to accurately describe things inevitably distorts and simplifies them, and (2) the critical distance which 'accurate' representation demands has the effect of turning that representation into a judgement. Instead of seeking to describe or judge, Deleuze avers that critical or philosophical writing should aim to *add* to the world, to make it more than it is, rather than less.

This expressive and creative principle of addition is echoed within some recent geographical writing. Thrift calls upon geographers to 'weave a poetic of the common practices and skills which produce people, selves and worlds' (2001: 216) while Dewsbury et al. argue that academic writing should aim to 'contribute to the stretch of expressions in the world' (2002: 439). Both of these citations are from advocates of what has rather clumsily been called 'non-representational theory'. One tenet of this nascent movement is that the reception of poststructural theory within geography has been flawed. First, there has been a persistent tendency to view it as an 'add-on' to structuralist and Marxist theories of power, society and identity; this is evident, for example, in the continuing framing of 'discourse' in terms of narratives of domination/resistance (e.g. see Pile and Keith, 1998). Second, there has been an 'idealist' tendency to use strategies such as discourse analysis and deconstruction within a movement that converts particular material flesh-and-blood actualities into an abstract realm of general symbols, texts, representations and so on, in which the 'meanings' of those actualities are held to inhere.

I concur with these criticisms. In this chapter my aim has been to outline the key principles and procedures of both Derrida's and Foucault's methodology, and to briefly describe their substantial use within human geographical research. In both cases, stress has been laid upon the fact that deconstruction and discourse analysis are *critical* methods – used in making critical assessments of social institutions, identities, cultural beliefs, political arrangements and so on. However, I have been concerned to stress that both are post-*structural* methods, in so far as they are concerned to go beyond 'top-down' structuralist conceptions of how power is exercised, of how identities are constructed. And I have also tried to stress that both deconstruction and discourse are processes already alive in the actuality of lived practice and performance. While a key part of critical geography has been to raise consciousness regarding the operation of cultural, political and economic ideologies, an unintended consequence has been a partial misreading of the scope and purchase of deconstruction and discourse analysis. These, I would suggest, are not just methods we can use to 'read' the world, nor are they means through which we can divide the world first into processes and events, and second into their 'inner' or 'wider' meaning. Instead, deconstruction *is* what happens and discourse *is* everyday practice. Using these methods is not so much about extracting the key themes and meanings of the text, image or situation being studied as about 'describing and transforming' it via experimental and expressive engagements.

References

Barnes, T. (1994) 'Probable writing: Derrida, deconstruction and the quantitative revolution in human geography', *Environment and Planning A*, 26: 1021–40.

Barnes, T.J. and Duncan, J.S. (1992) 'Introduction: writing worlds', in T.J. Barnes and J.S. Duncan (eds), *Writing Worlds*. London: Routledge.

Barnett, C. (1998) 'Impure and worldly geography: the Africanist discourse of the Royal Geographical Society, 1831–73', *Transactions of the Institute of British Geographers*, 23: 239–51.

Bell, D. and Valentine, G. (1995) 'The sexed self: strategies of performance, sites of resistance', in S. Pile and N. Thrift (eds), *Mapping the Subject: Geographies of Cultural Transformation*. London: Routledge.

Bhabha, H. (1994) *The Location of Culture*. London: Routledge.

Blunt, A. (1994) *Travel, Gender and Imperialism*. New York: Guilford.

Blunt, A. and McEwan, C. (eds) (2002) *Postcolonial Geographies*. Edinburgh: Cassell.

Butler, J. (1990) *Gender Trouble: Feminism and the Subversion of Identity*. London: Routledge.

Butler, J. (1993) *Bodies that Matter: On the Discursive Limits of 'Sex'*. London: Routledge.

Butler, R. and Parr, H. (1999) *Mind and Body Spaces*. London: Routledge.

Chouinard, V. (1997) 'Making space for disabling differences: challenging ableist geographies', *Environment and Planning D: Society and Space*,15: 379–87.

Clayton, D. (2000) *Islands of Truth*. Vancouver: University of British Columbia Press.

Crush, J. (1995) 'Introduction: imagining development', in J. Crush (ed.), *Power of Development*. London: Routledge.

Dalby, S. (1991) 'Critical geopolitics: discourse, difference, and dissent', *Environment and Planning D: Society and Space*, 9: 261–83.

Davis, M. (1990) *City of Quartz*. London: Verso.

Deleuze, G. (1990) *The Logic of Sense*. London: Athlone.

Deleuze, G. (1994) *Difference and Repetition*. London: Athlone.

Derrida, J. (1976) *Of Grammatology*. Baltimore, MD: Johns Hopkins University Press.

Derrida, J. (1978) *Writing and Difference*. Chicago, IL: University of Chicago Press.

Derrida, J. (1994) *Spectres of Marx*. London: Routledge.

Dewsbury, J.D., Wylie, J., Harrison, P. and Rose, M. (2002) 'Enacting geographies', *Geoforum*, 32: 437–41.

Dodds, K. (1996) 'The 1982 Falklands War and a critical geopolitical eye: Steve Bell and the *If...* cartoons', *Political Geography*, 15: 571–92.

Doel, M. (1992) 'In stalling deconstruction: striking out the postmodern', *Environment and Planning D: Society and Space*, 10: 163–80.

Doel, M. (1994) 'Deconstruction on the move: from libidinal economy to liminal materialism', *Environment and Planning A*, 26: 1041–59.

Doel, M. (1999) *Poststructuralist Geographies*. Edinburgh: Edinburgh University Press.

Domosh, M. (1991) 'Towards a feminist historiography of geography', *Transactions of the Institute of British Geographers*, 16: 95–104.

Dorn, M. and Laws, G. (1994) 'Social theory, body politics and medical geography', *Professional Geographer*, 46: 106–10.

Driver, F. (1985) 'Power, space and the body: a critical assessment of Foucault's *Discipline and Punish*', *Environment and Planning D: Society and Space*, 3: 425–47.

Driver, F. (1993) *Power and Pauperism: The Workhouse System, 1834–1884.* Cambridge: Cambridge University Press.

Driver, F. (2001) *Geography Militant.* Oxford: Blackwell.

Duncan, J. and Gregory, D. (eds) (1999) *Writes of Passage: Reading Travel Writing.* London: Routledge.

Eagleton, T. (1983) *Literary Theory: An Introduction.* Oxford: Blackwell.

Foucault, M. (1967) *Madness and Civilization: A History of Insanity in the Age of Reason.* London: Routledge.

Foucault, M. (1977) *Discipline and Punish: The Birth of the Prison.* London: Penguin.

Foucault, M. (1981) *The History of Sexuality, Vol. 1: An Introduction.* London: Allen Lane.

Gregory, D. (1994) *Geographical Imaginations.* Oxford: Blackwell.

Gregson, N. and Rose, G. (2000) 'Taking Butler elsewhere: performativities, spatialities and subjectivities', *Environment and Planning D: Society and Space*, 18: 433–52.

Harley, J.B. (1989) 'Deconstructing the map', *Cartographica*, 26: 1–20.

Jackson, P., Stevenson, N. and Brooks, K. (1999) 'Making sense of men's lifestyle magazines', *Environment and Planning D: Society and Space*, 17: 353–68.

Kearns, R.A. (1993) 'Place and health: towards a reformed medical geography', *Professional Geographer*, 45: 139–47.

Law, J. and Benchop, R. (1997) 'Resisting narrative Euclideanism: representation, distribution and ontological politics', in K. Hetherington and R. Munro (eds), *Ideas of Difference: Social Spaces and the Labour of Division.* Oxford: Blackwell.

Matless, D. (1998) *Landscape and Englishness.* London: Reaktion.

May, J. (1996) 'A little taste of something more exotic', *Geography*, 81: 57–64.

McDowell, L. (1991) 'The baby and the bathwater: diversity, deconstruction and feminist theory in geography', *Geoforum*, 23: 123–33.

Nash, C. (1996) 'Reclaiming vision: looking at landscape and the body', *Gender, Place and Culture*, 3: 149–69.

Olsson, G. (1991) *Lines of Power/Limits of Language.* Minneapolis: University of Minnesota Press.

O'Tuathail, G. (1996) 'An anti-geopolitical eye: Maggie O'Kane in Bosnia', *Gender, Place and Culture*, 3: 171–85.

O'Tuathail, G. (2000) 'Dis/placing the geo-politics which one cannot want', *Political Geography*, 19: 385–96.

Parr, H. (1999) 'Bodies and psychiatric medicine: interpreting different geographies of mental health', in R. Butler and H. Parr (eds), *Mind and Body Spaces.* London: Routledge.

Philo, C. (1995) 'The chaotic spaces of medieval madness', in M. Teich, R. Porter and B. Gustafson (eds), *Nature and Society in Historical Context.* Cambridge: Cambridge University Press.

Philo, C. and Parr, H. (1995) 'Mapping "mad" identities', in S. Pile and N. Thrift (eds), *Mapping the Subject: Geographies of Cultural Transformation.* London: Routledge.

Pile, S. and Keith, M. (eds) (1998), *Geographies of Resistance.* London: Routledge.

Plumwood, V. (1993) *Feminism and the Mastery of Nature.* London: Routledge.

Pollock, G. (1988) *Visions of Difference: Feminism, Feminity and the Histories of Art.* London: Routledge.

Pratt, M.L. (1992) *Imperial Eyes: Travel Writing and Transculturation.* London: Routledge.

Rabinow, P. (ed.) (1984) *A Foucault Reader.* London: Penguin.

Rose, G. (1993) *Feminism and Geography*. Cambridge: Polity.

Rose, G. (2004) 'Re-embracing metaphysics', *Environment and Planning A*, 36: 461–68.

Royle, N. (2003) *Derrida*. London: Routledge.

Ryan, S. (1996) *The Cartographic Eye: How Explorers Saw Australia*. Cambridge: Cambridge University Press.

Said, E. (1978) *Orientalism: Western Conceptions of the Orient*. London: Penguin.

Sharp, J.P. (1993) 'Publishing American identity: popular geopolitics, myth and the *Reader's Digest*', *Political Geography*, 12: 491–503.

Sidaway, J. (2000) 'Postcolonial geographies: an exploratory essay', *Progress in Human Geography*, 24: 591–612.

Spivak, G.C. (1988) 'Can the subaltern speak?', in C. Nelson and L. Grossberg (eds), *Marxism and the Interpretation of Culture*. London: Macmillan.

Thrift, N. (2001) 'Afterwords', *Environment and Planning D: Society and Space*, 18: 213–55.

28 RESEARCH IS THEFT: ENVIRONMENTAL INQUIRY IN A POSTCOLONIAL WORLD

Paul Robbins

One afternoon while I was doing research on forest cover outside a wildlife preserve in rural India, an old woman yelled at me over the thorn fence at the back of her house: 'You are writing down things about the forest so someone can come take it!'

Recording, interpreting, and analyzing the world is a way to appropriate it and control it. The terms and narratives used to describe social and environmental conditions serve to define the range of debate and to enclose its possibilities. If carried out by people with institutional authority and power (e.g. foreign researchers), such accounts can become the stuff of policy that dramatically impinges on local livelihoods and survival. Research on food has often made people hungrier. Research on forests has destroyed biodiversity. Research on poverty has made people poorer. In this way, even well intentioned research, especially when carried out by non-resident peoples, is a deeply structured part of a system of ongoing exploitation. So too, a researcher with explicitly normative and often lofty goals – more equitable distribution of resources, less exploitative labor relations, and defense of endangered species – is working not only to record how the forest is, but to imagine how the forest might be, an undeniable exercise of power. Finally, by making a living off the stories of other people and the records of environmental conditions in other places, research is unambiguously extractive. Researchers are paid for textual transcriptions of other people's histories and records of the condition of their plants, knowledge, technology, life, and land. In all these ways, research is theft.

The recent enclosure of the forest into a more restrictive wildlife preserve to which the old woman refers, however, is a move that has made access to the forest more difficult for subsistence producers who reside nearby, but one introduced and financed by a foreign state donor agency, which ushered in a new period of restriction and control. To record that fact and enumerate its deleterious impacts on the people of that neighborhood is a challenge to that control, and arguably represents an effort not to take the forest, but rather *to take it back*. Similarly, the research I was conducting that day was in cooperation with grassroots organizations, which explicitly do seek control of the forest, but with hopes of an argument based on scientifically recorded phenomena. So too, the research was directed at questions to which many local people want answers (How much forest cover is there? What are the effects of grazing?) but for which no funding is available from the state, from private firms, or from the very donor agencies that established the wildlife preserve in the first place. If research is theft, it is a theft in which many wish to participate, including local and marginal communities.

In this chapter, I seek to address this inevitable postcolonial contradiction (see also Chapter 12), and describe an ongoing research project that is in some ways colonial and in some ways an insurgent effort at anti-colonial science. In the process I hope to demonstrate the limits and potential of field-based investigation into issues of environmental change and sustainability, while relating a general sense of how this kind of research is done. But to do so requires answering a prior question: why do research at all?

Research as Expropriation

The argument that environmental research – objective, non-political, and truth seeking – is expropriative and often has normatively undesirable results, is one rooted in the global colonial experience. The great era of European expansion and domination of distant places was characterized not only by expressions of military force but also by the promulgation of expertise. Colonial states' activities, as Cohn observes, 'fostered official beliefs in how things are and how they ought to be', which depended upon systems of documentation

> that formed the basis of their capacity to govern. The reports and investigations of commissions, the compilation, storage, and publication of statistical data on finance, health, demography, crime, education, transportation, agriculture, and industry – these created data requiring as much exegetical and hermeneutical skill to interpret as an arcane Sanskrit text. (1996: 3)

In practical terms, this meant collecting a lot of data, arranged and defined in the categories of the colonizer. But it also entailed the establishment of an elite caste of people (e.g. statisticians, geographers, botanists, demographers) specially trained to interpret these data and make wise decisions on behalf of the colonized people. The social power, remuneration, and political authority of these people all depended on the consistent determination of what problems colonial regions faced, and on the creation of solutions. Such solutions commonly resulted in proscriptive and prescriptive policies with wide-ranging pernicious implications. In this way, theories of environment were invariably linked to theories of political domination. As environmental historian David Gilmartin puts it in the case of British colonial science:

> the definition of the environment as a natural field to be dominated for productive use, and the definition of the British as a distinctive colonial ruling class over alien peoples, went hand in hand. (1995: 211)

Conservation efforts in West African forestry, in an example relevant to later discussion here, are representative of this problem. French colonial authorities and scientists entering the savannas of Guinea in the 1800s saw a complex landscape of rotational fallows (land left to temporary regrowth), mixed with locally preserved forest, and open grazing land. As researchers James Fairhead and Melissa Leach (1994) record, the ecological complexity of the system and the difficult maintenance required from the local village residents, however, was absolutely lost on the colonial observers. They instead saw an area of great and increasing aridity, which they further suggested was a result of reckless local land use patterns leading to deforestation. Examination of aerial photographs and careful scrutiny of local records, however, suggest that in fact the reverse has been occurring: forests have been expanding in Guinea throughout and after the colonial period, precisely as a result of local land use practices.

These French scientists were not simply empirically wrong, seeing deforestation where in fact afforestation was occurring. More than this, they were led into their way of seeing, at least in part, as a result of their colonial relationship to local people. Knowing in advance that local accounts and practices were problematic, inevitably led to reading forest history backwards (Fairhead and Leach, 1995). And so too, their conclusions, that local people were destroying their ecosystem, had implications for power and control over resources for generations to come. The assumption that state conservation intervention is required to curb local behaviors invariably led to loss of control over traditional resources, a loss overseen not only by men with guns, but by those with notebooks, sketchpads, dog-eared botanical texts – the other tools of environmental expertise.

Thus, efforts at conservation appear, at least in retrospect, as exercises in control. What seemed perhaps at the time as enlightened efforts to preserve local ecologies and economies, were clearly efforts in expropriation; in the process, local people lost the rights to their land, the rights to their property, and the rights to govern their own affairs. That these colonial social and environmental science practitioners were well trained, well intentioned, and often very sympathetic to the situation of the colonized (though often they were not) matters little to such outcomes, and in some ways, is prerequisite to the confidence and zeal with which colonial science was conducted.

Nor was it unique to France (or Germany, Great Britain or the United States for that matter). This generalized expression of state hegemony in a form of research abstraction is arguably an inevitable product of state knowledge authority – what James Scott (1998) calls 'authoritarian high modernism', the simplification of ecology, economy, and society required to govern, and its inevitably pernicious effects.

Such research traditions are thus theft in two senses. First, they are instrumental to actually appropriating other people's stuff: forests, pastures, waterways, minerals, and knowledge. But more than this, they rob from other people the right to speak.

Colonialism Now: Environmental Science

The legacies of these research traditions are twofold. First, they have left the world with sets of historical notions that are empirically wrong, yet which hold tremendous influence over the global imagination. The idea that West African forests are receding under a tide of ignorant local populations, for example, is one as prevalent now as a century ago, despite local accounts and increasing historical evidence to the contrary (Fairhead and Leach, 1998).

But more than this, in the contemporary context, a tradition of colonial knowledge production has left a legacy of ongoing research in the underdeveloped world that is, put baldly, an *industry*. Hundreds of millions of dollars of national and international funds are poured into funding surveys, analyses, and examinations of a great range of environmental 'problems.' University professors, United States Agency for International Development workers, and arguably even reporters, trek across the world's poorest places, interviewing local people and recording their opinions, their resources, and their ideas, later bartering them for salary and prestige. All the while, such research is leveraged on providing 'authentic' accounts of local experience.

This last crisis, one of representation where local ideas are reframed in terms that make sense in the conceptual world of the expert observer, is perhaps the most pernicious, since it suggests the limits of emancipatory social and environmental science. As Gayatri Spivak (1990) explains, the impulse to 'save the poor' generally involves efforts to speak *on their behalf*, inevitably therefore in the language of the colonial expert, a self-defeating and internally contradictory effort. Rather, Spivak argues, scholars need not to learn more, but rather to *unlearn*:

> **There is an impulse among literary critics and other kinds of intellectuals to save the masses ... how about attempting to learn to speak in such a way that the masses will not regard as bullshit? When I think of the masses, I think of a woman belonging to 84% of women's work in India, which is unorganized peasant labor. Now if I could speak in such a way that such a person would actually listen to me and not dismiss me as yet another of those many colonial missionaries, that would embody the project of unlearning. (1990: 56)**

The extension of this line of criticism to the problem of dominant modes of colonial thinking in contemporary theory and research has come to be known as postocolonial theory. The term is quite contentious since it can be taken to mean:

1 the theorization of a historical period 'after' formal colonialism and the geographic spaces experiencing and negotiating decolonization, with their persistent experience of unequal power relations
2 a methodology to interrogate the colonial logics and practices of Euro-American cultural/scientific hegemony, both historically and contemporarily (Said, 1978; 1994; Mongia, 1997).

Despite their incredible heterogeneity, postcolonial theorists commonly share an interest in decolonizing the relations between north/south and exposing the contribution to global inequity and power relations of research, writing, and thinking on the 'Third World' (for lack of a better term) by First World scholars. In the process this work rewrites history and ecology from the point of view of the colonized subject and so inverts the privilege enjoyed at the expense of the world's most marginal people.

In this sense, contemporary environmental science conducted not only by foreign researchers but also by indigenous nationals funded by state organizations and private firms can be explored in terms of how it functions to reproduce colonial relationships while often ironically claiming to speak on behalf of marginal people. These practices appropriate local resources while simultaneously appropriating local voices: research is theft.

Ways Forward

All of this is to present a pretty formidable wall for researchers interested in environmental change and the power relations that flow from management and control of resources. Taking this critique seriously seems to suggest that the normal way of doing business is certainly not as progressive as one might think. So what are the directions forward for researchers seeking to do robust analysis of social and environmental processes, while admitting the political embeddedness of any claims about society and nature?

Ignore the critique?

One obvious option is to ignore the critique. Such postcolonial accounts, after all, only serve to muddy what is already a complex set of research tasks. And the fact that colonial accounts were perhaps incorrect, owing to the poor research technologies and ideologies held by researchers, is no guarantee that such failings are inevitable or inherent to contemporary research. Indeed *political* research may be the very problem in the colonial legacy; the key must be to create a less, rather than more, political ecology. Stay the course of positivist science, one might argue, perhaps getting 'feedback' from local people through participatory planning efforts.

While this is certainly the preference for most of the world's researchers, it is hard to proceed seriously with any normative projects (explicitly seeking more equitable and emancipatory ways of doing research, not just more accurate ones) if one takes the history of science seriously. To attempt to remove the 'politics' from science is precisely the failing of colonial science, which evacuated the normative moral responsibility from a practice laden with normative political categories and implications. As Said observes:

> **the general liberal consensus that 'true' knowledge is fundamentally non-political (and conversely, that overtly political knowledge is not 'true' knowledge) obscures the highly if obscurely organized political circumstances obtaining when knowledge is produced. No one is helped in understanding this today when the adjective political is used as a label to discredit any work for daring to violate the protocol of pretended suprapolitical objectivity. (1978: 10)**

As far as participation goes, while 'experts' have studiously recorded local opinions of problems, locals themselves have been largely written out of actual scientific practice. Even where local voices are allowed into the environmental planning process, as is increasingly the case, the actual practice of 'science', which precedes any community-based discussion, remains the untouched domain of experts.

Stay home?

Another possible response to the postcolonial critique, at least for privileged and First World researchers, is to stay home. Given the fraught nature of international research and the grossly unequal power relations between the research and the 'studied', the ethics of anti-colonial practice mandate that those daily geographies of colonialism not be reproduced.

Work at home, moreover, allows a great many important projects. The colonial nature of science can be explored historically, elites and their relations to the global south can be interrogated, and the political economy of practice 'at home', especially as it relates to processes 'far away', allows endless research projects in the researcher's own back yard. A deep historical reading of the colonial origins of the contemporary world, its imbalanced conceptual apparatuses, and its current implications, is a daunting and important enough project to justify staying home.

This seems a necessary but incomplete effort. First, there is nothing necessarily anti-colonial about physically doing work 'at home', even archival research, since all writing and interpretation are, by the definitions of the postcolonial critique, political acts. So too, the landscapes of home, whether urban Cleveland or rural West Virginia, are arguably themselves postcolonial landscapes, filled with unequal power relations, and embedded classed, raced, and gendered historical politics, all of which impinges on the conceptual worlds of the researcher and researched.

More than this, however, is the inevitable fact that though the critical researcher may choose to stay at home, the rest of the world most definitely will not. Increased global trade and foreign direct investment mean that multinational firms will not stay home. Global efforts at conservation mean that powerful environmental groups will not stay home. The United States armed forces will not stay home. The IMF will not stay home, nor will the World Bank, nor indeed the global labor force, whose migrations represent an endless set of complex movements with environmental portent. Indeed, if a (mis)reading of postcolonialism results in critical environmental researchers staying home, they will be numbered amongst a very few indeed.

Engage the critique: methodological implications

Obviously, therefore, a serious reading of postcolonialism suggests doing a different kind of environmental research, while at the same time engaging the colonial implications inherent in doing research. In such an approach, research revolves around exposing and interrogating the practice of scientific research and planning in the reproduction of colonial power relationships. This means engaging in ground-level, empirical study of how organizations operate, especially scientific and official ones. So too, this approach to research should examine how local people engage with agencies and knowledges under conditions of unequal power, exposing alliances, positions, and practices. This effort should work not only to understand the local politics of knowledge, but also to explain what kinds of social, cultural, and ecological work such alliances do, employing whatever ways of knowing we have at our disposal, and evaluating them in normative and emancipatory terms. In other words, a possible postcolonial research path is to explore and explain how landscapes and the knowledge of landscapes are produced through colonial practice.

Such an effort requires serious attention to methodology, since science attending to its own colonial implications must consider *how* research is performed, not simply the questions and answers that motivate inquiry. Research must retain methods that measure, describe, and explain, while linking them (or nesting them within) methodologies that seek to interpret, contextualize, and expose. The key is therefore to unite inquiry into scientific, environmental research questions with inquiry into the power of science. Or put another way, critical research may proceed by bringing together important and meaningful truth claims and questions (e.g. is soil being eroded, are groundwater levels changing, or is carbon being sequestered?) with explorations of the production of truth (e.g. who is doing soil science, how is groundwater categorized and defined, or who pays for global warming?).

The following case study from rural India represents my own efforts to follow this path. By inquiring into the politics of forestry science, while simultaneously seeking to measure alternative accounts of forest cover, the work seeks to critically assess the production of knowledge even while working to empirically explain forest cover change. Whether or not these goals are achieved is best left to the reader to decide.

How Much Forest Is There? A Postcolonial Inquiry

Global forest cover is on the decline. One much-cited estimate suggests that between 1700 and 1980, forests and woodlands declined by some 19 per cent, a loss of over 5 billion hectares of forest cover in less than 300 years. Of course, the trajectories of forest cover change are regionally divergent, with Latin America, tropical Africa, and South Asia recorded as suffering some of the greatest declines (Richards, 1990).

Certainly, there is much to say about these divergent trajectories of decline, especially in terms of the role of colonial science and modernization in their development and perpetuation. Many convincing critical explorations have been made into the driving forces of such change, relating current trends to colonial histories. The treatment of the forest periphery in Latin America as an extractive colony of the urban core, for example, shows that deforestation is commonly a result of explicit state policy for the appropriation of land from indigenous communities and other marginal groups (Hecht and Cockburn, 1989). The clearance of Asian forests in many countries, at a different scale, reflects the colonial dependency relationships with industrial neighbors, who accrue added value in the cutting and milling of products consumed in the First World (Kummer, 1992). Put simply, deforestation can be seen as an expression of power relationships.

Forest cover change in Godwar

Yet, as Fairhead and Leach have shown (see earlier), any measurement of forest cover change is itself inherently political. The first goal of a critical environmental science, therefore, is to place these statistics under critical scrutiny and explore the political stakes in enumerating forest cover in the first place. What is a forest? Who has the authority to say? And how are answers to these questions prefigured by the conceptual conditions within which they are asked? Under competing definitions, moreover, those enforced by officials versus those held by local producers, would rates of change be measured differently? In other words, by taking seriously the colonial character of environmental knowledge, what might an exploration of forest cover change look like?

To explore these conditions, my own research has focused on the savanna and grasslands regions of southern central Rajasthan in India. The Godwar region of the state, located along the spine of the highly eroded Aravalli hills, divide the arid northwest of the state from the humid southeast (Lodrick, 1994). This is an area of relatively good rainfall, receiving around 500 mm annually. The forests dominate the hills and contain a range of tree species, including *Anogeissus pendula*, *Butea monosperma*, and *Ziziphus nummularia* (Jain, 1992; Robbins, 2001a) (Figure 28.1). Between 1996 and 2000, I spent many months travelling in the region, examining satellite imagery, and having long discussions with foresters, farmers, and herders throughout the area, trying to answer the simple questions: is forest expanding or contracting, why, and at whose expense?

The expansion or contraction of forest in the region is a matter of some disagreement, and is of considerable political importance. In 1986, 562 square kilometres of the hills were enclosed to form the Kumbhalgarh Sanctuary, a wildlife park managed for panther, hyena, and sloth bear species (Chief Wildlife Warden, 1996). This enclosure, like many enforced throughout the region, is predicated on the explicit assumption that forests are declining throughout India and require immediate preservation. Indeed, the government's explicit goal, supported by that of the World Bank, is to put one-third of the subcontinent's land under forest cover: 'for achieving the target of the prescribed 33⅓ per cent area of the country to be under forests, we need about 35 million hectares to be planted and made into forests outside the traditional forests' (Maithani, 1988: v).

It is in this context, a government mandate nested within a global environmental crusade for forestry, that government statistics of regional land cover are created. These statistics tell an apparently optimistic story (Figure 28.2). Forest cover in Rajasthan, a poor and arid state on the margins of Indian industrialization, whose population growth rate is higher than the national average, has experienced more than a doubling of forest cover since 1965.

Contemporary Edenic imaginaries

Explorations of forestry texts and conversations with foresters in the field, however, open a window on a more complex process than a simple, value-neutral, land cover transformation. Specifically, this expansion of forests is commonly justified in terms of halting the encroachment of a vast desert. State officials insist that the crisis in forest cover is potentially catastrophic, a large-scale transformation of apocalyptic proportions. Generally, planting and protection efforts are extolled, therefore, as an effort to restore a lost garden that has been replaced by a desert. Each tree planted represents a step towards 'turning the desert region into a green belt full of vegetation and highly fertile land' (Bhalla, 1992: 284).

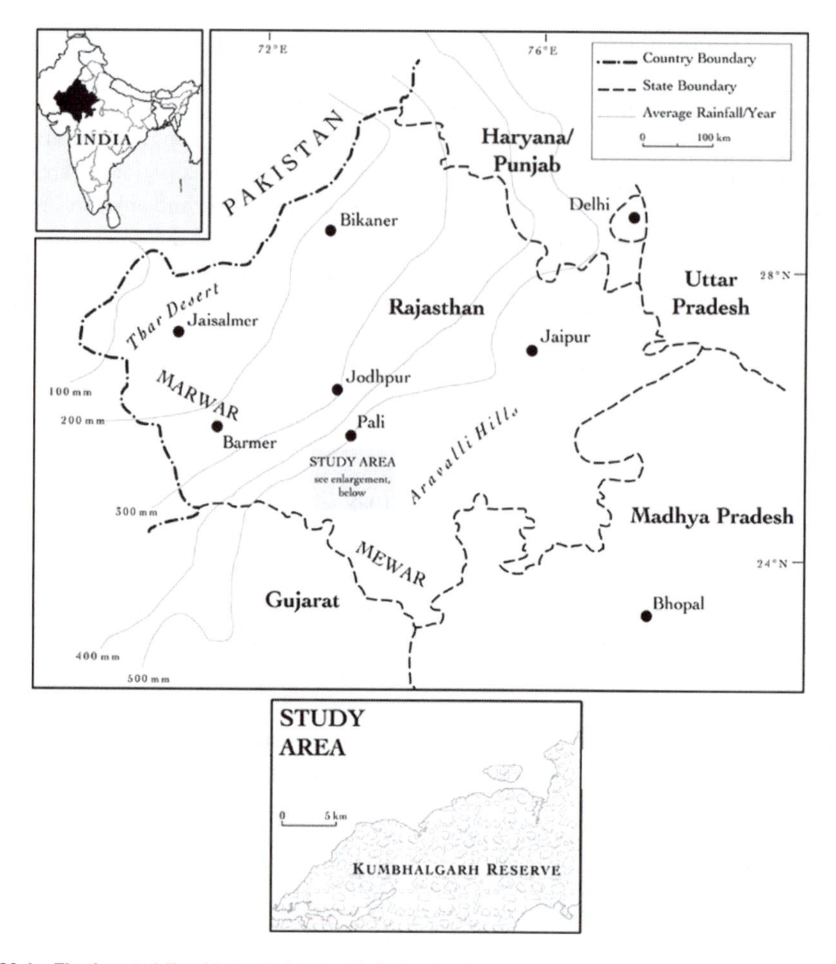

Figure 28.1 The forested Kumbhalgarh Reserve in Rajasthan, India

The quixotic conceptual system that such a discourse of land management implies is by no means unique to Indian forestry, but rather represents the 'Edenic imaginary' common to all modern tree-planting efforts. As Shaul Cohen (1999) has explained, forests and trees are powerful normative metaphors that drive a struggle to return to a pristine nature. Such a narrative supports the power of forestry agencies, private firms, and non-governmental organizations, all of whose efforts become suffused in a soft green glow by association with their planting of trees. With little or no specific ecological justification, our collective urge to restore a lost past props up the political and economic agendas of a range of powerful players.

So too, this approach to trees is underlain by two related assertions. As Cohen observes:

> **First, tree planters, regardless of their ideology, hold that the more trees planted the better, suggesting that the garden is improved with each tree. Second, trees and tree planting (and tree planters, by extension) are unmitigated goods. (1999: 429)**

Together this leads to a quantitative moral crusade, to plant as many trees as possible and to

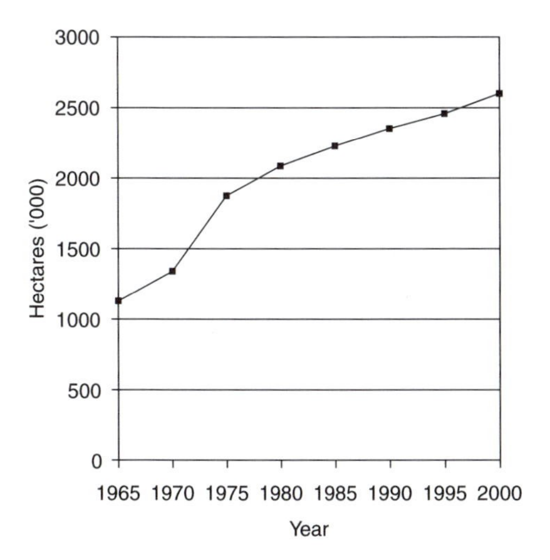

Figure 28.2 Forest cover in Rajasthan

assert successful land stewardship based on the number of trees on the ground. Such is clearly the case in Rajasthan. Targets are set quantitatively for tree plantation and survival, and professional promotions in the state Forest Department are fixed to quantitative assessment of success. So too, all tree plantations are described as unmitigated local goods, even though many of the species selected for the region are introduced from other countries and ecozones. Because of what has been 'lost', any efforts, no matter how ecologically questionable, represent 'gains'.

This metaphor of loss and recovery is obviously quite old, rooted in a specifically Judeo-Christian Edenic worldview, which may or may not be more general or universal. Whether forests, grasslands, or asphalt for that matter, characterize Marwari notions of Eden is of little relevance, however. The mobilization of the metaphor implies that local practice has erased a lost world, which only expert intervention can recover; an exercise in landscape control by a professional cadre, obscured in the assertion (never proven) of the normative good that will come from trees. Forestry texts and discourses, revealed in field-based exploration, show the Edenic imaginary of Rajasthani forestry to be in and of itself colonial.

Consistent colonial hegemony

A history of Marwari forests, moreover, suggests that the contemporary panic over deforestation is considerably less old than the Bible, and has its roots instead in a key historical moment of British hegemony. Surveying the colonial-era administrative reports for the Kumbhalgarh forest belt from the late 1800s reveals a clear linkage between forestry and colonial governance; as colonial authority expanded throughout the subcontinent during the period, facing resistance in many quarters, the imperative to maintain control over semi-independent regions and rulers, while avoiding the financial and military burden of widespread occupation, required the creation of experts.

Specifically, the princely states of Rajputana, which made up the geographical landscape of the region that would later become Rajasthan, were governed in the nineteenth century by semi-autonomous Rajas. These rulers, governing the states of Jaisalmer, Marwar, and Sirohi, were by no means 'free' agents. Rather, these states were linked to British authority through the offices of professionals, whose expertise was required in the conduct of modern governance.

Thus, in 1887 the first forestry survey of Marwar was published, under the authority of a Mr Lowrie, the Assistant Conservator of Forests of Ajmer, where the British enjoyed direct rule, who was 'loaned' to the kingdom for the task. Following Lowrie's tour of the region (and its tens of thousands of square miles) in a single brief month in 1884, the survey concluded that 'the soil of Marwar is generally poor, abounding with rock and sand, and ill-adapted for Forest growth' (Marwar, 1887: 26). As the *Gazetteer* from the region noted previously in 1871, 'the appointment of an Assistant Conservator of

Forests [is] more for the purpose of creating than of conserving forests' (Ajmeer, 1875: 9).

In administrative reports spanning the nearly two decades from 1887 to 1894, the tone and extent of forest control shifted dramatically. Enough valuable timber was found to justify conservation activity. Boundary pillars were built, forest laws were drawn up, and a number of traditional subsistence practices were criminalized, including 'grass cutting', 'grazing without permission', and lighting fires (Marwar, 1893: 19). As income began to flow into the state from forest revenue, forestry was simultaneously professionalized, with the first batch of English-speaking foresters sent for specialist training. Finally, the regional geography of the near-forest areas was divided up in favour of varying 'rights holders' with differing access and control over forest resources. All of this was accomplished through a set of complex negotiations where the princely owners of the forest were 'compensated' by the enclosure of the land by colonial authorities (Marwar, 1894). This was accompanied by an increasing number of experts and reports to codify the resources of the region. All of this is typical of colonial forestry more generally, but the Rajputana case is a remarkably clear example of the extension of colonial power through the establishment of expertise.

Yet simultaneous with this establishment of control was an explicit effort to 'denaturalize' the degradation of the region's forest, pointing to the hand of humanity, specifically non-colonials and historic Indian peoples, in the degradation of the forest. 'The reckless destruction by man' became a justification for action (Marwar, 1887). In particular, the Marattas – an earlier local imperial rival of the British – were blamed for destruction of what was once lush vegetation, along with local communities (Marwar, 1886).

Recalling the Edenic narrative promulgated in the contemporary period, a historical reading of the emergence of forestry in the region is equally marked with a regenerative crisis story. So too does the story coincide with the emergence of colonial technocratic control and the criminalization of the daily lives of the poor.

Living in someone else's forest

The upshot of this forest activity, including demarcation, institutionalization, and plantation, is the burgeoning expansion of tree cover in the Kumbhalgarh area. The energy and momentum described consistently from the colonial period to the present have resulted in dramatic landscape change, therefore (as suggested previously in Figure 28.2). So, how much forest is there?

To answer this question, my discussions with foresters and work in colonial archives were supplemented with analysis of satellite imagery and extensive interviews with local people living in and around these conservation areas. These two modes of inquiry produced a curiously contradictory picture of forest cover change, which sheds light on the postcolonial condition of regional forests.

Satellite imagery was unambiguous; as shown in Table 28.1, forest canopy cover in the 900 km^2 region facing the Kumbhalgarh reserve has increased dramatically between 1986 and 1999, of the order of 50 per cent in just a few years. A walking ground survey of these emergent forests, however, reveals complications. The trees in this emergent forest represent a relatively narrow range of species in plantation, including trees from the Americas (*Prosopis juliflora*) and the Near East (*Acacia tortilis*). Because they grow quickly and form a thick canopy, they often crowd out other important local species.

The problems encountered each day by people living in the material shadow of someone else's forest imaginary are manifold. The new thorn scrub discourages the growth of grassy ground cover for grazing. The leaves of the new trees are poor fodder. The new trees charcoal reasonably well but make poor materials for local construction. While the

Table 28.1 Net land cover change in Godwar, 1986 and 1999 (km^2)

Cover	*1986*	*1999*	*Change*	*Change as % of 1986*
Water/shadow	3.85	0.57	–3.27	–85.11
Urban/rocky	74.48	81.80	7.32	9.83
Grassy/fallow	336.50	185.29	–151.21	–44.94
Thorn scrub	161.69	161.20	–0.49	–0.31
Tree canopy	183.82	274.47	90.65	49.31
Agriculture/cultivated	18.12	75.39	57.27	315.95

Note: Figures exclude unclassified areas and represent roughly 775 km^2 of the study region.

Source: Robbins, 2001a

new forest cover is not exactly a nuisance, it by no means represents the return of 'forests' in any meaningful way for most locals (Robbins, 2001a). Not so for foresters or for state-level statistical reporting (Figure 28.2), which defines as forests all lands under the control of the Forest Department, who continue to insist that this land cover is successful forestry.

To test this notion, follow-up research mapped the varying estimates of forest cover by local farmers, herders, and foresters. The technique, utilizing photograph identification and satellite imagery, reveals that foresters consider a far larger expanse of land cover to be 'forest' than do their counterparts outside the bureaucracy (Robbins, 2001b). On the contrary, the areas considered by herders in the region to be forest, composed of savanna scrub and forage tree species, are in *decline*.

This is further confirmed in a survey of local land use and mapping records, called *jamabandi*, which are kept in each village and which record in local categorical terms the coverage of varying land uses. These records, which consist of large cloth cadastral maps and a geocoded record of land use for every village, are archived over long periods. Because of neglect, these records are in a regrettable condition of advanced physical decay (Figure 28.3). A sample of such records from villages in the region confirms the characterization of forest decline suggested by local farmers and herders. While 'forest' cover is indeed expanding in the region, on an average of 39 hectares per village between 1965 and 1992, there has been a concomitant decline in what the records describe as *oran* land. Surveyed villages in western Rajasthan lost on average 219 hectares (over 2 km^2) of *oran* each (Robbins, 1998). *Oran* lands, considered sacred by both the region's Hindus and Muslims, are typically covered in trees, especially important indigenous species (Gold, 1989). Thus, an expansion in forest in the region is accompanied by (and necessarily linked to) deforestation.

In sum, the local fact of deforestation is erased through a colonial practice of reforestation, which silences alternative notions of the landscape by forwarding not only a material practice of exogenous tree planting, but also a triumphant discourse of Edenic recovery, grounded in a colonial historical tradition of control. The more trees appear on the ground, moreover, the more physical vindication there is for the conceptual apparatus of expert authority. Trees are a colonial *fait accompli*, and the research and planning efforts of reforestation arguably represent a theft of local resources.

The Colonial Ambiguity of Research

Yet such pernicious processes have here been exposed and explained through *yet more research*. This empirical investigation, conducted to expose the practices and transformations born of technical imaginaries, thus itself represents a form of appropriation. It does not derive from local people themselves and it

Figure 28.3 Bundled *jamabandi* records for Marwari villages contain land use records in disaggregated and highly localized categorical vocabulary. Though a crucial archival resource, and an important alternative to technical records like satellite imagery, such records are commonly in a state of decay

does not (cannot?) give voice to the colonized subject, nor would it necessarily be understandable in any immediate way to that subject.

Put more directly, was the old lady who yelled at me across her back fence 'right' or 'wrong'? What gives this author the authority to say? Could I possibly speak for her in any case, since any such representation only reinscribes exactly the pernicious relationships the research seeks to undermine.

So too, the technical tools employed to build the case against colonial forestry, including satellite imagery and other equipment, actually served to reinforce the power of technical knowledge systems, which have historically silenced the voices of local practitioners and their ways of knowing. The apparent unambiguous 'truth' suggested in Table 28.1, as a specific example, is a privilege afforded to the researcher who uses the technical tools of colonial science, even where the results are used to challenge hegemonic authority. So too, it reinforces the very methods of 'abstraction' that James Scott so correctly criticizes (see earlier). Anti-colonial science is colonial too, at least to some degree.

This essential ambiguity is troubling and suggests the difficult methodological implications for environmental research in a postcolonial world. It does point a possible way forward, however, by critically examining the role of power in the production of environmental knowledge while going about the business of rigorously producing new knowledges.

The reconciliation of these projects will require more than has been suggested here, however. A necessary step for any such project is the establishment of politically viable alliances – those that might impact pernicious policy and practice – with other participants in the struggle over nature, whether these are farmers, herders, or foresters. By sharing results, by allowing research questions to be modified and diverted in consultation, and by arguing over the application of research to policy advocacy, the act of research becomes more complex, though really no more or less political. Admitting that the interests of these players, including the geographer, the forest range officer, and the *raika* pastoralist, do not and cannot entirely coincide, such alliances are necessarily fluid, temporary, and strategic. Even so, productive research to help reinvent the world will require robust and extensive research conducted in just such complex political networks – mutual exploitation, mutually agreed upon. Anything less is theft.

NOTE

This research described in this paper was made possible with support of the American Institute of

Indian Studies, The Ohio State University, and the National Science Foundation. Special thanks go to Hanwant Singh Rathore at the Lok Hit Pashu Palak Sansthan, Ilse Kohler-Rollefson at the League for Pastoral Peoples, and S.M. Mohnot at the School for Desert Sciences. Thanks also to the foresters and residents of Sadri and Mandigarh and to Anoop Banarjee and Sakka Ram Divasi. The author is deeply indebted to Joel Wainwright for comments on earlier drafts of the paper.

References

Ajmeer, State of (1875) *Gazetteer of Ajmeer–Merwara in Rajputana*. Calcutta: Office of the Superintendent of Government Printing.

Bhalla, L.R. (1992) 'Development of desert and waste land', in H.S. Sharma and M.L. Sharma (eds), *Geographical Facets of Rajasthan*. Ajmeer: Kuldeep, pp. 283–7.

Chief Wildlife Warden, F.D. (1996) *Management Plan: Kumbhalgarh Wildlife Sanctuary 1997–1998 to 2000–2001*. Udaipur: Rajasthan Forest Service.

Cohen, S. (1999) 'Promoting Eden: tree planting as the environmental panacea', *Ecumene*, 6: 424–46.

Cohn, B. (1996) *Colonialism and Its Forms of Knowledge*. Princeton, NJ: Princeton University Press.

Fairhead, J. and Leach, M. (1994) 'Contested forests', *African Affairs*, 93: 481–512.

Fairhead, J. and Leach, M. (1995) 'False forest history, complicit social analysis: rethinking some West African environmental narratives', *World Development*, 23: 1023–35.

Fairhead, J. and Leach, M. (1998) *Reframing Deforestation: Global Analysis and Local Realities: Studies in West Africa*. New York: Routledge.

Gilmartin, D. (1995) 'Models of the hydraulic environment: colonial irrigation, state power and community in the Indus Basin', in D. Arnold and R. Guha (eds), *Nature Culture Imperialism: Essays on the Environmental History of South Asia*. Bombay: Oxford University Press, pp. 210–36.

Gold, A. (1989) 'Of gods, trees, and boundaries: divine conservation in Rajasthan', *Asian Folklore Studies*, 48: 211–29.

Hecht, S. and Cockburn, A. (1989) *The Fate of the Forest: Developers, Destroyers and Defenders of the Amazon*. London: Verso.

Jain, R.R. (1992) 'Botanical and phytogenic resources of Rajasthan', in H.S. Sharma and M.L. Sharma (eds), *Geographical Facets of Rajasthan*. Chandranagar: Kuldeep, pp. 68–75.

Kummer, D.M. (1992) *Deforestation in the Postwar Philippines*. Chicago, IL: University of Chicago Press.

Lodrick, D.O. (1994) 'Rajasthan as a region: myth or reality', in K. Schomer, J.L. Erdman, D.O. Lodrick and L.I. Rudolph (eds), *The Idea of Rajasthan: Explorations in Regional Identity*. New Delhi: Manohar and the American Institute of Indian Studies, vol. I, pp. 1–44.

Maithani, G.P. (1988) 'Preface', *Wasteland Development for Fuelwood and Fodder Production* (*A Compilation*). Dehra Dun: Government of India, pp. v–vii.

Marwar, State of (1886) *Report of the Administration of the Marwar State for the Year 1885–86*. Jodhpur: Marwar State Press.

Marwar, State of (1887) *Report of the Administration of the Marwar State for the Year 1886–87*. Jodhpur: Marwar State Press.

Marwar, State of (1893) *Report of the Administration of the Marwar State for the Year 1891–92*. Jodhpur: Marwar State Press.

Marwar, State of (1894) *Report of the Administration of the Marwar State for the Year 1893–94*. Jodhpur: Marwar State Press.

Mongia, P. (1997) 'Introduction', in P. Mongia (ed.), *Contemporary Postcolonial Theory*. London: Arnold.

Richards, J.F. (1990) 'Land transformation', in B.L.T. Turner, W.C. Clark, R.W. Kates et al. (eds), *The Earth as Transformed by Human Action*. Cambridge: Cambridge University Press, pp. 163–78.

Robbins, P. (1998) 'Paper forests: imagining and deploying exogenous ecologies in Arid India', *Geoforum*, 29: 69–86.

Robbins, P. (2001a) 'Tracking invasive land covers in India: or, why our landscapes have never been modern', *Annals of the Association of American Geographers*, 91: 637–59.

Robbins, P. (2001b) 'Fixed categories in a portable landscape: the causes and consequences of land cover classification', *Environment and Planning A*, 33: 161–79.

Said, E. (1978) *Orientalism*. New York: Random House.

Said, E. (1994) *Culture and Imperialism*. London: Vintage.

Scott, J. (1998) *Seeing Like a State: How Certain Schemes to Improve the Human Condition Have Failed*. New Haven, CT: Yale University Press.

Spivak, G.C. (1990) *The Post-Colonial Critic: Interviews, Strategies, Dialogues*, (ed.) in Sarah Harasym. New York: Routledge.

29 CONTESTED GEOGRAPHIES: CULTURE WARS, PERSONAL CLASHES AND JOINING DEBATE

Gill Valentine and Stuart Aitken

'Geography is a social institution – it is made by human beings in social contexts – and as such its nature will always be contested' (Taylor and Overton, 1991: 1089). In this book so far the authors of each chapter have outlined very diverse ways of thinking about what constitutes geographical knowledge, the methods that should be used to collect data, and the politics and purpose of these endeavours. Sometimes implicitly and at other times explicitly, the chapters have touched on the conflicts between them in terms of geographical thought and practice and the implications of these for the direction and nature of the discipline. As Neil Smith argues:

> **The history of geography does not simply happen with the passing of time, but is an active creation, the result of struggle. There is a struggle over which ideas best explain the past, a struggle over concepts appropriate for current research, and insofar as scientific research is obliged to have some redeeming social importance, a struggle over the way in which the historical geographies of contemporary landscapes are to be fashioned. (1988: 160)**

This chapter provides some examples of the contested nature of geography. Henry Kissinger is credited with saying that debates amongst academics are the most vitriolic because there is so little at stake. This statement is as conflicted as the former US Secretary of State himself, who is lauded by some for his work towards peace and who is indicted by others as a war criminal. Yes, geographical debates can be vitriolic but the nature of the stakes is important because they are far from small.

This chapter picks out three examples – skirmishes about philosophy and relevance; skirmishes about institutions, people and the discipline; and skirmishes about philosophy and methodology – to demonstrate some of the ways that geography is being debated. What is at stake here are the ways that geographers see the world and practise research.

Skirmishes about Philosophy and Relevance

Our first example of a geographical skirmish took place within the pages of the journal *Progress in Human Geography*. In 2001 the geographer Ron Martin launched an attack on the failure of geography in the UK to make an effective contribution to the public policy agenda, at a time when it was being actively reworked and rethought. In doing so, he contrasted the discipline with others, such as sociology, whose leading figure Anthony Giddens is frequently invited by the Prime Minister to advise on policy-making.

Martin blamed the failure of geography to have an impact on the corrosive effects of postmodernism and the 'cultural turn' in Anglophone human geography. In mapping the process leading up to this, Martin recalled the publication of David Harvey's book *Social Justice and the City* in 1973. He argued that this critical Marxist take on capitalism, although largely theoretical, inspired a sense of social commitment and political engagement (see also Chapter 25) within the discipline. In the late 1970s as the postwar Keynesian economics and the welfare model of state intervention unravelled, geographers had the opportunity to be part of the political debate about new ways forward; or to challenge the free-market neoliberal policies of Thatcherism in the UK and Reaganism in the USA.

Instead of taking a 'policy turn' – with the notable exception of some individual geographers who have persisted with trying to make a contribution to policy – Martin (2001: 192) claims the discipline lost its relevance, fragmenting into a series of what he terms 'post-radical and post-Marxian movements', such as critical human geography, feminism, postmodernism and so on. In particular, he suggested that the contemporary emphasis on difference and particularism (see Chapters 10 and 12) had led to a fragmentation of geography's political project and a movement away from concerns with broader concepts such as social justice. Martin characterized 'post-radical and post-Marxian movements' as being preoccupied with theoretical and linguistic issues and losing their critical edge. He also lamented what he described as human geography's move away from doing rigorous empirical work; and its preoccupation with writing obtuse theory rather than using the kind of jargon-free language necessary to communicate with government. Rather, Martin suggested that human geographers had become dismissive of policy work, dubbing it atheoretical and descriptive and questioning the independence or integrity of its practitioners.

Calling for a 'policy turn' in the discipline, Martin argued that:

> **we need to temper our enthusiasm for seeking out the latest philosophical, theoretical or methodological fad, and develop a greater interest in practical social research, and as part of this reorientation, accord proper academic standing to policy studies ... we need to take detailed empirical work far more seriously: the drift towards 'thin empirics' needs to be reversed, and much greater attention directed to methodology and the quality of evidence. And ... human geographers need to decide how they, and the studies they undertake, are to be used: there is of course no such thing as neutral research and we need to be more explicit about, and more committed to, the political positions that inform and shape our work – whatever those political positions are. (2001: 202)**

Indeed, he went on to argue that geographers have a moral obligation to apply their research for the benefit of wider society (for example by exposing and explaining inequalities) rather than just their own careers.

Martin's article sparked a heated debate in the journal *Progress in Human Geography*. While he had blamed geography's failure to make an impact on government policy in terms of philosophical issues within the discipline, Doreen Massey (2001) located its difficulty in gaining any political influence, at least in part, on the unwillingness of government to listen to the radical implications of what geographers have to say. Massey warned against the dangers of abandoning theory in a bid to engage policy-makers, of separating theory from applied work, identifying instead the need for geographers to get their theoretical ideas across to wider audiences in a more accessible way. Indeed, she suggested that geographers needed to work through their own specific theoretical contribution to debates rather than being so quick to import ideas from other disciplines, calling for a more constructive dialogue between human and physical geography. And, she identified the need for geography to shake off its image as an intellectually dull subject associated with 'capes and bays' (2001: 13).

While Massey had responded to Martin in civil terms, Dorling and Shaw (2002) adopted a more aggressive tack, challenging both Martin's and Massey's arguments in brusque language. In particular, they took Martin and Massey to task for being preoccupied with geography, arguing that perhaps the reason why regional inequality is still on the policy agenda after 25 years is that geographers have spent too much time debating it with other geographers in academic journals, rather than getting involved with what is going on outside the discipline. They pointed out that most geographers 'are concerned with thinking about (and understanding and explaining) spatial relationships, not with changing them, and that is precisely why they are geographers' (2002: 632).

Dorling and Shaw then went on to argue that if geographers want to have an impact on policy, rather than waiting for a policy turn within the discipline, they must rethink how they get their message across outside the discipline. Here, they criticized geographers for their abstract conceptualizations of regional inequalities and the north–south divide, arguing that concrete arguments based on examples and statistical evidence are more convincing. Stressing the importance of quantitative methods (see Chapter 22) for 'showing how much things matter', Dorling and Shaw (2002: 633) accused critical geographers of losing sight of the effects of 'quantifying power'.

Like Martin, they too laid into the elitist jargon of poststructuralist theories, and called for geographers to articulate their message more effectively. Referring to a reader on poverty, inequality and health (Davey Smith et al., 2001), they noted that of the 30 seminal essays included, not one was written by a geographer. Controversially, they suggested that if geographers want to have any impact on policy they would be better off in another discipline, claiming that geography may not be well suited to political influence, characterizing it as 'an academic refugee camp – a place where academics can work on whatever they wish to work on and not be disturbed by the need to conform to the traditions of other disciplines' (2002: 638). Dorling and Shaw concluded by suggesting that if geographers wanted to be taken seriously outside the academy they needed to respect those that did take part in policy debates instead of dismissing them for being inadequately theoretical or not proper geographers.

Martin and Massey were given right of response to Dorling and Shaw's attack by the journal *Progress in Human Geography*. Martin (2002) took the opportunity to agree with Dorling and Shaw's (2002) criticism of Massey (2001) – that the cause of geography's lack of relevance is not politicians' failure to not listen but rather whether geographers will respond to calls for policy-oriented work or have anything informative/distinctive to say. He further picked up an underlying theme in Dorling and Shaw's paper, to argue that geography has a weak or inferior standing in wider educational and public domains, taking the opportunity to have another dig at poststructuralist theories and geography's dabblings in cultural and media studies, etc.

Massey entered back into the affray to defend herself against what she regarded as Dorling and Shaw's 'persistent misunderstanding (wilful?) [and] a scatter of insults (gratuitous)' (2002: 645). Pointing to her own involvement in policy arenas beyond the academy, Massey claimed that it was not necessary to sacrifice theory to be politically committed. Rather, she argued that the trick is to use appropriate language for appropriate audiences, characterizing her own role not as thinking of ideas and then trying to foist them onto politicians but as an 'endless moving between' in which she works with policy-makers and community groups in a range of forums, 'reflecting on these engagements, thinking, arguing, and writing' (2002: 645). In doing so, Massey took Dorling and Shaw to task for their conceptualization of policy, arguing it is not just a case of winning large government research contracts and providing technically correct answers to prespecified government questions. Rather, she suggested that policy work can also involve working with campaign groups or trying to influence wider public opinion and that academics have an intellectual responsibility to have a deeper, more difficult political engagement and debate about different understandings of the world.

Beyond this skirmish in *Progress in Human Geography*, the debate about geography's relevance and its relationship to different philosophical traditions has rumbled on. For Noel Castree (2002), if geographers want to make a difference beyond the academy they must first pay more attention to the institutions in which they are situated – to what Wills (1996) has dubbed the academic sweatshop or what Smith (2000) has described as the corporatization and commodification of universities. In focusing on the political economy of higher education, Castree (2002) argued for the need for more activism within universities to challenge and change what counts as valuable academic activity. While for Mitchell, 'to make a difference beyond the academy it is necessary to do good and important, and committed work, *within the academy*' (2004: 23). He cited the example of Karl Marx whose scholarship was driven by a commitment to a revolutionary project – to learn and explain how capital worked and how it might be changed – arguing that it was this commitment that has made Marx's work relevant to those outside academy for over a century and a half.

Skirmishes about Institutions, People and the Discipline of Geography

In contrast to the contemporary and UK-focused nature of our first example of the contested nature of geography, the second examines the history of the closure of geography in one of North America's most prestigious universities.

Following modest postwar expansion the geography program at Harvard University was abruptly closed in 1948, triggering a skirmish over the nature and future of the discipline of geography. This closure was symbolically important because of the eminent position of Harvard within the North American education system but was made even more significant by the President of Harvard's suggestion that geography was not an appropriate university subject.

Tracing the history of what happened, Neil Smith (1987) argued that the demise of Harvard geography was bound up with particular personalities. The story began optimistically: a wartime report on geography at Harvard (which at this time was a Department of Geology and Geography within the Division of Geological Sciences) highlighted a lack of trained geographers required for wartime operations, and therefore recommended expansion. Harvard took on several promising young geographers. In 1947 one of these, Ackerman, was recommended for promotion by the senior faculty to the Provost and Dean of the Faculty of Arts and Science, Paul Buck, who was the administrator directly in charge of geography. However, Marland Billings, a professor of geology and the Chair of the Division of Geological Sciences within which geography was located, was unhappy about this because Ackerman had originally been half in geology and half in geography. If Ackerman was promoted to associate professor in geography, geology would effectively lose half a post. Billings was concerned that geography's expansion threatened geology and chose the case of Ackerman's promotion to launch an attack on geography.

According to Smith's account, Billings lobbied Paul Buck to administratively separate geography and geology because the subjects were very different. His move was actually supported by the Head of Geography, Derwent Whittlesey, who welcomed the idea of autonomy for the emerging Geography Department.

This achieved, Billings wrote to Buck arguing that Ackerman's promotion had previously been supported by geology on the (mis)understanding that it would gain half a post if Ackerman's application was successful and he became a full-time geographer. Billings made the case that a half-post in geology was more valuable to the university than a new appointment in geography and questioned the importance of human geography. Paperwork flew back and forth. Whittlesey, and a number of prominent academics who acted as external independent referees, defended Ackerman's promotion and an Ad Hoc Committee on Geography, including outsiders, also recommended Ackerman's promotion.

But Buck had become convinced that geography should be closed. He refused to reappoint a teaching instructor who taught a number of core courses in the department, and the sophomore class was informed that there were insufficient courses for them to obtain a major in geography. With the exception of Whittlesey, the members of the Geography Department were to be fired. Despite a temporary stay of execution and the mobilization of support for the department on campus, geography closed.

In telling this story Neil Smith (1987) argued that there were three public factors in the decision to close geography: (1) the adverse financial circumstances of the university; (2) the efficacy of geography at Harvard; and (3) whether geography could be a university discipline. Behind the scenes, however, personalities also played a crucial part. A key figure was Isaiah Bowman, an eminent geographer and President of Johns Hopkins University, whom Smith identified as contributing to the elimination of geography at Harvard. Smith argued that although Bowman supported geography as a discipline, he did not support Harvard geography because he took a negative view, for personal reasons, of some of the staff. He found Whittlesey's homosexuality an anathema. In particular, he was critical of the way that Whittlesey had appointed Kemp – with whom he was having a gay relationship – as an instructor of geography, suggesting that Kemp was a mediocre scholar who probably would not have survived in Harvard without Whittlesey's patronage. He also had little time for Alexander Hamilton Rice, who ran the university's Institute for Geographical Exploration. According to Smith, Bowman saw Rice as a charlatan who had effectively purchased a professorship thanks to the money of his wife, a rich society figure.

However, the crucial factor in Bowman's reluctance to support the Harvard Geography Department was in Smith's (1987) analysis the fact that his personal antipathies were also bound up with an intellectual antipathy for Whittlesey. Bowman had trained within the Davisian paradigm, and regarded physical geography as the foundation stone of the discipline; as such he positioned geography within the sciences. He was critical of human geography, viewing it as descriptive, easy, and lacking in scientific character. His perception of the social sciences was further tainted by their association with leftwing radicals. Thus Whittlesey, who came from the Chicago School, and who believed that there could be a set of intellectual principles that could provide a foundation for human geography as a discipline, embodied Bowman's prejudices about the discipline. Smith argued that Bowman therefore let his judgement about the future of Harvard geography become clouded by his personal feelings, refusing to stand up in support of the department despite being encouraged to do so by other prominent figures in the discipline. He also suggests that Bowman was not kindly disposed to geography at Harvard because he felt intimidated by its wealth and elitism, coming as he did from a more modest background.

Smith (1987) therefore concludes that in the context of geography's institutional weakness at Harvard – emerging out of geology, it was numerically weak and had not yet identified a clear intellectual terrain or set of boundaries from other subjects for itself – it was vulnerable to the attack launched by Billings. Whittlesey was politically weak, and did not have the allies in the administration or among key faculty members to resist it.

Smith's explanation for the demise of geography at Harvard prompted a spate of responses from other geographers offering alternative readings of what had taken place. Martin (1988), for example, challenged Smith's account. He was more critical of the role played by Whittlesey, accusing him of having not developed a meaningful geography programme at Harvard (which was bound up with his alleged appointment of Kemp on the grounds of their relationship rather than his ability) and having failed to make his mark on the wider university. Concomitantly, Martin was more forgiving of Bowman. He suggested that he had failed to intervene because he respected the presidential prerogative, so once the President had made his decision to close geography Bowman would do nothing to undermine it, and did not want to engage in unseemly public squabbles about the discipline (although he suggested that he did make efforts in private to defend geography as a science). For Martin, Bowman's actions were shaped by his lack of respect for Whittlesey and Rice, regarding them both as liabilities, portraying Bowman as a good friend of geography.

Cohen (1988) took a different tack – identifying geography's vulnerability in terms of the department's failure to be aggressive at the first sign of crisis. He observed that the image of any department stems largely from the reputation in its faculty, and thus geography at Harvard was vulnerable because it was composed of poor teachers (such as Kemp) who lacked charisma. He also pointed out that the Geography Department did not work effectively as a team or come together. As a small community geography did not have the critical mass to work as a force within the university or to create a meaningful support network in the field, and the weaknesses of individuals were easily magnified.

Burghardt (1988) also waded into the debate, standing up for both Whittlesey and Bowman. He attacked Smith (1987), stating: 'I don't doubt that Whittlesey and Bowman did a poor job of defending the discipline. However this was at a time when human geography was just emerging from a physical cocoon. I feel that it is somewhat distasteful for contemporary geographers, with the richness of thirty years of intense discussion behind them, to blame the progenitors of the field for their lack of insight' (1988: 144).

Other commentators also defended different proponents in the story and criticized Smith (1987) for washing geography's dirty linen in public. In response to the commentaries on his paper, Smith (1988) struck back at those who argued that geographers should not be openly critical of each other in public. He challenged what he termed 'plodding histories' of the discipline that refused to acknowledge the intellectual and personal struggles that shaped it, stating that 'any outward appearance of unity and tranquillity fools only geographers' (1988: 160).

Skirmishes about Philosophy and Methodology

Our third example of an ongoing skirmish within geography focuses on the philosophical and methodological struggle for disciplinary supremacy between GIS practitioners and their critics in human geography.

GIS (see Chapter 23) is a collection of tools for quantitative data analysis that emerged out of the positivist tradition (see Chapter 2). In the early 1990s it was subject to hostile criticism on the basis of what was perceived to be its inherent positivism (Lake, 1993). Peter Taylor described the growth of GIS as a 'technocratic turn', characterized by a 'retreat from ideas to facts', 'trivial pursuit geography' and 'a return of the very worst sort of positivism, a naïve empiricism' (1990: 212).

In particular, critics attacked GIS researchers for unproblematically using scientific methods to study social phenomena, arguing that GIS 'did not accommodate less rational, more intuitive analyses of geographical issues and that its methodology, by definition, excluded a range of inquiry' (Schuurman, 2000: 577). GIS researchers were further lambasted for implying that all data are given (Taylor and Overton, 1991) rather than recognizing that data are always created and that there are social relations to its production (i.e. most data come from the state, and there is a contrast between the information at the disposal of rich and poor countries).

Critics of GIS also called its users to account for their claims of intellectual neutrality in the interpretation of data, and their lack of ethical concern for some of its applications. Neil Smith (1992: 257), for example, pointed out that the Iraq War of 1990–1 was the 'first full scale GIS war' in which advanced GIS and related technologies were used by pilots, tank commanders and geosmart bombs, altering the way modern warfare is fought. Moreover, he observed that a significant percentage of American geography graduates who study GIS end up in military-related jobs – noting that the US Defense Mapping Agency is the single largest employer of geography graduates. Denis Wood (1989) suggested that computer systems promote death not only through their military usages but also through their role in car manufacturing, observing that cars kill more people in the USA than most other causes of death combined.

Others critics railed at the under-representation of marginalized groups in the technology; challenged the commercial motivations behind the development of GIS; questioned its ethics, suggesting that the opportunities the technology provides for surveillance might threaten individuals' privacy and freedoms; and called for geographers to show responsibility for the ways that GIS is developed and applied (Pickles, 1993; Sheppard, 1993).

Advocates of GIS retaliated at various stages of the debate, arguing that it is a powerful tool that increases geographers' analytical abilities and that the discipline needs GIS (Goodchild, 1991). Dobson (1993) accused cultural geographers of being ignorant of GIS applications and contrasted its hostile reception inside the discipline with the positive ways it was being received outside the discipline. Openshaw (1991: 621) suggested that within geography there was 'genuine ignorance and wilfully misinformed prejudice' towards GIS. For Openshaw, GIS represents the essence of geography – the basis to hold the discipline together. He argued that GIS offers: 'an emerging all embracing implicit framework capable of integrating and linking all levels of past, present and possible future geographies' (1991: 628). He writes: 'A geographer of the impending new order may well be able to analyse river networks on Mars on Monday, study cancer in Bristol on Tuesday, map the underclass of London on Wednesday, analyse the groundwater flow in the Amazon basin on Thursday and end the week by modelling retail shoppers in Los Angeles on Friday' (1991: 624).

Like many academic skirmishes within the discipline, this one was characterized by both the critics and the advocates of GIS alike making excessive claims for their own positions while portraying the arguments of the other in derogatory terms (Schuurman, 2000). In the process of hyping the importance of GIS, Openshaw (1991) in particular was openly hostile towards its critics. He accused them of 'infecting' the younger generation; lambasted their 'fear and anxiety'; charged them with 'envy'; labelled them 'poor fools'; and most offensively of all described them in disablist terms as 'technical cripples' (1991: 621–4).

For Openshaw, the critics of GIS were motivated not just by concerns about epistemology but by a desire to protect and retain their disciplinary position. He suggested that with the growth of GIS 'the utility of soft and so-called intensive and squelchy-soft qualitative research paradigms could fade into insignificance' (1991: 622). Smith (1992: 258) struck back, ridiculing the 'outlandish disciplinary ambitions' of GIS users and their obsession with a technical agenda at the exclusion of intellectual pursuits and other perspectives.

These intense and often hostile skirmishes that characterized the debate about the role of GIS in geography in the first half of the 1990s gave way to a calmer discussion about the social effects of the technology and its impact on the discipline (Schuurman, 2000). By now there was a growing demand for GIS academics (and a corresponding falloff in positions for cultural geographers); and the technology was making inroads into other disciplines as well as the commercial and public sectors. As such, Schuurman argues that GIS had become a legitimate and permanent feature of geography, and so its critics began to recognize that the technology was non-negotiable and to focus instead on opportunities to rework GIS in postpositivist ways. This marked a period of negotiation between users of GIS and social theorists, although Schuurman observes that this process was undercut by communication problems as a result of ignorance of each other's fields. Rather than making their criticisms of GIS relevant to its users, its critics, unfamiliar with the language of technology, used the language of social theory to talk about the epistemological and ethical shortcomings of GIS. In turn, GIS researchers, unused to the language of social theory, employed the vocabulary of technology, written in the language of computational algebra and laws of physics, to extol its virtues. Schuurman argued that if social theorists wanted to influence the use of GIS they must learn to communicate in its own language.

Some GIS researchers, concerned by the attack on GIS in terms of positivism, began to draw attention to the fact that neither the technology nor its users were inherently positivist. Rather GIS could be compatible with a wide range of philosophic positions (Schuurman and Pratt, 2002). Indeed, Schuurman and Pratt (2002) argued that critiques from within the GIS research community might be more effective at shaping its use than those offered by social theorists operating outside its community who have no stake in its future.

Mei-Po Kwan (2002), for example, a feminist GIS user, observed that the crude oppositional polemic of the GIS debate – positivist/quantitative methods versus critical/qualitative methods; GIS/spatial analysis versus social/critical theory – had marginalized the contribution of feminist GIS researchers and the potential to develop feminist GIS practices.

She claimed that GIS can disrupt the quantitative/qualitative methods division in geography because qualitative data like video or voice clips and photographs, hand drawn maps, or sketches can be incorporated into the technology; and GIS/spatial analysis can also be informed and contextualized by data generated by qualitative methods such as interviews. Moreover, she showed that GIS can be employed in alternative ways that subvert dominant GIS practices and are compatible with feminist epistemologies and politics. For example, Mei-Po Kwan (1999a; 1999b; 2000; 2002) has used GIS to trace and visualize women's life paths and the impact of space–time constraints on their mobility and employment; and to examine the spatially constrained life spaces of African Americans. GIS software and data do not predetermine the ways that technologies are used; rather, GIS can create many different products. As another feminist geographer Sarah Elwood (2000) has shown, the outcome of GIS depends less on the technology itself than on the critical agency of its users. In her study, GIS technologies, by legitimizing a neighbourhood association, shifted the power dynamics between a community group and the state.

Reflecting on how GIS can be used in feminist research, Schuurman and Pratt (2002) commented that feminist geographers can bring reflexivity to GIS in relation to the creation, use and interpretation of the visual data and so challenge the detached, disembodied mode of knowing that characterizes conventional GIS practices. They suggested that feminist geographers need to think about what knowledges are excluded from GIS representations; to consider the relevance of the knowledge produced by the technology to those participating in the research; to identify which groups are empowered or disempowered by GIS; and to be sensitive to the impact of it on the lives of vulnerable groups. They concluded by arguing that feminist geographers need to destabilize masculinist computer cultures and GIS practices, and find ways of using GIS technologies to advance social justice.

Up Close and Personal

As all the previous skirmishes (about philosophy and relevance; people and institutions; and philosophy and methodology) have shown, the terms of geographical debate are rarely confined to scholarly remarks. Professional differences can become personal antagonisms; while personal antipathy can be projected on to professional positions. Rather than academic argument moving smoothly from an exchange of views to a negotiated consensus, in the whole process there is often a tendency for individuals to resort to varying degrees of provocation, irony, and even insult.

In the mid 1990s *The Canadian Geographer* published a keynote conference speech given by Peter Gould. In it he rounded on radical feminist geography, claiming that it had 'been elevated rapidly to the sacred, where faith and belief rule, but reason is not relevant' (Gould, 1994: 10). He attacked radical feminists' concern with sexist language, accusing them of being ignorant of the historical origins of words and linguistic structures. Implicitly, Gould argued that feminist geography was constructing men like himself as 'other' (Peake, 1994). And in defence of his position he referred to himself as a lover of Geographia – invoking images from his book *The Geographer at Work* (1985), which depicts the discipline as a naked woman. In his published speech Gould also took postcolonialism to task, reappraising colonialists for spreading the advantages of civilization such as 'scrupulous honesty in all matters of accounting … and a sense of service' (1994: 14), while accusing black African leaders of being corrupt despots. At the same time, Gould (1994: 10) attempted to deflect potential criticisms of his position, by referring to some of his best friends as black.

Feminist geographers Linda Peake (1994) and Janet Momsen (1994) hit back. Peake (1994) questioned Gould's (1994) representation of women, and the political effects of the symbols he invoked. Likewise, Peake pointed out that Gould's reappraisal of colonialism could be read very differently, drawing attention to the dishonesty of colonial regimes such as the UK government in British Guyana – which, she pointed out, rigged elections for 29 years and left the country with one of the lowest levels of GNP per capita in the world. Peake lambasted Gould for implicitly constituting himself as an objective detached viewer in what she termed a 'dangerously nostalgic vein of masculinist rationality' (1994: 203). She concluded by arguing that:

> **Professor Gould suggests that it is time we all grew up … The problem is that the Young Turks, the (exclusively male) intellectual vanguard of academic geography to whom he makes reference several times, have a habit of growing up**

> **into Old Turkeys. Maybe it is time not only to grow up, but for some Old Turkeys to come down off their roosts. That would be just fine by me. (1994: 206)**

While this skirmish in *The Canadian Geographer* might be seen as a rather personalized contest framed through academic discourse, on other occasions conflict can go beyond this rough and tumble of academic debate. Michael Dear (2001) argues that conflicts between geographers can descend into hate. He takes the dictionary definition of hate as having 'feelings of hostility or antipathy towards' (2001: 2). Using this definition, he recounts how he has been the subject of hate in geography, in terms of being branded with labels used pejoratively such as liberal, Marxist and postmodernist, as well as more blatant insults such as 'freak', 'zealot', 'un-American trash' and 'subversive scum'. Referring to responses to his essay on postmodernism and urbanism (Dear and Flusty, 1998), he recalls how his promotion of postmodernism was referred to by one commentator as a form of 'academic AIDS' (2001: 4). Other commentators have frequently used animalistic and excretory references to describe his work as well as making threats about 'unpleasant actions' that 'will be taken' against him (2001:5).

Dear argues that these experiences of what he dubs 'rabid referees' and 'combatants in the culture wars' – as well as the public cases of hate incidents and intimidation directed at Michael Storper and Gill Valentine – have made him aware of the role of the personal in disciplinary politics. Dear suggests that personal hatred is often a product of professional antagonisms, such as jealousies over differences in reputation, or the distribution of resources, responsibilities, power, and even popularity with students within departments and the discipline. He questions the silence in geography surrounding such issues, and the role of professional networks in generating hate. Rather, Dear argues for the need for geographers to teach students the value of civility in academic discourse and to develop a culture of respectful criticism. He suggests that academics should spend less time looking for each other's faults and more time looking for strengths, and that they should embrace diversity (in terms of intellectual commitments and projects as well as personal identities) rather than attack it. Dear concludes his paper with the claim that geographers have a moral responsibility to treat colleagues with civil collegiality.

In a response Wolfgang Natter (2001) makes three practical suggestions for how this might be achieved. First, he suggests that if geographers overhear conversations where colleagues are being unfairly criticized they should demonstrate zero tolerance for character assassination by challenging unacceptable personal characterizations. Second, as referees and editors geographers should encourage scholarly writing and promote an ethics of discussion. Third, the discipline needs to develop, and extend, codes of professional conduct.

In such ways, geographers might keep a bit of civility in their everyday skirmishes over the content, boundaries and directions of the discipline without losing any of the passion or excitement of the debate.

References

Burghardt, A.F. (1988) 'On "academic war over the field of geography": the elimination of geography at Harvard, 1947–51', *Annals of the Association of American Geographers*, 78: 144.

Castree, N. (2002) 'Border geography', *Area*, 34: 103–12.

Cohen, S. (1988) 'Reflections on the elimination of geography at Harvard, 1947–51', *Annals of the Association of American Geographers*, 78: 148–51.

Davey Smith, G., Dorling, D. and Shaw, M. (2001) *Poverty, Inequality and Health 1800–2000: A Reader.* Bristol: Policy.

Dear, M. (2001) 'The politics of geography: hate mail, rabid referees, and culture wars', *Political Geography*, 20:1–12.

Dear, M. and Flusty, S. (1998) 'Postmodernism urbanism', *Annals of the Association of American Geographers*, 88: 50–72.

Dobson, M. (1993) 'Automated geography', *Professional Geographer*, 35: 135–43.

Dorling, D. and Shaw, M. (2002) 'Geographies of the agenda: public policy, the discipline and its (re)turns', *Progress in Human Geography*, 26: 629–46.

Elwood, S. (2000) 'Information for change: the social and political implications of geographic information technologies'. Unpublished PhD dissertation, University of Minnesota, Minneapolis.

Goodchild, M. (1991) 'Just the facts', *Political Geography Quarterly*, 10: 335–7.

Gould, P. (1985) *The Geographer at Work.* London: Routledge, Kegan Paul.

Gould, P. (1994) 'Sharing a tradition: geographies from the enlightenment', *The Canadian Geographer*, 38: 194–202.

Harvey, D. (1973) *Social Justice and the City.* London: Arnold.

Kwan, M.P. (1999a) 'Gender, the home–work link, and space–time patterns of non-employment activities', *Economic Geography*, 75: 370–94.

Kwan, M.P. (1999b) 'Gender and individual access to urban opportunities: a study using space–time measures', *Professional Geographer*, 51: 210–27.

Kwan, M.P. (2000) 'Interactive geovisualisation of activity-travel patterns using three dimensional geographical information systems: a methodological exploration with a large data set', *Transportation Research C*, 8: 185–203.

Kwan, M.P. (2002) 'Is GIS for women? Reflections on the critical discourse in the 1990s', *Gender, Place and Culture*, 9: 271–9.

Lake, R.W. (1993) 'Planning and applied geography: positivism, ethics and geographic information systems', *Progress in Human Geography*, 17: 404–13.

Martin, G.J. (1988) 'On Whittlesey, Bowman and Harvard', *Annals of the Association of American Geographers*, 78: 152–8.

Martin, R. (2001) 'Geography and public policy: the case of the missing agenda', *Progress in Human Geography*, 25: 189–210.

Martin, R. (2002) 'A geography for policy, or a policy for geography? A response to Dorling and Shaw', *Progress in Human Geography*, 26: 642–4.

Massey, D. (2001) 'Geography on the agenda', *Progress in Human Geography*, 25: 5–17.

Massey, D. (2002) 'Geography, policy and politics: a response to Dorling and Shaw', *Progress in Human Geography*, 26: 645–6.

Momsen, J. (1994) 'Response to Peter Gould', *The Canadian Geographer,* 38: 202–4.

Natter, W. (2001) 'From hate to antagonism: toward an ethics of emotion, discussion and the political', *Political Geography*, 20: 25–34.

Openshaw, S. (1991) 'A view on the GIS crisis in geography: or, using GIS to put Humpty-Dumpty back together again', *Environment and Planning A*, 23: 621–8.

Peake, L. (1994) '"Proper words in proper places ..." or, of young Turks and old turkeys', *The Canadian Geographer*, 38: 204–6.

Pickles, J. (1993) 'Discourse on methods and the history of the discipline: reflections on Jerome Dobson's 1993 "Automated geography"' *Professional Geographer*, 45: 451–5.

Schuurman, N. (2000) 'Trouble in the heartland: GIS and its critics in the 1990s', *Progress in Human Geography*, 24: 569–90.

Schuurman, N. and Pratt, G. (2002) 'Care of the subject: feminism and critiques of GIS', *Gender, Place and Culture*, 9: 291–9.

Sheppard, E. (1993) 'Automated geography: what kind of geography for what kind of society?', *Professional Geographer*, 45: 457–60.

Smith, N. (1987) 'Academic war over the field of geography: the elimination of geography at Harvard, 1947–51', *Annals of the Association of American Geographers*, 77: 155–72.

Smith, N. (1988) 'For a history of geography: a response to comments', *Annals of the Association of American Geographers*, 78: 159–63.

Smith, N. (1992) 'History and philosophy of geography: real wars and theory wars', *Progress in Human Geography*, 16: 257–71.

Smith, N. (2000) 'Who rules this sausage factory?', *Antipode*, 32: 330–9.

Taylor, P.J. (1990) 'GKS', *Political Geography Quarterly*, 9: 211–12.

Taylor, P.J. and Overton, M. (1991) 'Further thoughts on geography and GIS', *Environment and Planning A*, 23: 1087–94.

Wills, J. (1996) 'Labouring for love?', *Antipode*, 28: 292–303.

Wood, D. (1989) 'Commentary', *Cartographica*, 26: 117–19.

EXERCISES

1. Define the following terms and briefly state their significance for human geography: dialectic, patriarchy, epistemology, ontology, phenomenology, gender.
2. Structuration theories are inherently attractive as an ontology, but using them as a basis for research is highly problematic. Discuss.
3. By focusing solely on the thoughts and actions of 'ordinary folk', humanistic geography is romantic, superficial and empirical. Critically evaluate this assessment of humanistic geography.
4. Human geography has moved from the absolutes of positivism to the relativities of postmodernism. Discuss this statement and identify what has been gained and lost on the way.
5. Outline the main consequences for human geography of the quantitative revolution.
6. What was radical about the development of Marxist approaches to human geography in the 1960s and 1970s?
7. 'We need to contemplate the human world less in terms of "grand theories" and more in terms of humble, eclectic and empirically grounded materials' (P. Cloke, C. Philo and D. Sadler, eds, 2003, *Approaching Human Geography*, London: Chapman). To what extent do you agree with this approach to the study of human geography?
8. Critically evaluate the role of agency in actor-network theory.
9. Outline postcolonial critiques of representation. What are the implications of this for doing human geography research?
10. Choose a well-known human geographer, and research the development of their career and writing. How does their work reflect particular philosophical or methodological approaches to geography?
11. The topic of your research project is immigration. For each of the philosophies outlined in Part 1 of this book, identify how a geographer adopting this approach might go about researching this topic. What are the similarities and differences between them?

GLOSSARY

Actor-network theory A theoretical approach that holds to the indivisibility of human and non-human agents, exploring the ways that different materials are enrolled in networks. Originally developed in debates about the production of scientific knowledge, actor-network theory (often abbreviated to ANT) opens up the 'black boxes' of action to explore the way that heterogeneous materials are continually assembled to allow actions to occur.

Agency Literally, the ability to act. Commonly used to refer to the ability of people to make choices or decisions which shape their own lives. This notion of self-determination has formed an important part of the humanistic critique of structuralist approaches to geography.

Behaviouralism An outlook or system of thought that believes that human activity can best be explained by studying the human decision-making processes that shape that activity. Originally developed in psychology, largely as a reaction to the mechanistic excesses of experimental psychology, behaviouralism – and more particularly cognitive behaviouralism – came to prominence in the human geography of the 1960s and 1970s. Primarily based on methods of quantification, behavioural geography has been criticized for its adherence to positivist principles, as well as its unwillingness to explore the role of the unconscious mind, although it still underpins many research projects, particularly those based on survey research.

Capital accumulation The use or investment of capital to produce more capital. This is the aim of, or driving force in, a capitalist society. It results in patterns of uneven development.

Capitalism The political-economic system in which the organization of society is structured in relation to a mode of production that prioritizes the generation of profit for those who own the means of production. Such a structuring sees a clear division in status, wealth and living conditions between those few who own or control the means of production (bourgeois) and those who work for them (proletariat).

Citizenship The relationship between individuals and a political body (i.e. the nation-state). Usually conceptualized in terms of the rights/privileges that individuals can expect in return for fulfilling certain obligations to the state.

Class A system of social stratification based on people's economic position (specifically the social relations of property and work). Understandings and definitions of class are highly contested.

Commodification The processes through which people, ideas or things are converted into commodities that can be bought and sold. As such it is a manifestation of capitalism. Such is the extensiveness of commodification that it is alleged by some writers to be contributing to the creation of a global culture.

Critical geography Though diverse in its epistemology, ontology and methodology, and hence lacking a distinctive theoretical identity, critical geography nonetheless brings together those working with different approaches (e.g. Marxist, feminist, postcolonial, poststructural) through a shared commitment to expose the sociospatial processes that (re)produce inequalities between people and places. In other words, critical geographers are united in general terms by their *ideological* stance and their desire to study and engender a more just world. This interest in studying and changing the social, cultural, economic or political relations that create unequal, uneven, unjust and exploitative geographies is manifest in engagements with questions of moral philosophy and social and environmental justice as well as in attempts to bridge the divide between research and praxis.

Cultural turn A trend in the late twentieth and early twenty-first centuries which has seen the social sciences and humanities increasingly focus on culture (and specifically the construction, negotiation and contestation of meanings). Linked to postmodernist philosophies.

Deconstruction A method of analysis that seeks to critique and destabilize apparently stable systems of meaning in discourses by illustrating their contradictions, paradoxes, and contingent nature.

Dialectics A form of explanation and representation that emphasizes the resolution of binary oppositions. Rather than understanding the relationship between two elements as a one-way cause and effect, dialectical thinking understands them to be part of, and inherent in, each other. A dialectic approach has been an important part of structuralist accounts that seek to understand the interplay between individuals and society.

Difference Poststructuralist theory has emphasized the need to recognize the complexity of human social differences associated with culturally constructed notions of gender, race, sexuality, age, disability, etc. This means providing an analysis that is sensitive to the differences between individuals and avoids overgeneralizations.

Discourse Sets of connected ideas, meanings and practices through which we talk about or represent the world.

Dualism Where two factors (e.g. home/work; body/mind; nature/culture; private/public) are assumed to be distinct and mutually exclusive and to have incompatible characteristics.

Empiricism A philosophy of science that emphasizes empirical observation over theory. In other words it assumes 'facts speak for themselves'.

Essentialism The belief that social differences (such as gender, race, etc.) are determined by biology and that bodies therefore have fixed properties or 'essences'.

Feminism A set of perspectives that seek to explore the way that gender relations are played out in favour of men rather than women. In human geography, such perspectives have suggested that space is crucial in the maintenance of patriarchy – the structure by which women are exploited in the private and public sphere.

Humanistic A theoretical approach to geography which is characterized by an emphasis on human agency, consciousness and meanings and values. It developed in the 1970s partially as a critique of the spatial science of positivism.

Ideology A set of meanings, ideas or values which (re)produce relations of domination and subordination.

Marxism A set of theories developed from the writing of Karl Marx, a nineteenth-century German philosopher. Marxist approaches to geography use these insights to examine geographies of capitalism, challenging the processes that produce patterns of uneven development.

Mode of production A Marxist-derived term that denotes the way that relations of production are organized in specific periods. Currently, it is accepted that the world is organized so as to reproduce and maintain a capitalist mode of production, though it is emphasized that feudal, socialist and communist modes of production have been (and in some cases still remain) dominant in some nations.

Non-representational theory A theory that seeks to move the emphasis of analysis from representation and interpretation to practice and mobility. Emphasis is placed on studying processes of becoming, recognizing that the world is always in the making, and that such becoming is not always discursively formed (framed within, or arising out of, discourse). Here, society consists of a set of heterogeneous actants who produce space and time through embodied action that often lacks reason and purpose. To understand how the world is becoming involves observant participation – a self-directed analysis of how people interact and produce space through their movement and practice.

Objectivity The assumption that knowledge is produced by individuals who can detach themselves from their own experiences, positions and values and therefore approach the object being researched in a neutral or disinterested way.

Other/othering The 'other' refers to the person that is different or opposite to the self. Othering is the process through which the other is often defined in relation to the self in negative ways: for example, woman is often constructed as other to man; black as the other of white; and so on.

Paradigm The assumptions and ideas that define a particular way of thinking about and undertaking research that become the dominant way of theorizing a discipline over a period of time until challenged and replaced by a new paradigm.

Political economy Theoretical approaches that stress the importance of the political organization of economic reproduction in structuring social, economic and political life. Associated in human geography with the influence of Marxist thinking, political economy perspectives in fact encompass a variety of approaches that explore the workings of market economies.

Positionality Refers to the way that our own experiences, beliefs and social location affect the way we understand the world and go about researching it.

Positivism A theoretical approach to human geography, characterized by the adoption of a scientific approach in which theories/models derived from observations are empirically verified through scientific methods to produce spatial laws. Positivism came to the fore in the 1950s and 1960s in what was known as the quantitative revolution.

Postcolonialism A set of approaches that seek to expose the ongoing legacy of the colonial era for those nations that were subject to occupation by white, European colonizers. Emphasizing both the material and the symbolic effects of colonialism, postcolonial perspectives are particularly concerned with the ways that notions of inferiority and otherness are mapped onto the global south by the north, though postcolonial perspectives have also been utilized to explore the race and ethnic relations played out on different scales.

Postmodernism A theoretical approach to human geography which rejects the claims of grand theories or metanarratives. Instead it recognizes that all knowledge is partial, fluid and contingent and emphasizes a sensitivity to difference and an openness to a range of voices. Deconstruction is a postmodern method. Postmodernism is also a style, associated with a particular form of architecture and aesthetics.

Poststructuralism A broad set of theoretical positions that problematize the role of language in the construction of knowledge. Contrary to structural approaches, which see the world as constructed through fixed forms of language, such approaches emphasize the slipperiness of language and the instability of text. A wide-ranging set of assertions follow from this key argument, including the assertion that subjects are made through language; the idea that life is essentially unstable, and only given stability through language; the irrelevance of distinctions between realities and simulacra; ultimately, that there is nothing 'beyond the text'. In human geography, poststructural thought has provoked attempts to deconstruct a wide variety of texts (including maps) and has encouraged geographers to reject totalizing and foundationalist discourses (especially those associated with structural Marxism).

Qualitative method In human geography, this denotes those methods that accept words and text as legitimate forms of data, including discourse analysis, ethnography, interviewing, and numerous methods of visual analysis. Mainly tracing their roots to the arts and humanities, such methods have often been depicted as 'soft' methods, and hence described as feminist in orientation. Latterly, however, such simplistic assertions have been dismissed, and qualitative methods proliferate across the discipline in areas including economic and political geography.

Quantitative method In human geography, this denotes those methods that prioritize numerical data, including survey techniques, use of secondary statistics, numerous forms of experimentation, and many forms of content analysis. Mainly derived from the natural sciences, such methods are often depicted as 'hard' methods, deriving their analytical rigour and validity by association with masculine modes of science and exploration. However, numerous critiques have exposed the subjectivity of quantitative methods, and suggested their techniques cannot be understood as objective ways of looking at and understanding the world. This has led to a reappraisal of the quantitative method in areas of the discipline such as social and cultural geography where it has long been anathema.

Realism A theoretical perspective that seeks to transcend many of the problems associated with positivism and structuralism by seeking to isolate the causal properties of things that cause other things to happen in given situations. Based on a methodological distinction between extensive and intensive research, this approach was widely embraced in human geography in the 1980s as a way of distinguishing between spurious associations and meaningful relations.

Reflexivity Refers to a process of reflection about who we are, what we know, and how we come to know it.

Situated knowledge In a challenge to objectivity, a situated knowledge is one where theorization and empirical research are framed within the context in which they were formulated. Here, it is posited, knowledge is not simply 'out there' waiting to be collected but is rather made by actors who are situated within particular contexts. Research is not a neutral or an objective activity but is shaped by a host of influences ranging from personal beliefs to the culture of academia, to the conditions of funding, to individual relationships between researcher and researched, and so on. This situatedness of knowledge production needs to be reflexively documented to allow other researchers to understand the positionality of the researcher and the findings of a study.

Social justice Refers to the distribution of income and other forms of material benefits within society.

Spatial science An approach to understanding human geography that holds to the idea that there can be a search for general laws that will explain the distribution of human activity across the world's surface. Associated with the precepts of positivism, and mainly reliant on quantitative method, spatial science signalled geography's transition from an atheoretical discipline to one concerned with explanation rather than mere description. Emerging in the 1950s, and bolstered by the quantitative revolution of the 1960s, spatial science continues to be dominant in many areas of the discipline, though in others its philosophical underpinnings and theoretical conceits have long discredited it.

Structuralism A theoretical approach to human geography which is characterized by a belief that in order to understand the surface patterns of human behaviour it is necessary to understand the structures underlying them which produce or shape human actions.

Subject/subjective Subject refers to the individual human agent (includes both physical embodiment and thought/emotional dimensions). Subjective research is that which acknowledges the personal judgements, experiences, tastes, values and so on of the researcher.

INDEX

A page number followed by an asterisk (*) refers to a mention in the glossary.